T0181697

Lie Groups and Geometric Aspects
of Isometric Actions

Marcos M. Alexandrino • Renato G. Bettiol

Lie Groups and Geometric Aspects of Isometric Actions

 Springer

Marcos M. Alexandrino
Departamento de Matemática
Instituto de Matemática e Estatística
Universidade de São Paulo
São Paulo, Brazil

Renato G. Bettiol
Department of Mathematics
University of Pennsylvania
Philadelphia, PA, USA

ISBN 978-3-319-38627-0   ISBN 978-3-319-16613-1   (eBook)
DOI 10.1007/978-3-319-16613-1

Springer Cham Heidelberg New York Dordrecht London
© Springer International Publishing Switzerland 2015
Softcover reprint of the hardcover 1st edition 2015

Printed on acid-free paper

Springer International Publishing AG Switzerland is part of Springer Science+Business Media (www.springer.com)

Dedicated to our families.

Preface

This book is intended for advanced undergraduates, graduate students, and young researchers in geometry. It was written with two main goals in mind. First, we give a gentle introduction to the classical theory of Lie groups, using a concise geometric approach. Second, we provide an overview of topics related to isometric actions, exploring their relations with the research areas of the authors and giving the main ideas of proofs. We discuss recent applications to active research areas, such as isoparametric submanifolds, polar actions and polar foliations, cohomogeneity one actions, and positive curvature via symmetries. In this way, the text is naturally divided in two interrelated parts.

Let us give a more precise description of such parts. The goal of the first part (Chaps. 1 and 2) is to introduce the concepts of Lie groups, Lie algebras and adjoint representation, relating these objects. Moreover, we give basic results on closed subgroups, bi-invariant metrics, Killing forms, and splitting of Lie algebras in simple ideals. This is done concisely due to the use of Riemannian geometry, whose fundamental techniques are also quickly reviewed.

The second part (Chaps. 3–6) is slightly more advanced. We begin with some results on proper and isometric actions in Chap. 3, presenting a few research comments. In Chap. 4, classical results on adjoint and conjugation actions are presented, especially regarding maximal tori, roots of compact Lie groups, and Dynkin diagrams. In addition, the connection with isoparametric submanifolds and polar actions is explored. In Chap. 5, we survey on the theory of polar foliations, which generalizes some of the objects studied in the previous chapter. Finally, Chap. 6 briefly discusses basic aspects of homogeneous spaces and builds on all the previous material to explore the geometry of low cohomogeneity actions and its interplay with manifolds with positive (and nonnegative) sectional curvature.

Prerequisites expected from the reader are a good knowledge of advanced calculus and linear algebra, together with rudiments of calculus on manifolds. Nevertheless, a brief review of the main definitions and essential results is given in the Appendix A.

This book can be used for a one-semester graduate course (of around 3 h per week) or an individual study, as it was written to be as self-contained as possible.

Part of the material in Chap. 3, as well as Chaps. 5 and 6, may be skipped by students in a first reading. Most sections in the book are illustrated with several examples, designed to convey a geometric intuition on the material. These are complemented by exercises that are usually accompanied by a hint. Some exercises are labeled with a star ($\star$), indicating that they are slightly more involved than the others. We encourage the reader to think about them, in an effort to develop a good working knowledge of the material and practice active reading.

The present book evolved from several lecture notes that we used to teach graduate courses and minicourses. In 2007, 2009, and 2010, graduate courses on Lie groups and proper actions were taught at the University of São Paulo, Brazil, exploring mostly the first four chapters of the text. Graduate students working in various fields followed these courses, with very positive results. During this period, the same material was used in a graduate course at the University of Parma, Italy. Relevant contributions also originated from short courses given by the authors during the XV Brazilian School of Differential Geometry (Fortaleza, Brazil, July 2008), the Second São Paulo Geometry Meeting (São Carlos, Brazil, February 2009), and the Rey Pastor Seminar at the University of Murcia (Murcia, Spain, July 2009). In 2009, a preliminary draft of this text was posted on the arXiv (0901.2374), which prompted instructors in various universities to list it as complementary study material. Since then, we have substantially improved and updated the text, particularly the last chapters, featuring many recent advances in the research areas discussed.

There are several important research areas related to the content of this book that are not treated here. We would like to point out two of these, for which we hope to give the necessary background: first, *representation theory* and *harmonic analysis*, for which we recommend Bröken and Tom Dieck [56], Deitmar [78], Fulton and Harris [90], Gangolli and Varadarajan [94], Helgason [125], Katznelson [136], Knapp [144], and Varadarajan [217], and, second, *symmetries in differential equations* and *integrable systems*, for which we recommend Bryant [57], Fehér and Pusztai [86, 87], Guest [119], Noumi [175], and Olver [176].

Acknowledgements We acknowledge financial support from CNPq and FAPESP (Brazil) and NSF (USA). The authors are very grateful to Alexander Lytchak, Gudlaugur Thorbergsson, Karsten Grove, Paolo Piccione, and Wolfgang Ziller for their constant support and several valuable discussions. It is also a pleasure to thank Augusto Ritter Stoffel, Daniel Victor Tausk, Dirk Töben, Fábio Simas, Flausino Lucas Spindola, Francisco Carlos Junior, Ion Moutinho, Ivan Struchiner, John Harvey, László Fehér, Leandro Lichtenfelz, Leonardo Biliotti, Martin Kerin, Martin Weilandt, Rafael Briquet, Renée Abib, and Ricardo Mendes for their helpful comments and suggestions during the elaboration of this book.

São Paulo   Marcos M. Alexandrino
Philadelphia   Renato G. Bettiol
June 2015

Contents

Part I Lie Groups

1 Basic Results on Lie Groups 3
   1.1 Lie Groups and Lie Algebras 3
   1.2 Lie Subgroups and Lie Homomorphisms 7
   1.3 Exponential Map and Adjoint Representation 13
   1.4 Closed Subgroups and More Examples 18

2 Lie Groups with Bi-invariant Metrics 27
   2.1 Basic Facts of Riemannian Geometry 27
   2.2 Bi-invariant Metrics 38
   2.3 Killing Form and Semisimple Lie Algebras 41
   2.4 Splitting Lie Groups with Bi-invariant Metrics 45

Part II Isometric Actions

3 Proper and Isometric Actions 51
   3.1 Proper Actions and Fiber Bundles 51
   3.2 Slices and Tubular Neighborhoods 64
   3.3 Isometric Actions 69
   3.4 Principal Orbits 73
   3.5 Orbit Types 76

4 Adjoint and Conjugation Actions 85
   4.1 Maximal Tori and Polar Actions 85
   4.2 Normal Slices of Conjugation Actions 92
   4.3 Roots of a Compact Lie Group 93
   4.4 Weyl Group 99
   4.5 Dynkin Diagrams 102

5 Polar Foliations 109
   5.1 Definitions and First Examples 109
   5.2 Holonomy and Orbifolds 111

5.3   Surgery and Suspension of Homomorphisms 116
5.4   Differential and Geometric Aspects of Polar Foliations 117
5.5   Transnormal and Isoparametric Maps 126
5.6   Perspectives 135

6 Low Cohomogeneity Actions and Positive Curvature 139
6.1   Cheeger Deformation 139
6.2   Compact Homogeneous Spaces 145
6.3   Cohomogeneity One Actions 157
6.4   Positive and Nonnegative Curvature via Symmetries................. 171

A Rudiments of Smooth Manifolds 185
A.1   Smooth Manifolds.. 185
A.2   Vector Fields... 187
A.3   Foliations and the Frobenius Theorem 190
A.4   Differential Forms, Integration, and de Rham Cohomology 192

References.. 199

Index.. 209

Part I
Lie Groups

Chapter 1
Basic Results on Lie Groups

This chapter gives an introduction to Lie group theory, presenting the main concepts and giving detailed proofs of basic results. Some knowledge of group theory, linear algebra and advanced calculus is assumed. However, as a service to the reader, a few facts about differentiable manifolds are recalled in Appendix A, which can be used as preliminary reading.

The following references, which inspire our approach to the subject, can be used for further reading material on the contents of this chapter: Bump [58], Duistermaat and Kolk [79], Fegan [85], Gilmore [97], Gorbatsevich et al. [88, 98–100], Helgason [126], Hilgert and Neeb [128], Hsiang [129], Knapp [145], Onishchik [179], Spivak [198], Varadarajan [216] and Warner [227].

1.1 Lie Groups and Lie Algebras

Definition 1.1. A smooth (respectively, analytic) manifold G is said to be a *smooth* (respectively, *analytic*) *Lie group*[1] if G is a group and the maps

$$G \times G \ni (x,y) \longmapsto xy \in G \tag{1.1}$$

$$G \ni x \longmapsto x^{-1} \in G \tag{1.2}$$

are smooth (respectively, analytic).

---

[1] Sophus Lie was a nineteenth century Norwegian mathematician, who laid the foundations of continuous symmetry groups and transformation groups.

© Springer International Publishing Switzerland 2015
M.M. Alexandrino, R.G. Bettiol, *Lie Groups and Geometric Aspects of Isometric Actions*, DOI 10.1007/978-3-319-16613-1_1

Remark 1.2. The requirements that (1.1) and (1.2) be smooth, or analytic, often appear as a requirement that the map $G \times G \ni (x, y) \longmapsto xy^{-1} \in G$ be smooth, or analytic. It is easy to prove that these conditions are equivalent. See also Exercise 1.50.

In this book, we only deal with *smooth* Lie groups. Nevertheless, we mention the following important result, which is explored in detail in Duistermaat and Kolk [79].

Theorem 1.3. *Each C^2 Lie group admits a unique* analytic *structure, turning G into an analytic Lie group.*

We stress that every result proved in this chapter on (smooth) Lie groups is hence automatically valid on analytic Lie groups. Henceforth, G denotes a smooth Lie group, and the word *smooth* is omitted.

Remark 1.4. Historically, an important problem was to determine if a connected locally Euclidean topological group has a smooth structure. This problem was known as *5th Hilbert's problem*, posed by Hilbert at the International Congress of Mathematicians in 1900, and solved by von Neumann in 1933 in the compact case. Only in 1952 the general case was solved, by Gleason, Montgomery and Zippen [168].

Some of the most basic examples of Lie groups are $(\mathbb{R}^n, +)$, $(S^1, \cdot)$ with group operation $e^{i\theta} \cdot e^{i\eta} = e^{i(\theta+\eta)}$, and the *n*-torus $T^n = S^1 \times \cdots \times S^1$ as a product group.

Example 1.5. More interesting examples are the so-called *classical Lie groups*, which form four families of matrix Lie groups, closely related to symmetries of Euclidean spaces. The term *classical* appeared in 1940 in Weyl's monograph, probably referring to the *classical* geometries in the spirit of Klein's Erlangen program.

We begin with $GL(n, \mathbb{R})$, the *general linear group* of nonsingular (i.e., invertible) $n \times n$ real matrices. Similarly, $GL(n, \mathbb{C})$ and $GL(n, \mathbb{H})$ are respectively the groups of nonsingular $n \times n$ matrices over the complex numbers and the quaternions.[2] Furthermore, the following complete the list of classical Lie groups, where I denotes the identity matrix:

(i) $SL(n, \mathbb{R}) = \{M \in GL(n, \mathbb{R}) : \det M = 1\}$, $SL(n, \mathbb{C})$ and $SL(n, \mathbb{H})$, the *special linear groups*;

(ii) $O(n) = \{M \in GL(n, \mathbb{R}) : M^t M = I\}$, the *orthogonal group*;

(iii) $SO(n) = O(n) \cap SL(n, \mathbb{R})$, the *special orthogonal group*;

---

[2]The *quaternions* form a real normed division algebra, which provides a noncommutative extension of complex (and real) numbers. Quaternions were discovered by Sir William Hamilton, after whom the usual notation is $\mathbb{H}$. The algebra $\mathbb{H}$ is a 4-dimensional real vector space, endowed with the quaternion multiplication. Its orthonormal basis is denoted $(1, i, j, k)$, and the product of basis elements is given by $i^2 = j^2 = k^2 = ijk = -1$. Any elements of $\mathbb{H}$ are of the form $a + bi + cj + dk \in \mathbb{H}$, and their product is determined by the equations above and the distributive law. Conjugation, norm and division can be defined as natural extensions of those in $\mathbb{C}$.

(iv) $U(n) = \{M \in GL(n,\mathbb{C}) : M^*M = I\}$, the *unitary group*;
(v) $SU(n) = U(n) \cap SL(n,\mathbb{C})$, the *special unitary group*;
(vi) $Sp(n) = \{M \in GL(n,\mathbb{H}) : M^*M = I\}$, the *symplectic group*.

In order to verify that those are indeed Lie groups, see Exercise 1.51. For now, we encourage the reader to keep them in mind as important examples of Lie groups.

Remark 1.6. Another class of examples of Lie groups is constructed by quotients of Lie groups by their normal and closed subgroups (see Corollary 3.38). In this class of examples, there are Lie groups that are *not* matrix groups. In fact, consider

$$G := \left\{ \begin{pmatrix} 1 & a & b \\ 0 & 1 & c \\ 0 & 0 & 1 \end{pmatrix} : a,b,c \in \mathbb{R} \right\} \quad \text{and} \quad N := \left\{ \begin{pmatrix} 1 & 0 & n \\ 0 & 1 & 0 \\ 0 & 0 & 1 \end{pmatrix} : n \in \mathbb{Z} \right\}.$$

Then G/N is a Lie group. It is possible to prove that there are no injective homomorphisms $\varphi \colon G/N \to \mathrm{Aut}(V)$, where $\mathrm{Aut}(V)$ denotes the group of linear automorphisms of a finite-dimensional vector space V (see Carter, Segal and MacDonald [66]).

Definition 1.7. A *Lie algebra* $\mathfrak{g}$ is a real vector space endowed with a bilinear map $[\cdot,\cdot] \colon \mathfrak{g} \times \mathfrak{g} \to \mathfrak{g}$, called the *Lie bracket*, satisfying for all $X,Y,Z \in \mathfrak{g}$,

(i) *Skew-symmetry*: $[X,Y] = -[Y,X]$;
(ii) *Jacobi identity*: $[[X,Y],Z] + [[Y,Z],X] + [[Z,X],Y] = 0$.

Example 1.8. Basic examples of Lie algebras are the vector spaces of $n \times n$ square matrices over $\mathbb{R}$ and $\mathbb{C}$, respectively denoted $\mathfrak{gl}(n,\mathbb{R})$ and $\mathfrak{gl}(n,\mathbb{C})$, endowed with the Lie bracket given by the matrix commutator $[A,B] = AB - BA$.

Exercise 1.9. Let $\mathfrak{so}(3) = \{A \in \mathfrak{gl}(3,\mathbb{R}) : A^t + A = 0\}$.

(i) Verify $\mathfrak{so}(3)$ is a Lie algebra, with Lie bracket given by the matrix commutator;
(ii) Let $A_X = \begin{pmatrix} 0 & -x_3 & x_2 \\ x_3 & 0 & -x_1 \\ -x_2 & x_1 & 0 \end{pmatrix}$. Prove that $A_X v = X \times v$.
(iii) Verify that $A_{X \times Y} = [A_X, A_Y] = A_X A_Y - A_Y A_X$. Using the fact that $(\mathbb{R}^3, \times)$ is a Lie algebra endowed with the cross product of vectors, conclude that the map $(\mathbb{R}^3, \times) \ni X \mapsto A_X \in \mathfrak{so}(3)$ is a *Lie algebra isomorphism*, i.e., a linear isomorphism that preserves Lie brackets.

We now proceed to associate to each Lie group a Lie algebra, by considering left-invariant vector fields. For each $g \in G$, denote by L_g and R_g the *left* and *right translation* maps on G, that is,

$$L_g(x) := gx \quad \text{and} \quad R_g(x) := xg. \tag{1.3}$$

These smooth maps are easily verified to be diffeomorphisms, since they have simultaneous left and right smooth inverses $L_{g^{-1}}$ and $R_{g^{-1}}$, respectively.

A (not necessarily smooth) vector field X on G is said to be *left-invariant* if X is L_g-related to itself for all $g \in G$, i.e., $\mathrm{d}L_g \circ X = X \circ L_g$. This means that $X(gh) = \mathrm{d}(L_g)_h X(h)$, or shortly $X = \mathrm{d}L_g X$, for all $g \in G$. Similarly, a vector field is *right-invariant* if it is R_g-related to itself for all $g \in G$, meaning that $X = \mathrm{d}R_g X$ for all $g \in G$. A simultaneously left- and right-invariant vector field is said to be *bi-invariant*.

Lemma 1.10. *Left-invariant vector fields are smooth.*

Proof. Let X be a left-invariant vector field on G, and consider the group operation

$$\mu: G \times G \to G, \quad \mu(g,h) = gh.$$

Differentiating μ, we obtain $\mathrm{d}\mu: T(G \times G) \cong TG \times TG \to TG$, a smooth map. Define $s: G \to TG \times TG$ by $s(g) := (0_g, X(e))$, where $g \mapsto 0_g$ is the null section of TG. Since $X = \mathrm{d}\mu \circ s$, it follows that X is smooth. $\square$

Remark 1.11. Clearly, the above result also holds for *right-invariant* vector fields.

Given two Lie algebras $\mathfrak{g}_1$ and $\mathfrak{g}_2$, a linear map $\psi: \mathfrak{g}_1 \to \mathfrak{g}_2$ is called *Lie algebra homomorphism* if

$$[\psi(X), \psi(Y)] = \psi([X,Y]), \quad \text{for all } X, Y \in \mathfrak{g}_1.$$

Theorem 1.12. *Let $\mathfrak{g}$ be the set of left-invariant vector fields on the Lie group G. Then the following hold:*

(i) $\mathfrak{g}$ is a Lie algebra, endowed with the Lie bracket of vector fields;
(ii) Consider the tangent space T_eG with the bracket defined as follows. If $X^1, X^2 \in T_eG$, set $[X^1, X^2] := \widetilde{[X^1, X^2]}_e$ where $\widetilde{X_g^i} = \mathrm{d}(L_g)_e X^i$. Define $\psi: \mathfrak{g} \to G$ by $\psi(X) := X_e$. Then ψ is a Lie algebra isomorphism, where $\mathfrak{g}$ is endowed with the Lie bracket of vector fields and T_eG with the bracket defined above.

Proof. First, note that $\mathfrak{g}$ has a (real) vector space structure, by the linearity of $\mathrm{d}(L_g)_e$. It is not difficult to see that the Lie bracket of vector fields is a Lie bracket, i.e., it is skew-symmetric and satisfies the Jacobi identity. Equation (A.3) implies that the Lie bracket of left-invariant vector fields is still left-invariant. Hence $\mathfrak{g}$ is a Lie algebra, proving (i).

To prove (ii), we first claim that ψ is injective. Indeed, if $\psi(X) = \psi(Y)$, for each $g \in G$, $X(g) = \mathrm{d}L_g(X(e)) = \mathrm{d}L_g(Y(e)) = Y(g)$. Furthermore, it is also surjective, since for each $v \in T_eG$, $X(g) := \mathrm{d}L_g(v)$ is clearly left-invariant and satisfies $\psi(X) = v$. Therefore, ψ is a linear bijection between two vector spaces, hence an isomorphism. From the definition of Lie bracket on T_eG, we have $[\psi(X), \psi(Y)] = [X,Y]_e = \psi([X,Y])$. Thus ψ is a Lie algebra isomorphism. $\square$

Definition 1.13. The *Lie algebra of the Lie group G* is the Lie algebra $\mathfrak{g}$ of left-invariant vector fields on G.

According to the above theorem, $\mathfrak{g}$ could be equivalently defined as the tangent space T_eG, with the bracket defined as in (ii). In this way, a Lie group G gives rise to a canonically determined Lie algebra $\mathfrak{g}$. A converse is given by the following foundational result, see e.g., Duistermaat and Kolk [79].

Lie's Third Theorem 1.14. *Let $\mathfrak{g}$ be a (finite-dimensional) Lie algebra. There exists a unique connected and simply-connected Lie group G with Lie algebra isomorphic to $\mathfrak{g}$.*

We do not prove the existence of G, but its uniqueness follows from Corollary 1.27 below. We conclude this section with an exercise that complements Exercise 1.9, verifying that $\mathfrak{so}(3)$ is the Lie algebra of SO(3).

Exercise 1.15. Assume the following result to be seen in Exercise 1.38: *The tangent space at the identity to a Lie subgroup of $\mathrm{GL}(n,\mathbb{R})$, endowed with the matrix commutator, is isomorphic to its Lie algebra.* Consider $\mathscr{S} \subset \mathrm{GL}(3,\mathbb{R})$ the subset of symmetric matrices, and define $\varphi \colon \mathrm{GL}(3,\mathbb{R}) \to \mathscr{S}$, $\varphi(A) = AA^{\mathrm{t}}$.

(i) Verify that the kernel of $d\varphi(I) \colon \mathfrak{gl}(3,\mathbb{R}) \to \mathscr{S}$ is the subspace of skew-symmetric matrices in $\mathfrak{gl}(3,\mathbb{R})$;

(ii) Prove that, for all $A \in \mathrm{GL}(3,\mathbb{R})$,

$$\ker d\varphi(A) = \left\{ H \in \mathfrak{gl}(3,\mathbb{R}) : A^{-1}H \in \ker d\varphi(I) \right\}.$$

Conclude that $\dim \ker d\varphi(A) = 3$, for all $A \in \mathrm{GL}(3,\mathbb{R})$;

(iii) Prove that the identity matrix I is a regular value of φ;

(iv) Conclude that the Lie group O(3) may be written as $\varphi^{-1}(I)$, and compute its dimension.

(v) Recall that SO(3) is the subgroup of O(3) determined by the connected component of the identity I. Prove that $T_I\mathrm{SO}(3) = \ker d\varphi(I) = \mathfrak{so}(3)$. Finally, conclude that $\mathfrak{so}(3)$ is the Lie algebra of SO(3).

Analogous statements hold for SO(n) and $\mathfrak{so}(n)$, with identical proofs.

1.2 Lie Subgroups and Lie Homomorphisms

The aim of this section is to establish the basic relations between Lie algebras and Lie groups, together with their natural maps and subobjects.

A group homomorphism between Lie groups $\varphi \colon G_1 \to G_2$ is called a *Lie group homomorphism* if it is smooth. In Corollary 1.49, we prove that continuity is actually sufficient for a group homomorphism between Lie groups to be smooth. Recall that, given two Lie algebras $\mathfrak{g}_1, \mathfrak{g}_2$, a linear map $\psi \colon \mathfrak{g}_1 \to \mathfrak{g}_2$ is a *Lie algebra homomorphism* if $\psi([X,Y]) = [\psi(X), \psi(Y)]$, for all $X,Y \in \mathfrak{g}_1$. A *Lie subgroup H* of a Lie group G is an abstract subgroup, such that H is an immersed submanifold of G and

$$H \times H \ni (x,y) \longmapsto xy^{-1} \in H \tag{1.4}$$

is smooth. In addition, if $\mathfrak{g}$ is a Lie algebra, a subspace $\mathfrak{h}$ is a *Lie subalgebra* if it is closed with respect to the Lie bracket.

Proposition 1.16. *Let G be a Lie group and $H \subset G$ an embedded submanifold of G that is also a group with respect to the group operation of G. Then H is a closed Lie subgroup of G.*

Proof. Consider the map $f: H \times H \to G$ given by (1.4). Then f is smooth and $f(H \times H) \subset H$. Since H is embedded in G, from Proposition A.7, $f: H \times H \to H$ is smooth. Hence H is a Lie subgroup of G.

It remains to prove that H is a closed subgroup of G. Since H is an embedded submanifold, there exists a neighborhood W of the identity $e \in G$, and a submanifold chart $\varphi = (x_1, \ldots, x_k): W \to \mathbb{R}^{\dim G}$, such that

$$H \cap W = \{g \in G \cap W : x_i(g) = 0, \ i = 1, \ldots \dim H\}. \tag{1.5}$$

Consider a sequence $\{h_n\}$ in H that converges to $h \in G$. Continuity of (1.4) implies the existence of a neighborhood U of $e \in G$ such that $UU^{-1} \subset W$. As h_n converges to h, for n sufficiently large, $h_n h^{-1} \in U$. In particular, $h_n h_m^{-1} \in H \cap W$ for n sufficiently large. From (1.5), this means that $x_i(h_n h_m^{-1}) = 0$ for $i = 1, \ldots, \dim H$. Fixing m and letting $n \to +\infty$, we have that $x_i(h h_m^{-1}) = 0$ for $i = 1, \ldots, \dim H$. Therefore, $h h_m^{-1} \in H \cap W$, hence $h \in H$, concluding the proof. $\qquad\square$

A related result to Proposition 1.16 is proved in Sect. 1.4. We now investigate the nature of the Lie algebra of a Lie subgroup.

Lemma 1.17. *Let G_1 and G_2 be Lie groups and $\varphi: G_1 \to G_2$ be a Lie group homomorphism. Given any left-invariant vector field $X \in \mathfrak{g}_1$, there exists a unique left-invariant vector field $Y \in \mathfrak{g}_2$ that is φ-related to X.*

Proof. First, if $Y \in \mathfrak{g}_2$ is φ-related to X, since φ is a Lie group homomorphism, $Y_e = d\varphi_e X_e$. Hence, uniqueness follows from left-invariance of Y.

Define $Y := d(L_g)_e (d\varphi_e(X_e))$. It remains to prove that Y is φ-related to X. Observing that φ is a Lie group homomorphism, $\varphi \circ L_g = L_{\varphi(g)} \circ \varphi$, for all $g \in G_1$. Therefore, for each $g \in G_1$,

$$d\varphi_g(X_g) = d\varphi_g(d(L_g)_e X_e)$$
$$= d(\varphi \circ L_g)_e X_e$$
$$= d(L_{\varphi(g)} \circ \varphi)_e X_e$$

$$= d(L_{\varphi(g)})_e (d\varphi_e X_e)$$

$$= Y(\varphi(g)).$$

This shows that Y is φ-related to X, completing the proof. □

Proposition 1.18. *Let G_1 and G_2 be Lie groups and $\varphi\colon G_1 \to G_2$ be a Lie group homomorphism. Then $d\varphi_e\colon \mathfrak{g}_1 \to \mathfrak{g}_2$ is a Lie algebra homomorphism.*

Proof. We want to prove that $d\varphi_e\colon (T_e G_1, [\cdot, \cdot]) \to (T_e G_2, [\cdot, \cdot])$ is a Lie algebra homomorphism, where the Lie bracket $[\cdot, \cdot]$ on $T_e G_i$ was defined in Theorem 1.12. For each $X^1, X^2 \in T_e G_1$, define the vectors $Y^i := d\varphi_e X^i \in T_e G_2$ and extend them to left-invariant vector fields $\widetilde{X}^i \in \mathfrak{X}(G_1)$ and $\widetilde{Y}^i \in \mathfrak{X}(G_2)$, by setting $\widetilde{X}^i_g := dL_g X^i$ and $\widetilde{Y}^i_g := dL_g Y^i$.

On the one hand, it follows from the definition of the Lie bracket in $T_e G_2$ that

$$[d\varphi_e X^1, d\varphi_e X^2] = [Y^1, Y^2] = [\widetilde{Y}^1, \widetilde{Y}^2]_e.$$

On the other hand, it follows from Lemma 1.17 that $\widetilde{X}^i$ and $\widetilde{Y}^i$ are φ-related, and from (A.3), $[\widetilde{X}^1, \widetilde{X}^2]$ and $[\widetilde{Y}^1, \widetilde{Y}^2]$ are φ-related. Therefore

$$[\widetilde{Y}^1, \widetilde{Y}^2]_e = d\varphi_e [\widetilde{X}^1, \widetilde{X}^2]_e = d\varphi_e [X^1, X^2].$$

The above imply that $[d\varphi_e X^1, d\varphi_e X^2] = d\varphi_e [X^1, X^2]$, concluding the proof. □

Corollary 1.19. *Let G be a Lie group and $H \subset G$ a Lie subgroup. Then the inclusion map $i\colon H \hookrightarrow G$ induces an isomorphism di_e between the Lie algebra $\mathfrak{h}$ of H and a Lie subalgebra $di_e(\mathfrak{h})$ of $\mathfrak{g}$.*

A converse result is given below, on conditions under which a Lie subalgebra gives rise to a Lie subgroup. Before that, however, we need a result on how an open neighborhood of the identity *generates* the entire Lie group.

Proposition 1.20. *Let G be a Lie group and G_0 be the connected component of G containing the identity $e \in G$. Then G_0 is a normal Lie subgroup of G and connected components of G are of the form gG_0, for some $g \in G$. Moreover, for any open neighborhood U of e, $G_0 = \bigcup_{n \in \mathbb{N}} U^n$, where $U^n := \{g_1^{\pm 1} \cdots g_n^{\pm 1} : g_i \in U\}$.*

Proof. Since G_0 is a connected component, it is an open and closed subset of G. In order to verify that it is also a Lie subgroup, let $g_0 \in G_0$ and consider $g_0 G_0 = L_{g_0}(G_0)$. Note that $g_0 G_0$ is a connected component of G, as L_{g_0} is a diffeomorphism. Since $g_0 \in G_0 \cap g_0 G_0$, it follows from the maximality of the connected component that $g_0 G_0 = G_0$. Similarly, as the inversion map is also a diffeomorphism, the subset $G_0^{-1} := \{g_0^{-1} : g_0 \in G_0\}$ is connected, with $e \in G_0^{-1}$. Hence, $G_0^{-1} = G_0$, using the same argument. Therefore, G_0 is a subgroup of G and an embedded submanifold of G. From Proposition 1.16, it follows that G_0 is a Lie subgroup of G.

For each $g \in G$, consider the diffeomorphism given by conjugation $x \mapsto gxg^{-1}$. Using the same argument of maximality of the connected component, one may conclude that $gG_0g^{-1} = G_0$ for all $g \in G$, hence G_0 is normal. The proof that gG_0 is the connected component of G containing g is completely analogous.

Finally, since G_0 connected, to show that $G_0 = \bigcup_{n \in \mathbb{N}} U^n$ it suffices to check that $\bigcup_{n \in \mathbb{N}} U^n$ is open and closed in G_0. It is clearly open, since U (hence U^n) is open. To verify that it is also closed, let $h \in G_0$ be the limit of a sequence $\{h_j\}$ in $\bigcup_{n \in \mathbb{N}} U^n$, i.e., $\lim h_j = h$. Since $U^{-1} = \{u^{-1} : u \in U\}$ is an open neighborhood of $e \in G$, hU^{-1} is an open neighborhood of h. From the convergence of the sequence $\{h_j\}$, there exists $j_0 \in \mathbb{N}$ such that $h_{j_0} \in hU^{-1}$, that is, there exists $u \in U$ such that $h_{j_0} = hu^{-1}$. Hence $h = h_{j_0} u \in \bigcup_{n \in \mathbb{N}} U^n$. Therefore, this set is closed in G_0.   □

Theorem 1.21. *Let G be a Lie group with Lie algebra $\mathfrak{g}$, and $\mathfrak{h}$ be a Lie subalgebra of $\mathfrak{g}$. There exists a unique connected Lie subgroup $H \subset G$ with Lie algebra $\mathfrak{h}$.*

Proof. Define the distribution $D_q = \{X_q : X_q = dL_q X \text{ for } X \in \mathfrak{h}\}$. Since $\mathfrak{h}$ is a Lie algebra, D is involutive. It follows from the Frobenius Theorem (see Theorem A.19) that there exists a unique foliation $\mathscr{F} = \{F_q : q \in G\}$ tangent to this distribution, i.e., $D_q = T_q F_q$, for all $q \in G$. Define $H := F_e$ to be the leaf passing through the identity.

Note that L_g maps leaves to leaves, since $dL_g D_a = D_{ga}$. Furthermore, for each $h \in H$, $L_{h^{-1}}(H)$ is the leaf passing through the identity. Therefore $L_{h^{-1}}(H) = H$, which means that H is a group. It also follows from the Frobenius Theorem that H is quasi-embedded. Consider $\psi : H \times H \ni (x,y) \mapsto x^{-1}y \in H$. Since the inclusion $i : H \hookrightarrow G$ and $i \circ \psi$ are smooth, ψ is also smooth. Thus, H is a connected Lie subgroup of G with Lie algebra $\mathfrak{h}$. Uniqueness follows from the Frobenius Theorem and Proposition 1.20.   □

Definition 1.22. A smooth surjective map $\pi : E \to B$ is a *covering* map if, for each $p \in B$, there exists an open neighborhood U of p such that $\pi^{-1}(U)$ is a disjoint union of open sets $U_\alpha \subset E$ mapped diffeomorphically onto U by π. In other words, $\pi|_{U_\alpha} : U_\alpha \to U$ is a diffeomorphism for each α.

Theorem 1.23. *Let G be a connected Lie group. There exist a unique simply-connected Lie group $\widetilde{G}$ and a Lie group homomorphism $\pi : \widetilde{G} \to G$, which is also a covering map.*

A proof of this theorem can be found in Boothby [46] or Duistermaat and Kolk [79]. An example of covering map that is also a Lie group homomorphism is the usual covering $\pi : \mathbb{R}^n \to T^n$ of the n-torus by Euclidean space. Other examples of Lie group coverings are discussed in Exercise 1.55 and Remark 1.56.

Proposition 1.24. *Let G_1 and G_2 be connected Lie groups and $\pi : G_1 \to G_2$ be a Lie group homomorphism. Then π is a covering map if and only if $d\pi_e$ is an isomorphism.*

Proof. Suppose that $d\pi_{e_1} : T_{e_1} G_1 \to T_{e_2} G_2$ is an isomorphism, where $e_i \in G_i$ denotes the identity element of G_i. We claim that π is surjective.

Indeed, since $d\pi_{e_1}$ is an isomorphism, from the Inverse Function Theorem, there exist open neighborhoods U of the identity $e_1 \in G_1$ and V of the identity $e_2 \in G_2$, such that $\pi(U) = V$ and $\pi|_U$ is a diffeomorphism. Let $h \in G_2$. From Proposition 1.20, there exist $h_i \in V$ such that $h = h_1^{\pm 1} \cdots h_n^{\pm 1}$. Since $\pi|_U$ is a diffeomorphism, for each $1 \le i \le n$, there exists a unique $g_i \in U$ such that $\pi(g_i^{\pm 1}) = h_i^{\pm 1}$. Thus,

$$\pi(g_1^{\pm 1} \cdots g_n^{\pm 1}) = \pi(g_1^{\pm 1}) \cdots \pi(g_n^{\pm 1}) = h_1^{\pm 1} \cdots h_n^{\pm 1} = h.$$

Therefore π is a surjective Lie group homomorphism.

Let $\{g_\alpha\} = \pi^{-1}(e_2)$. Using the fact that π is a Lie group homomorphism and $d\pi_{e_1}$ is an isomorphism, one can prove that $\{g_\alpha\}$ is discrete. Thus, π is a covering map, since:

(i) $\pi^{-1}(qV) = \bigcup_\alpha g_\alpha pU$;
(ii) $(g_\alpha pU) \cap (pU) = \emptyset$, if $g_\alpha \ne e_1$;
(iii) $\pi|_{g_\alpha pU} : g_\alpha pU \to qV$ is a diffeomorphism.

Conversely, if π is a covering map, it is locally a diffeomorphism, hence $d\pi_{e_1}$ is an isomorphism. □

Let G_1 and G_2 be Lie groups and $\theta : \mathfrak{g}_1 \to \mathfrak{g}_2$ be a Lie algebra homomorphism. We prove that if G_1 is connected and simply-connected, then θ induces a Lie group homomorphism. We begin by proving uniqueness in the following lemma.

Lemma 1.25. *Let G_1 and G_2 be Lie groups, with identities e_1 and e_2 respectively, and $\theta : \mathfrak{g}_1 \to \mathfrak{g}_2$ be a fixed Lie algebra homomorphism. If G_1 is connected, and $\varphi, \psi : G_1 \to G_2$ are Lie group homomorphisms with $d\varphi_{e_1} = d\psi_{e_1} = \theta$, then $\varphi = \psi$.*

Proof. It is easy to see that the direct sum $\mathfrak{g}_1 \oplus \mathfrak{g}_2$ of Lie algebras has a natural Lie algebra structure, and the direct product $G_1 \times G_2$ of Lie groups has a natural Lie group structure. Consider the Lie subalgebra of $\mathfrak{g}_1 \oplus \mathfrak{g}_2$ given by

$$\mathfrak{h} := \{(X, \theta(X)) : X \in \mathfrak{g}_1\}, \tag{1.6}$$

i.e., the graph of $\theta : \mathfrak{g}_1 \to \mathfrak{g}_2$. It follows from Theorem 1.21 that there exists a unique connected Lie subgroup $H \subset G_1 \times G_2$ with Lie algebra $\mathfrak{h}$.

Suppose that $\varphi : G_1 \to G_2$ is a Lie group homomorphism with $d\varphi_{e_1} = \theta$. Then

$$\sigma : G_1 \to G_1 \times G_2, \quad \sigma(g) := (g, \varphi(g)),$$

is a Lie group homomorphism, and

$$d\sigma_{e_1} : \mathfrak{g}_1 \to \mathfrak{g}_1 \oplus \mathfrak{g}_2, \quad d\sigma_{e_1} X = (X, \theta(X))$$

is a Lie algebra homomorphism. Note that $\sigma(G_1)$ is the graph of φ, hence embedded in $G_1 \times G_2$. From Proposition 1.16, $\sigma(G_1)$ is a Lie subgroup of $G_1 \times G_2$, with Lie algebra $\mathfrak{h} = d\sigma_{e_1}(\mathfrak{g}_1)$. Therefore, from Theorem 1.21, $\sigma(G_1) = H$. In other words, H is the graph of φ. If $\psi: G_1 \to G_2$ is another Lie group homomorphism with $d\psi_{e_1} = \theta$, following the same construction above, the graphs of ψ and φ would both be equal to H, hence $\varphi = \psi$. $\square$

Theorem 1.26. *Let G_1 and G_2 be Lie groups and $\theta: \mathfrak{g}_1 \to \mathfrak{g}_2$ be a Lie algebra homomorphism. There exist an open neighborhood V of e_1 and a smooth map $\varphi: V \to G_2$ that is a local homomorphism[3] with $d\varphi_{e_1} = \theta$. In addition, if G_1 is connected and simply-connected, there exists a unique Lie group homomorphism $\varphi: G_1 \to G_2$ with $d\varphi_{e_1} = \theta$.*

Proof. Let $\mathfrak{h}$ be defined as in (1.6). From Theorem 1.21, there exists a unique connected Lie subgroup $H \subset G_1 \times G_2$ with Lie algebra $\mathfrak{h}$, whose identity element we denote $\widetilde{e} \in H$. Consider the inclusion $i: H \hookrightarrow G_1 \times G_2$ and the projections $\pi_j: G_1 \times G_2 \to G_j$, $j = 1,2$. The map $(\pi_1 \circ i): H \to G_1$ is a Lie group homomorphism, such that $d(\pi_1 \circ i)_{\widetilde{e}}(X, \theta(X)) = X$, for all $X \in T_{e_1}G_1$. It follows from the Inverse Function Theorem that there exist open neighborhoods U of $\widetilde{e}$ in H and V of e_1 in G_1, such that $(\pi_1 \circ i)|_U: U \to V$ is a diffeomorphism.

Define $\varphi = \pi_2 \circ (\pi_1 \circ i)^{-1}: V \to G_2$. Then φ is a local homomorphism with $d\varphi_{e_1} = \theta$. In fact, for each $X \in T_{e_1}G_1$,

$$d\varphi_{e_1}(X) = d\left(\pi_2 \circ (\pi_1 \circ i)^{-1}\right)_{e_1}(X)$$

$$= d(\pi_2)_{\widetilde{e}}\, d\left((\pi_1 \circ i)^{-1}\right)_{e_1} X$$

$$= d(\pi_2)_{\widetilde{e}}(X, \theta(X))$$

$$= \theta(X).$$

Furthermore, $(\pi_1 \circ i)$ is a Lie group homomorphism and $d(\pi_1 \circ i)_{\widetilde{e}}$ is an isomorphism. From Proposition 1.24, $(\pi_1 \circ i): H \to G_1$ is a covering map. Assuming G_1 simply-connected, it follows that $(\pi_1 \circ i)$ is a diffeomorphism, since a covering map onto a simply-connected space is a diffeomorphism. Thus, $(\pi_1 \circ i)$ can be globally inverted, and we obtain a global homomorphism $\varphi = \pi_2 \circ (\pi_1 \circ i)^{-1}: G_1 \to G_2$ with $d\varphi_{e_1} = \theta$. The uniqueness of φ follows from Lemma 1.25, under the assumption that G_1 is simply-connected. $\square$

Corollary 1.27. *If G_1 and G_2 are connected and simply-connected and $\theta: \mathfrak{g}_1 \to \mathfrak{g}_2$ is an isomorphism, there exists a unique Lie group isomorphism $\varphi: G_1 \to G_2$ with $d\varphi_{e_1} = \theta$. In other words, if G_1 and G_2 are as above, then G_1 and G_2 are isomorphic if and only if $\mathfrak{g}_1$ and $\mathfrak{g}_2$ are isomorphic.*

Proof. By Theorem 1.26, there exists a unique Lie group homomorphism $\varphi: G_1 \to G_2$ with $d\varphi_{e_1} = \theta$. By Proposition 1.24, since $d\varphi_{e_1} = \theta$ is an isomorphism, φ is a covering map. Since G_2 is simply-connected, φ is a diffeomorphism. Thus, φ is a Lie group homomorphism and a diffeomorphism, therefore an isomorphism. $\square$

---

[3]This simply means that $\varphi(ab) = \varphi(a)\varphi(b)$, for all $a, b \in V$ such that $ab \in V$.

1.3 Exponential Map and Adjoint Representation

Let G be a Lie group and $\mathfrak{g}$ its Lie algebra. We recall that a Lie group homomorphism $\varphi \colon \mathbb{R} \to G$ is called a 1-*parameter subgroup* of G. Let $X \in \mathfrak{g}$ and consider the Lie algebra homomorphism

$$\theta \colon \mathbb{R} \to \mathfrak{g}, \quad \theta(t) := tX.$$

From Theorems 1.21 and 1.26, there is a unique 1-parameter subgroup $\lambda_X \colon \mathbb{R} \to G$, such that $\lambda_X'(0) = X$.

Remark 1.28. Note that λ_X is an integral curve of the left-invariant vector field X passing through e. In fact,

$$\lambda_X'(t) = \tfrac{\mathrm{d}}{\mathrm{d}s}\lambda_X(t+s)\big|_{s=0} = \mathrm{d}L_{\lambda_X(t)}\lambda_X'(0) = \mathrm{d}L_{\lambda_X(t)}X = X(\lambda_X(t)).$$

Definition 1.29. The *(Lie) exponential map* of G is defined as

$$\exp \colon \mathfrak{g} \to G, \quad \exp(X) := \lambda_X(1),$$

where λ_X is the unique 1-parameter subgroup of G such that $\lambda_X'(0) = X$.

Proposition 1.30. *The exponential map satisfies the following properties, for all* $X \in \mathfrak{g}$ *and* $t \in \mathbb{R}$,

  (i) $\exp(tX) = \lambda_X(t)$;
  (ii) $\exp(-tX) = \exp(tX)^{-1}$;
  (iii) $\exp(t_1 X + t_2 X) = \exp(t_1 X)\exp(t_2 X)$;
  (iv) $\exp \colon T_e G \to G$ *is smooth and* $\mathrm{d}(\exp)_0 = \mathrm{id}$, *hence* $\exp$ *is a diffeomorphism from an open neighborhood of the origin of* $T_e G$ *onto an open neighborhood of the identity* $e \in G$.

Proof. We claim that $\lambda_X(t) = \lambda_{tX}(1)$. Consider the 1-parameter subgroup $\lambda(s) = \lambda_X(st)$. Differentiating at $s = 0$, it follows that

$$\lambda'(0) = \tfrac{\mathrm{d}}{\mathrm{d}s}\lambda_X(st)\big|_{s=0} = t\lambda_X'(0) = tX.$$

Hence, from uniqueness of the 1-parameter subgroup in Definition 1.29, $\lambda_X(st) = \lambda_{tX}(s)$. Choosing $s = 1$, we obtain the expression in (i). Items (ii) and (iii) are immediate consequences of (i), since λ_X is a Lie group homomorphism.

In order to prove item (iv), we construct a vector field V on the tangent bundle TG. This tangent bundle can be identified with $G \times T_e G$, since G is parallelizable. The construction is such that the projection of the integral curve of V passing through (e, X) coincides with the curve $t \mapsto \exp(tX)$. Thus, from Theorem A.14, the flow of V depends smoothly on the initial conditions, hence its projection (i.e., the exponential map) is also smooth.

Consider $G \times T_e G \simeq TG$. Note that for all $(g, X) \in G \times T_e G$, the tangent space $T_{(g,X)}(G \times T_e G)$ can be identified with $T_g G \oplus T_e G$. Define the vector field

$$V \in \mathfrak{X}(G \times T_e G), \quad V(g, X) := \widetilde{X}(g) \oplus \{0\} \in T_g G \oplus T_e G,$$

where $\widetilde{X}(g) = dL_g X$. It is not difficult to see that V is a smooth vector field. Since $t \mapsto \exp(tX)$ is the unique integral curve of $\widetilde{X}$ for which $\lambda_X(0) = e$, as $\widetilde{X}$ is left-invariant, $L_g \circ \lambda_X$ is the unique integral curve of $\widetilde{X}$ that takes value g at $t = 0$. Hence, the integral curve of V through (g, X) is $t \mapsto (g \exp(tX), X)$. In other words, the flow of V is given by $\varphi_t^V(g, X) = (g \exp(tX), X)$ and, in particular, V is complete. Let $\pi_1 : G \times T_e G \to G$ be the projection onto G. Then $\exp(X) = \pi_1 \circ \varphi_1^V(e, X)$, so $\exp$ is given by composition of smooth maps and is hence smooth. Finally, (i) and Remark 1.28 imply that $d(\exp)_0 = \mathrm{id}$. $\qquad \square$

The exponential map, in general, may not be surjective. The classical example of this situation is given by $SL(2, \mathbb{R})$, see Duistermaat and Kolk [79].

Considering Lie groups of matrices $GL(n, \mathbb{K})$, for $\mathbb{K} = \mathbb{C}$ or $\mathbb{K} = \mathbb{R}$, it is natural to inquire whether the Lie exponential map $\exp : \mathfrak{gl}(n, \mathbb{K}) \to GL(n, \mathbb{K})$ coincides with the usual exponentiation of matrices, given for each $A \in \mathfrak{gl}(n, \mathbb{K})$ by

$$e^A := \sum_{k=0}^{\infty} \frac{A^k}{k!}. \tag{1.7}$$

We now prove that these notions indeed coincide.

To this aim, we recall two well-known properties of the exponentiation of matrices. First, the right-hand side of (1.7) converges uniformly for A in a bounded region of $\mathfrak{gl}(n, \mathbb{K})$. This can be easily verified using the Weierstrass M-test. In addition, given $A, B \in GL(n, \mathbb{K})$, it is true that $e^{A+B} = e^A e^B$ if and only if A and B commute (this fact is generalized in Remark 1.40).

Consider the map $\mathbb{R} \ni t \mapsto e^{tA} \in GL(n, \mathbb{K})$. Since each entry of e^{tA} is a power series in t with infinite radius of convergence, it follows that this map is smooth. Differentiating the power series term by term, it is easy to see that its tangent vector at the origin of $\mathfrak{gl}(n, \mathbb{K})$ is A, and from the properties above, this map is also a Lie group homomorphism, hence a 1-parameter subgroup of $GL(n, \mathbb{K})$. Since $\exp(A)$ is the *unique* 1-parameter subgroup of $GL(n, \mathbb{K})$ whose tangent vector at the origin is A, it follows that $e^A = \exp(A)$, for all $A \in \mathfrak{gl}(n, \mathbb{K})$.

Remark 1.31. The above holds, more generally, for the exponential of endomorphisms of any real or complex vector space V. Namely, $\exp : \mathrm{End}(V) \to \mathrm{Aut}(V)$ is given by the exponentiation of endomorphisms, defined exactly as in (1.7).

Proposition 1.32. *Let G_1 and G_2 be Lie groups and $\varphi : G_1 \to G_2$ a Lie group homomorphism. Then $\varphi \circ \exp^1 = \exp^2 \circ d\varphi_e$, i.e., the following diagram commutes:*

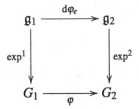

Proof. Consider the 1-parameter subgroups of G_2 given by $\alpha(t) = \varphi \circ \exp^1(tX)$ and $\beta(t) = \exp^2 \circ d\varphi_e(tX)$. Then $\alpha'(0) = \beta'(0) = d\varphi_e X$, hence, it follows from Theorem 1.26 that $\alpha = \beta$, that is, the diagram above commutes. $\qquad\square$

Remark 1.33. Using Proposition 1.32, we now show that if H is a Lie subgroup of G, then the exponential map $\exp^H$ of H coincides with the restriction to H of the exponential map $\exp^G$ of G. Consider the inclusion $i \colon H \hookrightarrow G$, which is a Lie group homomorphism, and its differential $di_e \colon \mathfrak{h} \hookrightarrow \mathfrak{g}$. According to Corollary 1.19, this is an isomorphism between the Lie algebra $\mathfrak{h}$ and a Lie subalgebra of $\mathfrak{g}$. Then the following diagram is commutative

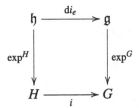

Thus, with the appropriate identifications, $\exp^H(X) = i(\exp^H(X)) = \exp^G(di_e(X)) = \exp^G(X)$, for all $X \in \mathfrak{h}$, which proves the assertion.

We proceed by stating three important identities known as the *Campbell formulas*, or *Campbell-Baker-Hausdorff formulas*. A proof can be found in Spivak [198].

Campbell formulas 1.34. *Let G be a Lie group and $X,Y \in \mathfrak{g}$. There exists $\varepsilon > 0$ such that, for all $|t| < \varepsilon$, the following hold:*

(i) $\exp(tX)\exp(tY) = \exp\left(t(X+Y) + \frac{t^2}{2}[X,Y] + O(t^3)\right)$;
(ii) $\exp(tX)\exp(tY)\exp(-tX) = \exp\left(tY + t^2[X,Y] + O(t^3)\right)$;
(iii) $\exp(-tX)\exp(-tY)\exp(tX)\exp(tY) = \exp\left(t^2[X,Y] + O(t^3)\right)$;

where $\frac{O(t^3)}{t^3}$ is bounded.

We now pass to the second part of this section, discussing some properties of the adjoint representation, or adjoint action.

Definition 1.35. Let G be any group and V a vector space. A *(linear) representation* of G on V is a group homomorphism $\varphi \colon G \to \mathrm{Aut}(V)$, where $\mathrm{Aut}(V)$ is the group of vector space isomorphisms of V.

Consider the action of G on itself by conjugation, i.e.,

$$a: G \times G \to G, \quad a(g,x) := a_g(x) = gxg^{-1}.$$

Definition 1.36. Let G be a Lie group, and $\mathfrak{g}$ its Lie algebra. The representation

$$\text{Ad}: G \to \text{Aut}(\mathfrak{g}), \quad \text{Ad}(g) := d(a_g)_e = d(L_g)_{g^{-1}} \circ d(R_{g^{-1}})_e,$$

is called the *adjoint representation* (or *adjoint action*) of G.

It follows from the above definition that

$$\text{Ad}(g)X = \tfrac{d}{dt}\left(g\exp(tX)g^{-1}\right)\big|_{t=0}. \tag{1.8}$$

Applying Proposition 1.32 to the automorphism a_g, it follows that

$$\exp(t\,\text{Ad}(g)X) = a_g(\exp(tX)) = g\exp(tX)g^{-1}, \tag{1.9}$$

and, in particular, for $t = 1$,

$$g\exp(X)g^{-1} = \exp(\text{Ad}(g)X). \tag{1.10}$$

The differential of the adjoint representation Ad is denoted ad, and given by

$$\text{ad}: \mathfrak{g} \to \text{End}(\mathfrak{g}), \quad \text{ad}(X)Y := d\text{Ad}_e(X)(Y).$$

It follows from the above definition that

$$\text{ad}(X)Y = \tfrac{d}{dt}\left(\text{Ad}(\exp(tX))\,Y\right)\big|_{t=0}. \tag{1.11}$$

Using Proposition 1.32 once more, we have

$$\text{Ad}(\exp(tX)) = \exp(t\,\text{ad}(X))$$

i.e., the following diagram is commutative:

$$
\begin{array}{ccc}
\mathfrak{g} & \xrightarrow{\ \text{ad}\ } & \text{End}(\mathfrak{g}) \\
{\scriptstyle \exp}\big\downarrow & & \big\downarrow{\scriptstyle \exp} \\
G & \xrightarrow[\text{Ad}]{} & \text{Aut}(\mathfrak{g})
\end{array}
$$

In particular, for $t = 1$ we obtain

$$\text{Ad}(\exp(X)) = \exp(\text{ad}(X)) \tag{1.12}$$

Proposition 1.37. *If $X, Y \in \mathfrak{g}$, then $\mathrm{ad}(X)Y = [X, Y]$.*

Proof. From the Campbell formulas (Theorem 1.34), it follows that

$$\exp(tX)\exp(tY)\exp(-tX) = \exp\left(tY + t^2[X,Y] + O(t^3)\right).$$

Using (1.10) with $g = \exp(tX)$, $\exp\left(\mathrm{Ad}(\exp(tX))tY\right) = \exp(tY + t^2[X,Y] + O(t^3))$. Since $\exp$ is locally injective near zero, we have $\mathrm{Ad}(\exp(tX))tY = tY + t^2[X,Y] + O(t^3)$, for sufficiently small t. Dividing by t, differentiating at $t = 0$, and applying (1.11), we conclude that $\mathrm{ad}(X)Y = [X, Y]$. $\qquad\square$

Exercise 1.38. Let G be a Lie subgroup of $\mathrm{GL}(n, \mathbb{R})$. For each $g \in G$ and $X, Y \in \mathfrak{g}$, verify the following properties:

(i) $\mathrm{d}L_g X = gX$ and $\mathrm{d}R_g X = X g$ are given by matrix multiplication;
(ii) $\mathrm{Ad}(g)Y = gY g^{-1}$ is given by matrix conjugation;
(iii) Use Proposition 1.37 to prove that $[X, Y] = XY - YX$ is the matrix commutator.

Let us now prove a result relating commutativity and the Lie bracket. Recall that vector fields $X, Y \in \mathfrak{X}(M)$ are said to *commute* if $[X, Y] = 0$.

Proposition 1.39. *Let G be a connected Lie group with Lie algebra $\mathfrak{g}$. The Lie algebra $\mathfrak{g}$ is abelian if and only if G is abelian.*

Proof. Fix $X, Y \in \mathfrak{g}$ and assume that $[X, Y] = 0$. From (1.12) and Proposition 1.37,

$$\mathrm{Ad}(\exp(X))Y = \exp(\mathrm{ad}(X))Y = \sum_{k=0}^{\infty} \frac{\mathrm{ad}(X)^k}{k!} Y = Y.$$

Thus, from (1.10), $\exp(X)\exp(Y)\exp(-X) = \exp(Y)$, and hence $\exp(X)\exp(Y) = \exp(Y)\exp(X)$. This means that there exists an open neighborhood U of e such that if $g_1, g_2 \in U$, then $g_1 g_2 = g_2 g_1$. It follows from Proposition 1.20, that $G = \bigcup_{n \in \mathbb{N}} U^n$, where $U^n = \{g_1^{\pm 1} \cdots g_n^{\pm 1} : g_i \in U\}$. Therefore G is abelian.[4]

Conversely, suppose G abelian. In particular, for all $s, t \in \mathbb{R}$ and $X, Y \in \mathfrak{g}$, $\exp(sX)\exp(tY)\exp(-sX) = \exp(tY)$. Differentiating at $t = 0$,

$$\tfrac{\mathrm{d}}{\mathrm{d}t}\exp(sX)\exp(tY)\exp(-sX)\big|_{t=0} = Y.$$

From (1.8), it follows that $\mathrm{Ad}(\exp(sX))Y = Y$. Hence, differentiating at $s = 0$, it follows from (1.11) and Proposition 1.37 that $\mathrm{ad}(X)Y = [X, Y] = 0$, which means that $\mathfrak{g}$ is abelian. $\qquad\square$

Remark 1.40. If $X, Y \in \mathfrak{g}$ commute, that is, $[X, Y] = 0$, then

$$\exp(X + Y) = \exp(X)\exp(Y).$$

---

[4]Note that one cannot infer that G is abelian directly from the commutativity of $\exp$, since $\exp$ might not be surjective.

Note that this does not hold in general. To verify this identity, set $\alpha\colon \mathbb{R} \to G$, $\alpha(t) := \exp(tX)\exp(tY)$. From Proposition 1.39, α is a 1-parameter subgroup, and differentiating α at $t = 0$, we have $\alpha'(0) = X + Y$. Hence $\alpha(t) = \exp(t(X+Y))$, and we get the desired equation setting $t = 1$.

We conclude this section with a characterization of connected abelian Lie groups.

Theorem 1.41. *Let G be a connected n-dimensional abelian Lie group. Then G is isomorphic to $T^k \times \mathbb{R}^{n-k}$, where $T^k = S^1 \times \cdots \times S^1$ is a k-torus. In particular, an abelian connected and compact Lie group is isomorphic to a torus.*

Proof. Using Proposition 1.39, since G is abelian, $\mathfrak{g}$ is also abelian. Thus $\mathfrak{g}$ can be identified with $\mathbb{R}^n$. Since G is connected, it follows from Remark 1.40 that $\exp\colon \mathfrak{g} \to G$ is a Lie group homomorphism. From Proposition 1.24, it is a covering map.

Consider the normal subgroup given by $\Gamma = \ker \exp$. We must now use two results whose proof only appears later. The first (Theorem 1.42) asserts that any closed subgroup of a Lie group is a Lie subgroup. As $\exp$ continuous and Γ is closed, Γ is a Lie subgroup of $\mathbb{R}^n$. The second (Corollary 3.38) asserts that the quotient of a Lie group by a normal Lie subgroup is also a Lie group, hence $\mathbb{R}^n/\Gamma$ is a Lie group. Note that G is isomorphic to $\mathbb{R}^n/\Gamma$, because $\exp\colon \mathbb{R}^n \to G$ is a surjective Lie group homomorphism and $\Gamma = \ker \exp$.

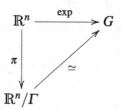

Using the fact that $\exp$ is a covering map, it is possible to prove that the isomorphism between $\mathbb{R}^n/\Gamma$ and G defined above is in fact smooth, i.e., is a Lie group isomorphism (this also follows from Corollary 1.49).

Finally, it is well-known that the only nontrivial discrete subgroups of $\mathbb{R}^n$ are integral lattices. In other words, there exists a positive integer $k \leq n$ and linearly independent vectors $e_1, \ldots, e_k \in \mathbb{R}^n$ such that $\Gamma = \{\sum_{i=1}^{k} n_i e_i : n_i \in \mathbb{Z}\}$. Therefore, G is isomorphic to $\mathbb{R}^n/\Gamma = T^k \times \mathbb{R}^{n-k}$, where $T^k = S^1 \times \cdots \times S^1$ is a k-torus. $\square$

1.4 Closed Subgroups and More Examples

The goal of this section is to prove that closed subgroups of a Lie group are Lie subgroups. This fact is a very useful tool to prove that a subgroup is a *Lie subgroup*. For instance, it applies to the subgroups of $\mathrm{GL}(n, \mathbb{K})$, for $\mathbb{K} = \mathbb{C}$ and $\mathbb{K} = \mathbb{R}$, defined in Sect. 1.1. We then briefly explore some corollaries, and discuss a few important examples that complement the material of this chapter.

Theorem 1.42. *Let G be a Lie group and $H \subset G$ a closed subgroup of G. Then H is an embedded Lie subgroup of G.*

Proof. We prove this result through a sequence of five claims, following closely the approach in Spivak [198, pp. 530–534]. The central idea of the proof is to reconstruct the Lie algebra of H as a Lie subalgebra $\mathfrak{h} \subset \mathfrak{g}$. The natural candidate is

$$\mathfrak{h} := \{X \in T_eG : \exp(tX) \in H, \text{ for all } t \in \mathbb{R}\}.$$

Claim 1.43. *Let $\{X_i\}$ be a sequence in T_eG with $\lim X_i = X$, and $\{t_i\}$ a sequence of real numbers, with $\lim t_i = 0$. If $\exp(t_iX_i) \in H$, for all $i \in \mathbb{N}$, then $X \in \mathfrak{h}$.*

Since $\exp(-t_iX_i) = \big(\exp(t_iX_i)\big)^{-1}$, without loss of generality, assume $t_i > 0$. Define $R_i(t)$ to be the largest integer $\leq \frac{t}{t_i}$. Then $\frac{t}{t_i} - 1 < R_i(t) \leq \frac{t}{t_i}$, so $\lim t_iR_i(t) = t$. Therefore $\lim t_iR_i(t)X_i = tX$. On the one hand, it follows from continuity of exp that $\lim \exp(t_iR_i(t)X_i) = \exp(tX)$. On the other hand, $\exp(t_iR_i(t)X_i) = [\exp(t_iX_i)]^{R_i(t)} \in H$. Since H is closed, $\exp(tX) \in H$. Therefore $X \in \mathfrak{h}$.

Claim 1.44. $\mathfrak{h} \subset T_eG$ *is a vector subspace of $\mathfrak{g}$.*

Let $X, Y \in \mathfrak{h}$. It is clear that $sX \in \mathfrak{h}$ for all $s \in \mathbb{R}$. From the Campbell formulas (Theorem 1.34),

$$\exp\left(t_i(X+Y) + \tfrac{t_i^2}{2}[X,Y] + O(t_i^3)\right) = \exp(t_iX)\exp(t_iY) \in H,$$

so $\exp\left(t_i(X+Y+\tfrac{t_i}{2}[X,Y]+O(t_i^2))\right) \in H$ and $\left(X+Y+\tfrac{t_i}{2}[X,Y]+O(t_i^2)\right)$ converges to $X+Y$ when t_i goes to 0. From Claim 1.43, $X+Y \in \mathfrak{h}$.

Claim 1.45. *Let $\mathfrak{k}$ be a vector space such that $T_eG = \mathfrak{h} \oplus \mathfrak{k}$, and*

$$\psi: \mathfrak{h} \oplus \mathfrak{k} \to G, \quad \psi(X,Y) := \exp(X)\exp(Y).$$

There exists an open neighborhood U of the origin $(0,0) \in \mathfrak{h} \oplus \mathfrak{k}$, such that $\psi|_U$ is a diffeomorphism.

Differentiating ψ with respect to each component, it follows that

$$d\psi_{(0,0)}(X,0) = d(\exp)_0X = X,$$
$$d\psi_{(0,0)}(0,Y) = d(\exp)_0Y = Y.$$

Thus, $d\psi_0 = \mathrm{id}$. From the Inverse Function Theorem, there exists U an open neighborhood of the origin, such that $\psi|_U$ is a diffeomorphism.

Claim 1.46. *There exists an open neighborhood V of the origin of $\mathfrak{k}$, such that $\exp(Y) \notin H$ for $Y \in V \setminus \{0\}$.*

Suppose that there exists a sequence $\{Y_i\}$ with $Y_i \in \mathfrak{k}$, $\lim Y_i = 0$ and $\exp(Y_i) \in H$. Choose an inner product on $\mathfrak{k}$ and define $t_i = \|Y_i\|$ and $X_i = \frac{1}{t_i}Y_i$. Since $\{X_i\}$ is a sequence in the unit sphere of $\mathfrak{k}$, which is compact, it converges up to passing to a subsequence. Hence $\lim X_i = X$, $\lim t_i = 0$ and $\exp(t_i X_i) \in H$. Therefore, from Claim 1.43, $X \in \mathfrak{h}$. This contradicts $\mathfrak{h} \cap \mathfrak{k} = \{0\}$.

Claim 1.47. There exists an open neighborhood W of the origin in $T_e G$ such that $H \cap \exp(W) = \exp(\mathfrak{h} \cap W)$.

It follows from the construction of $\mathfrak{h}$ that $H \cap \exp(W) \supset \exp(\mathfrak{h} \cap W)$. According to Claims 1.45 and 1.46, there exists a sufficiently small open neighborhood W of the origin of $T_e G$ such that $\exp|_W$ and $\psi|_W$ are diffeomorphisms, and $(W \cap \mathfrak{k}) \subset V$.

Let $a \in H \cap \exp(W)$. As $\psi|_W$ is a diffeomorphism, there exist a unique $X \in \mathfrak{h}$ and a unique $Y \in \mathfrak{k}$ such that $a = \exp(X)\exp(Y)$. Hence $\exp(Y) = (\exp(X))^{-1} a \in H$. From Claim 1.46, $Y = 0$, that is, $a = \exp(X)$, with $X \in \mathfrak{h}$. Therefore $H \cap \exp(W) \subset \exp(\mathfrak{h} \cap W)$. From Claim 1.47, H is an embedded submanifold of G in a neighborhood of the identity $e \in G$. Thus, since H is a group, it is an embedded submanifold. Finally, from Proposition 1.16, H is an embedded Lie subgroup of G. $\square$

In order to explore some consequences of the above result, we recall the *Constant Rank Theorem*. It states that if a smooth map $f: M \to N$ is such that df_x has constant rank, then for each $x_0 \in M$ there exists a neighborhood U of x_0 such that:

(i) $f(U)$ is an embedded submanifold of N;
(ii) The partition $\{f^{-1}(y) \cap U\}_{y \in f(U)}$ is a foliation of U;
(iii) For each $y \in f(U)$, $\ker df_x = T_x(f^{-1}(y))$, for all $x \in f^{-1}(y)$.

Lemma 1.48. *Let G_1 and G_2 be Lie groups and $\varphi: G_1 \to G_2$ be a Lie group homomorphism. Then the following hold:*

(i) $d\varphi_g$ *has constant rank;*
(ii) $\ker \varphi$ *is a normal Lie subgroup of G_1;*
(iii) $\ker d\varphi_e = T_e(\ker \varphi)$.

Proof. Since φ is a Lie group homomorphism, $\varphi \circ L_g^1 = L_{\varphi(g)}^2 \circ \varphi$, where L_g^i denotes the left multiplication by g on G_i. Hence, for all $X \in T_g G_1$,

$$d\varphi_g X = d\varphi_g \, d(L_g^1)_e \, d(L_{g^{-1}}^1)_g X = dL_{\varphi(g)}^2 \, d\varphi_e \, d(L_{g^{-1}}^1)_g X.$$

Since $L_{\varphi(g)}^2$ is a diffeomorphism, $d\varphi_g X = 0$ if and only if $d\varphi_e d(L_{g^{-1}}^1)X = 0$. Hence $\dim \ker d\varphi_g = \dim \ker d\varphi_e$, therefore $d\varphi_g$ has constant rank, proving (i). Item (ii) follows from Theorem 1.42, since $\ker \varphi = \varphi^{-1}(e)$ is a closed normal subgroup. Finally, (iii) follows directly from the Constant Rank Theorem. $\square$

Corollary 1.49. *Let G_1 and G_2 be Lie groups and $\varphi \colon G_1 \to G_2$ a continuous homomorphism. Then φ is smooth.*

Proof. Let $R = \{(g, \varphi(g)) : g \in G_1\}$ be the graph of φ. Then R is a closed subgroup of $G_1 \times G_2$, hence, from Theorem 1.42, R is an embedded Lie subgroup of $G_1 \times G_2$. Consider $i \colon R \hookrightarrow G_1 \times G_2$ the inclusion and the projections $\pi_j \colon G \times G \to G_j$, $j = 1, 2$. Then $(\pi_1 \circ i)$ is a Lie group homomorphism, and from Lemma 1.48, $d(\pi_1 \circ i)_g$ has constant rank. On the other hand, R is a graph, hence, by the Constant Rank Theorem, $(\pi_1 \circ i)$ is an immersion.

In addition, $\dim R = \dim G_1$, otherwise $(\pi_1 \circ i)(R)$ would have measure zero, contradicting $(\pi_1 \circ i)(R) = G_1$. From the Inverse Function Theorem, $(\pi_1 \circ i)$ is a local diffeomorphism. Since it is also bijective, it is a global diffeomorphism, therefore $\varphi = \pi_2 \circ (\pi_1 \circ i)^{-1}$ is smooth. $\qquad\square$

Exercise 1.50 ($\star$). Let G be a Lie group and denote by $\mu \colon G \times G \to G$ the multiplication map $\mu(g, h) = gh$.

(i) Show that the differential $d\mu_{(g,h)} \colon T_g G \oplus T_h G \to T_{gh} G$ is given by

$$d\mu_{(g,h)}(X, Y) = dL_g Y + dR_h X, \quad X \in T_g G, Y \in T_h G, \tag{1.13}$$

and conclude that $\mu \colon G \times G \to G$ is a submersion (see Definition A.4);

(ii) Show that $\mu^{-1}(e) = \{(g, g^{-1}) : g \in G\}$ is a submanifold of $G \times G$, and describe its tangent space $T_{(g,h)}\big(\mu^{-1}(e)\big) \subset T_g G \oplus T_h G$;

(iii) Denote by $i \colon \mu^{-1}(e) \hookrightarrow G \times G$ the inclusion and by $\pi_j \colon G \times G \to G$, $j = 1, 2$, the projection maps. Show that the restriction $\pi_1 \colon \mu^{-1}(e) \to G$ is a diffeomorphism. Conclude that smoothness of the inversion map (1.2), which can be written as $\pi_2 \circ (\pi_1 \circ i)^{-1} \colon G \to G$, is a consequence of the smoothness of μ.

Exercise 1.51. Prove that $GL(n, \mathbb{R})$, $GL(n, \mathbb{C})$, $SL(n, \mathbb{R})$, $SL(n, \mathbb{C})$, $O(n)$, $SO(n)$, $U(n)$, $SU(n)$ and $Sp(n)$ are Lie groups (recall definitions in Example 1.5). Verify that their Lie algebras are, respectively,

(i) $\mathfrak{gl}(n, \mathbb{R})$, the space of $n \times n$ square matrices over $\mathbb{R}$;
(ii) $\mathfrak{gl}(n, \mathbb{C})$, the space of $n \times n$ square matrices over $\mathbb{C}$;
(iii) $\mathfrak{sl}(n, \mathbb{R}) = \{X \in \mathfrak{gl}(n, \mathbb{R}) : \operatorname{tr} X = 0\}$;
(iv) $\mathfrak{sl}(n, \mathbb{C}) = \{X \in \mathfrak{gl}(n, \mathbb{C}) : \operatorname{tr} X = 0\}$;
(v) $\mathfrak{o}(n) = \mathfrak{so}(n) = \{X \in \mathfrak{gl}(n, \mathbb{R}) : X^t + X = 0\}$;
(vi) $\mathfrak{u}(n) = \{X \in \mathfrak{gl}(n, \mathbb{C}) : X^* + X = 0\}$;
(vii) $\mathfrak{su}(n) = \mathfrak{u}(n) \cap \mathfrak{sl}(n, \mathbb{C})$;
(viii) $\mathfrak{sp}(n) = \{X \in \mathfrak{gl}(n, \mathbb{H}) : X^* + X = 0\}$.

Compare (v) with Exercise 1.15.

Hint: Prove directly that $GL(n, \mathbb{R})$ and $GL(n, \mathbb{C})$ are Lie subgroups and verify that the others are closed subgroups of these. Use Remark 1.33 to prove that if $X \in \mathfrak{g}$ and $\mathfrak{h} \subset \mathfrak{g}$, then $X \in \mathfrak{h}$ if and only if $\exp(tX) \in H$, for all $t \in \mathbb{R}$. Then, use Lemma 1.48, combined with the identity $\det e^A = e^{\operatorname{tr} A}$ for any $A \in \mathfrak{gl}(n, \mathbb{K})$, to assist with the computation of the above Lie algebras.

Exercise 1.52. Let $S^3 = \{(z_1, z_2) \in \mathbb{C}^2 : |z_1|^2 + |z_2|^2 = 1\}$ be the unit 3-sphere. Verify that S^3 is a Lie group when endowed with the multiplication

$$(z_1, z_2) \cdot (w_1, w_2) := (z_1 w_1 - \overline{z_2} w_2, \ w_1 z_2 + \overline{z_1} w_2).$$

(i) Consider the map $\varphi \colon S^3 \to \mathrm{SU}(2)$, given by

$$\varphi(z_1, z_2) = \begin{pmatrix} z_1 & -\overline{z_2} \\ z_2 & \overline{z_1} \end{pmatrix};$$

Verify that φ is a continuous homomorphism, hence a Lie group homomorphism. Verify that $\ker \varphi$ is trivial, so φ is an isomorphism;

(ii) Consider the group of unit quaternions

$$\mathrm{Sp}(1) = \{a + bi + cj + dk \in \mathbb{H} : a^2 + b^2 + c^2 + d^2 = 1\}$$

Find a continuous (hence smooth) group homomorphism $\psi \colon \mathrm{Sp}(1) \to S^3$;

(iii) Conclude that S^3, $\mathrm{Sp}(1)$ and $\mathrm{SU}(2)$ are isomorphic Lie groups.

Exercise 1.53. Let G be a Lie group and consider $\rho \colon \widetilde{G} \to G$ its universal covering. Prove that:

(i) $H := \rho^{-1}(e)$ is a normal discrete closed subgroup, and $gh = hg$, for all $h \in H, g \in \widetilde{G}$;
(ii) $G \cong \widetilde{G}/H$;
(iii) $\pi_1(G)$ is abelian.

Hint: Item (i) follows directly from ρ being a continuous homomorphism. The fact that $gh = hg$ can be proved defining $f \colon \widetilde{G} \to H$ as $f(x) := xhx^{-1}$, noticing that $\{h\}$ is an open and closed subset of H and hence concluding that $f^{-1}(\{h\})$ is an open and closed subset of $\widetilde{G}$. In order to prove (iii), consider two loops α_i in G such that $\alpha_i(0) = e$, and their lifts $\widetilde{\alpha}_i$ with $\widetilde{\alpha}_i(0) = e$. Set $h_i = \widetilde{\alpha}_i(1)$ and use (i) to conclude that $\widetilde{\alpha_2 * \alpha_1} = h_2 h_1 = h_1 h_2 = \widetilde{\alpha_1 * \alpha_2}$ where $*$ denotes concatenation of curves.

Further details on the smooth structure of quotients of Lie groups, such as $\widetilde{G}/H$ above, are given in Chap. 3, see Corollary 3.38.

Remark 1.54. The above exercise gives a glimpse of the structure of homotopy groups of Lie groups. By using Morse theory on the space of paths on a compact Lie group G, it is possible to prove that, in addition to $\pi_1(G)$ being abelian, $\pi_2(G)$ is trivial and $\pi_3(G)$ is torsion-free. These techniques also led to the discovery that homotopy groups of classical Lie groups are periodic, which is a foundational result known as *Bott periodicity*. For further details, we refer the reader to Milnor [158].

Exercise 1.55 ($\star$). Prove that $SU(2)$ is the universal covering of $SO(3)$, via the following steps:

(i) Let $g \in SU(2) \cong S^3$, $u \in S^2 \subset \mathbb{R}^3$ and $\theta \in [0, 2\pi]$ be such that $g = \cos(\theta) + \sin(\theta)u$. Define $T_g(v) = gvg^{-1}$ for all $v \in \mathbb{R}^3$. Prove that $T_g \colon \mathbb{R}^3 \to \mathbb{R}^3$ is a linear orthogonal transformation, and $T_g = e^{A_{2\theta u}}$, where A_X is as in Exercise 1.9;

(ii) Prove that $\varphi \colon S^3 \ni g \mapsto T_g \in SO(3)$ is a double covering map, concluding $\pi_1(SO(3)) \cong \mathbb{Z}_2$.

Hint: Recall that S^3 is isomorphic to $SU(2)$, see Exercise 1.52. Also, recall that the product of quaternions corresponds to $(t_1 + u_1) \cdot (t_2 + u_2) = (t_1 t_2 - \langle u_1, u_2 \rangle) + (t_2 u_1 + t_1 u_2 + u_1 \times u_2)$, and verify that $\frac{d}{dt} T_{\cos(t\theta) + \sin(t\theta)u}(v)|_{t=0} = 2\theta\, u \times v$. Note that under the identifications $SU(2) \cong S^3$ and $SO(3) \cong \mathbb{R}P^3$, the double covering $SU(2) \to SO(3)$ is precisely the double covering $S^3 \to \mathbb{R}P^3$.

Remark 1.56. Using results from basic topology, it is not difficult to show that $\pi_1(SO(n)) \cong \mathbb{Z}_2$ for all $n \geq 3$. The universal covering of $SO(n)$, $n \geq 3$, is called the *spin group* and denoted $\mathrm{Spin}(n)$, recall Theorem 1.23. Algebraic manipulations slightly more elaborate than those in Exercise 1.55 show that $\mathrm{Sp}(1) \times \mathrm{Sp}(1) \cong SU(2) \times SU(2)$ is a double covering of $SO(4)$, $\mathrm{Sp}(2)$ is a double covering of $SO(5)$, and $SU(4)$ is a double covering of $SO(6)$. In particular, we have the isomorphisms:

$$\mathrm{Spin}(3) \cong \mathrm{Sp}(1), \quad \mathrm{Spin}(4) \cong \mathrm{Sp}(1) \times \mathrm{Sp}(1), \quad \mathrm{Spin}(5) \cong \mathrm{Sp}(2), \quad \mathrm{Spin}(6) \cong SU(4).$$

The *center* of a Lie group G is the subgroup given by

$$Z(G) := \left\{ g \in G : ghg^{-1} = h, \text{ for all } h \in G \right\}, \tag{1.14}$$

and the *center* of a Lie algebra $\mathfrak{g}$ is defined as

$$Z(\mathfrak{g}) := \left\{ X \in \mathfrak{g} : [X, Y] = 0, \text{ for all } Y \in \mathfrak{g} \right\}. \tag{1.15}$$

The following result relates the centers of a Lie group and of its Lie algebra.

Corollary 1.57. *Let G be a connected Lie group. Then the following hold:*

(i) $Z(G) = \ker \mathrm{Ad}$;
(ii) $Z(G)$ *is a normal Lie subgroup of G;*
(iii) $Z(\mathfrak{g}) = \ker \mathrm{ad}$;
(iv) $Z(\mathfrak{g})$ *is the Lie algebra of $Z(G)$.*

Proof. First, we verify that $Z(G) = \ker \mathrm{Ad}$. If $g \in Z(G)$, clearly $\mathrm{Ad}(g) = \mathrm{id}$. Conversely, let $g \in \ker \mathrm{Ad}$. It follows from (1.10) that $g \exp(tX) g^{-1} = \exp(tX)$, for all $X \in \mathfrak{g}$. Hence g commutes with the elements of a neighborhood of the identity $e \in G$. From Proposition 1.20, we conclude that $g \in Z(G)$, proving (i).

Item (ii) follows from (i) and Lemma 1.48. Item (iii) follows directly from Proposition 1.37. Since $d(\mathrm{Ad})_e = \mathrm{ad}$, (iv) follows directly from (iii) in Lemma 1.48. $\qquad\square$

Remark 1.58. The above result gives an alternative proof of an assertion in Proposition 1.39, that G is abelian if $\mathfrak{g}$ is abelian. Indeed, if $[X,Y] = 0$ for all $X,Y \in \mathfrak{g}$, then $Z(\mathfrak{g}) = \mathfrak{g}$. Thus, from (iv) in Corollary 1.57, $Z(G)$ is open in G and it follows from Proposition 1.20 that $G = Z(G)$, hence G is abelian.

Exercise 1.59 ($\star$). In this exercise, we generalize the concept of *center* of a Lie group G and Lie algebra $\mathfrak{g}$, to the notion *centralizer* of subsets of G and $\mathfrak{g}$. Define the *centralizer* of a subset S in G as

$$Z_G(S) := \left\{ g \in G : ghg^{-1} = h, \text{ for all } h \in S \right\},$$

and the *centralizer* of a subset $\mathfrak{s}$ in $\mathfrak{g}$ as

$$Z_{\mathfrak{g}}(\mathfrak{s}) := \left\{ X \in \mathfrak{g} : [X,Y] = 0, \text{ for all } Y \in \mathfrak{s} \right\}.$$

Note that $Z_G(G) = Z(G)$ and $Z_{\mathfrak{g}}(\mathfrak{g}) = Z(\mathfrak{g})$.

Prove that $Z_G(S)$ is a Lie subgroup of G and $Z_{\mathfrak{g}}(\mathfrak{s})$ is a Lie subalgebra of $\mathfrak{g}$. Moreover, if G is connected, H is a Lie subgroup of G and $\mathfrak{h}$ is the corresponding Lie subalgebra of $\mathfrak{g}$, then the following hold (cf. Corollary 1.57):

(i) $Z_G(H) = \{g \in G : \mathrm{Ad}(g)X = X \text{ for all } X \in \mathfrak{h}\}$;
(ii) $Z_{\mathfrak{g}}(\mathfrak{h}) = \{X \in \mathfrak{g} : [X,Y] = 0 \text{ for all } Y \in \mathfrak{h}\}$;
(iii) $Z_{\mathfrak{g}}(\mathfrak{h})$ is the Lie algebra of $Z_G(H)$.

Exercise 1.60 ($\star$). In this exercise, we discuss relations between $U(n)$ and $SU(n)$.

(i) Show that the centers of $U(n)$ and $SU(n)$ are respectively given by

$$Z(U(n)) = \{z\,\mathrm{id} : z \in \mathbb{C}, |z| = 1\} \cong S^1 \text{ and } Z(SU(n)) = \{z\,\mathrm{id} : z^n = 1\} \cong \mathbb{Z}_n.$$

In particular, conclude that the Lie groups $U(n)$ and $SU(n) \times S^1$ are not isomorphic, since their centers are not isomorphic;

(ii) Show that the multiplication map $Z(U(n)) \times SU(n) \ni (z\,\mathrm{id}, A) \mapsto zA \in U(n)$ is a Lie group homomorphism, and also an n-fold covering $S^1 \times SU(n) \to U(n)$;

(iii) Find a Lie group homomorphism $\varphi : U(n) \to S^1$ such that $\ker \varphi = SU(n)$ and $\varphi \circ \iota = \mathrm{id}$, where $\iota : S^1 \to U(n)$ is the inclusion with image consisting of diagonal matrices with $e^{i\theta}$ in the upper left corner and 1 in the rest of the diagonal. Conclude that $U(n) \cong SU(n) \rtimes S^1$ is a *semidirect product* of $SU(n)$ by S^1.

Exercise 1.61 ($\star$). Let G be a Lie group and H a closed Lie subgroup. Define the *normalizer* of H in G as

$$N(H) := \left\{ g \in G : gHg^{-1} = H \right\}, \tag{1.16}$$

which is also sometimes denoted $N_G(H)$. Prove that $N(H)$ is a closed Lie subgroup of G. Assuming that H is connected and denoting its Lie algebra by $\mathfrak{h}$, prove that the following hold:

(i) $N(H) = \{g \in G : \mathrm{Ad}(g)X \in \mathfrak{h} \text{ for all } X \in \mathfrak{h}\}$;

(ii) The Lie algebra of $N(H)$ is $\mathfrak{n} = \{X \in \mathfrak{g} : [X,Y] \in \mathfrak{h} \text{ for all } Y \in \mathfrak{h}\}$.

Hint: Use (1.8), (1.10) and (1.12). For a solution, see the proof of Proposition 2.37.

Chapter 2
Lie Groups with Bi-invariant Metrics

This chapter deals with Lie groups with special types of Riemannian metrics: *bi-invariant* metrics. Every compact Lie group admits one such metric (see Proposition 2.24), which plays a very important role in the study of its geometry. In what follows, we use tools from Riemannian geometry to give concise proofs of several classical results on compact Lie groups.

We begin by reviewing some auxiliary facts of Riemannian geometry. Basic results on bi-invariant metrics and Killing forms are then discussed, e.g., we prove that the exponential map of a compact Lie group is surjective (Theorem 2.27), and we show that a semisimple Lie group is compact if and only if its Killing form is negative-definite (Theorem 2.35). We also prove that a simply-connected Lie group admits a bi-invariant metric if and only if it is a product of a compact Lie group with a vector space (Theorem 2.45). Finally, we prove that if the Lie algebra of a compact Lie group G is simple, then the bi-invariant metric on G is unique up to rescaling (Proposition 2.48).

Further suggested readings on the contents of this chapter are Fegan [85], Grove [106], Ise and Takeuchi [133], Milnor [158, 159], Onishchik [179] and Petersen [183], which inspired the present text.

2.1 Basic Facts of Riemannian Geometry

The main objective of this section is to review basic results of Riemannian geometry that are used in the following chapters. Proofs of most results in this section are omitted, and can be found in any standard textbook on Riemannian geometry, such as do Carmo [61], Petersen [183] or Lee [152].

A *Riemannian manifold* is a smooth manifold M endowed with a *(Riemannian) metric*, i.e., a $(0,2)$-tensor field g on M that is

© Springer International Publishing Switzerland 2015
M.M. Alexandrino, R.G. Bettiol, *Lie Groups and Geometric Aspects of Isometric Actions*, DOI 10.1007/978-3-319-16613-1_2

(i) Symmetric: $g(X,Y) = g(Y,X)$, for all $X,Y \in TM$;
(ii) Positive-definite: $g(X,X) > 0$, if $X \neq 0$.

This means that a metric determines an inner product g_p on each tangent space T_pM. For this reason, we sometimes write $\langle X,Y \rangle = g(X,Y)$ and $\|X\|^2 = g(X,X)$, if the metric g is evident from the context. It is not difficult to prove that every manifold admits a Riemannian metric, using partitions of unity. As indicated below, Riemannian metrics provide a way to measure distances, angles, curvature, and other geometric properties.

It is possible to associate to each metric a so-called *connection*. Such map allows to *parallel translate* vectors along curves (see Proposition 2.7), *connecting* tangent spaces of M at different points. It is actually possible to define connections on *any* vector bundles over a manifold.

Definition 2.1. A *linear connection* on a Riemannian manifold (M,g) is a map

$$\nabla \colon \mathfrak{X}(M) \times \mathfrak{X}(M) \ni (X,Y) \longmapsto \nabla_X Y \in \mathfrak{X}(M),$$

satisfying the following properties:

(i) $\nabla_X Y$ is $C^\infty(M)$-linear in X, i.e., for all $f,g \in C^\infty(M)$,

$$\nabla_{fX_1 + gX_2} Y = f\nabla_{X_1} Y + g\nabla_{X_2} Y;$$

(ii) $\nabla_X Y$ is $\mathbb{R}$-linear in Y, i.e., for all $a,b \in \mathbb{R}$,

$$\nabla_X(aY_1 + bY_2) = a\nabla_X Y_1 + b\nabla_X Y_2;$$

(iii) ∇ satisfies the Leibniz rule, i.e., for all $f \in C^\infty(M)$,

$$\nabla_X(fY) = f\nabla_X Y + (Xf)Y.$$

Moreover, a linear connection is said to be *compatible with the metric* g if

$$Xg(Y,Z) = g(\nabla_X Y,Z) + g(Y,\nabla_X Z), \quad X,Y,Z \in \mathfrak{X}(M).$$

It turns out that requiring a connection to be compatible with the metric does not determine a unique connection on (M,g). To this purpose, define the *torsion tensor* of ∇ to be the $(1,2)$-tensor field given by $T(X,Y) = \nabla_X Y - \nabla_Y X - [X,Y]$. A connection ∇ is said to be *symmetric* if its torsion vanishes identically, that is, if $[X,Y] = \nabla_X Y - \nabla_Y X$ for all $X,Y \in \mathfrak{X}(M)$.

Exercise 2.2. Let M be an embedded surface in $\mathbb{R}^3$ with the induced metric, i.e., $g = i^* g_0$ where g_0 is the Euclidean metric and $i \colon M \to \mathbb{R}^3$ is the inclusion. Let $\overline{\nabla}$ be the Euclidean derivative. Set $\nabla_X Y := (\overline{\nabla}_{\overline{X}} \overline{Y})^\top$, i.e., the tangent part of $\overline{\nabla}_{\overline{X}} \overline{Y}$ where $\overline{X}$ and $\overline{Y}$ are local extension of the vector field X and Y. Show that ∇ is a symmetric connection compatible with the metric g.

Hint: One can prove that the connection is symmetric using Remark A.11.

Levi-Civita Theorem 2.3. *On a Riemannian manifold* $(M, \mathbf{g})$, *there exists a unique linear connection* ∇ *that is compatible with* $\mathbf{g}$ *and symmetric.*

The key fact on the proof of this theorem is the equation known as *connection formula*, or *Koszul formula*. It exhibits the natural candidate to the desired connection, and shows that it is uniquely determined by the metric:

$$\langle \nabla_Y X, Z \rangle = \tfrac{1}{2} \Big(X \langle Y, Z \rangle - Z \langle X, Y \rangle + Y \langle Z, X \rangle$$
$$- \langle [X, Y], Z \rangle - \langle [X, Z], Y \rangle - \langle [Y, Z], X \rangle \Big). \tag{2.1}$$

The unique symmetric linear connection compatible with the metric is called the *Levi-Civita connection*, or *Riemannian connection*, and we refer to it simply as *connection*. Using this connection, one can differentiate vector fields on a Riemannian manifold $(M, \mathbf{g})$ as we describe next.

Proposition 2.4. *Let M be a manifold with linear connection ∇ and $\gamma: I \to M$ be a smooth curve. Let $\Gamma(\gamma^*TM)$ denote the set of smooth vector fields along γ. There exists a unique correspondence that, to each $X \in \Gamma(\gamma^*TM)$ associates $\frac{D}{dt} X \in \Gamma(\gamma^*TM)$, called the covariant derivative of X along γ, satisfying:*

(i) *Linearity, i.e., for all $X, Y \in \Gamma(\gamma^*TM)$,*

$$\frac{D}{dt}(X + Y) = \frac{D}{dt}X + \frac{D}{dt}Y;$$

(ii) *Leibniz rule, i.e., for all $X \in \Gamma(\gamma^*TM), f \in C^\infty(I)$,*

$$\frac{D}{dt}(fX) = \frac{df}{dt}X + f\frac{D}{dt}X;$$

(iii) *If X is induced from a vector field $\widetilde{X} \in \mathfrak{X}(M)$, that is $X(t) = \widetilde{X}(\gamma(t))$, then $\frac{D}{dt}X = \nabla_{\gamma'}\widetilde{X}$.*

Note that to each linear connection on M, the proposition above gives a covariant derivative operator for vector fields along γ. As mentioned before, we only consider the Levi-Civita connection, hence the covariant derivative is uniquely defined.

Equipped with this notion, it is possible to define the *acceleration* of a curve as the covariant derivative of its tangent vector field, and *geodesics* as curves with null acceleration. More precisely, $\gamma: I \to M$ is a *geodesic* if $\frac{D}{dt}\gamma' = 0$. It is often convenient to reparametrize a geodesic γ to become a *unit speed geodesic* $\widetilde{\gamma}$, that is, $\|\widetilde{\gamma}'\| = 1$ and $\widetilde{\gamma}$ has the same image as γ.

Writing a local expression for the covariant derivative, it is easy to see that a curve is a geodesic if and only if it satisfies a second-order system of ODEs, called the *geodesic equation*. Hence, applying the classical ODE result that guarantees existence and uniqueness of solutions, one can prove the following.

Theorem 2.5. *For any $p \in M$, $t_0 \in \mathbb{R}$ and $v \in T_pM$, there exist an open interval $I \subset \mathbb{R}$ containing t_0 and a geodesic $\gamma: I \to M$ satisfying the initial conditions $\gamma(t_0) = p$ and $\gamma'(t_0) = v$. In addition, any two geodesics with those initial conditions agree on their common domain.*

From uniqueness of the solution, it is possible to obtain the existence of a geodesic with *maximal domain* for a prescribed initial data. The restriction of a geodesic to a bounded subset of its maximal domain is called a *geodesic segment*.

Exercise 2.6. Let M be an embedded surface in $\mathbb{R}^3$, with the induced connection (defined in Exercise 2.2). Let V be a plane, and assume that, for all $p \in M \cap V$, the normal vector $\xi(p)$ is tangent to V. Show that the intersection $M \cap V$ can be locally parametrized by a geodesic segment on M. Conclude that the profile curve of a surface of revolution is a geodesic (up to reparametrization). In particular, conclude that great circles are the geodesics of the round sphere $S^2 \subset \mathbb{R}^3$.

Hint: Consider $\gamma \in V \cap M$ such that $\|\gamma'(t)\| = 1$. Note that $\langle \gamma''(t), \gamma'(t) \rangle = 0$, and hence $\gamma''(t)$ is normal to γ and contained in V. Since ξ is also normal to γ and contained in V, conclude that $\gamma'' = \lambda \xi$ and therefore $\frac{D}{dt}\gamma' = 0$.

Another construction that involves covariant differentiation along curves is parallel translation. A vector field $X \in \Gamma(\gamma^*TM)$ is said to be *parallel along γ* if $\frac{D}{dt}X = 0$. Thus, a geodesic γ can be characterized as a curve whose tangent field γ' is parallel along γ. A vector field is called *parallel* if it is parallel along every curve.

Proposition 2.7. *Let $\gamma: I \to M$ be a curve, $t_0 \in I$ and $v_0 \in T_{\gamma(t_0)}M$. There exists a unique parallel vector field X along γ such that $X(t_0) = v_0$.*

This vector field is called the *parallel translate* of v_0 along γ. Once more, the proof depends on basic ODE results.

An important question after having existence and uniqueness of geodesics is how geodesics change under perturbations of the initial data, which leads to the definition of the *geodesic flow* of a Riemannian manifold. This is the flow

$$\varphi^{\mathscr{G}} : U \subset \mathbb{R} \times TM \to TM$$

defined in an open subset U of $\mathbb{R} \times TM$ that contains $\{0\} \times TM$, of the unique vector field $\mathscr{G}$ on the tangent bundle, i.e., $\mathscr{G} \in \mathfrak{X}(TM)$, whose integral curves are of the form $t \mapsto (\gamma(t), \gamma'(t))$, where γ is a geodesic in M. This means that:

(i) $\gamma(t) = \pi \circ \varphi^{\mathscr{G}}(t, (p, v))$ is the geodesic with initial conditions $\gamma(0) = p$ and $\gamma'(0) = v$, where $\pi: TM \to M$ is the bundle projection;

(ii) $\varphi^{\mathscr{G}}(t, (p, cv)) = \varphi^{\mathscr{G}}(ct, (p, v))$, for all $c \in \mathbb{R}$ such that this equation makes sense.

In fact, supposing that such vector field $\mathscr{G}$ exists, one can obtain conditions in local coordinates that this field must satisfy (corresponding to the geodesic equation). Defining the vector field as its solutions, one may use Theorem A.14 to guarantee existence and smoothness of $\varphi^{\mathscr{G}}$.

We now proceed to define a similar concept to the Lie exponential map, which coincides with it when considering appropriate (bi-invariant) Riemannian metrics on compact Lie groups, see Theorem 2.27.

Definition 2.8. The *(Riemannian) exponential map* at $p \in M$ is the map

$$\exp_p \colon B_\varepsilon(0) \subset T_p M \to M, \quad \exp_p(v) := \pi \circ \varphi^{\mathscr{G}}(1, (p, v)).$$

Smoothness of $\exp_p$ follows immediately from Theorem A.14. Using the Inverse Function Theorem, one can verify that for any $p \in M$, there exist a neighborhood V of the origin in $T_p M$ and a neighborhood U of $p \in M$, such that $\exp_p|_V \colon V \to U$ is a diffeomorphism. Such neighborhood U is called a *normal neighborhood* of p.

Define the *(Riemannian) distance* $\mathrm{dist}(p, q)$ for any pair of points $p, q \in M$ to be the infimum of lengths of all piecewise smooth curve segments joining p and q, where the *length* of a curve segment $\gamma \colon [a, b] \to M$ is defined as

$$\ell(\gamma) := \int_a^b \sqrt{g(\gamma'(t), \gamma'(t))} \, dt.$$

Then (M, dist) is a metric space, and the topology induced by this distance coincides with the original topology from the atlas of M. It is a very important fact that geodesics *locally minimize* ℓ among piecewise smooth curves. Geodesic segments $\gamma \colon [a, b] \to M$ that globally minimize distance, i.e., such that $\mathrm{dist}(\gamma(t_1), \gamma(t_2)) = |t_1 - t_2|$ for all $t_1, t_2 \in [a, b]$, are called *minimal*.

A Riemannian manifold is called *geodesically complete* if the maximal domain of definition of all geodesics is $\mathbb{R}$. It is not difficult to see that compactness is a sufficient condition for a manifold to be complete. The following important result states that all completeness notions for a Riemannian manifold are equivalent.

Hopf-Rinow Theorem 2.9. *Let (M, g) be a connected Riemannian manifold and $p \in M$. The following statements are equivalent:*

(i) $\exp_p$ is globally defined, that is, $\exp_p \colon T_p M \to M$;
(ii) M is geodesically complete;
(iii) (M, dist) is a complete metric space;
(iv) Every closed bounded set in M is compact.

If M satisfies any (hence all) of the above items, each two points of M can be joined by a minimal geodesic segment. In particular, for each $x \in M$ the exponential map $\exp_x \colon T_x M \to M$ is surjective.

Exercise 2.10. Give an example of a Riemannian manifold which is not complete, but such that any pair of points can be joined by a minimal geodesic segment.

Consider M and N manifolds and $f \colon M \to N$ a smooth map. Recall that any $(0, s)$-tensor τ on N may be *pulled back* by f, the result being a $(0, s)$-tensor $f^* \tau$ on M, see (A.6). A diffeomorphism $f \colon (M, g^M) \to (N, g^N)$ satisfying $f^* g^N = g^M$, that is, $g_p^M(X,Y) = g_{f(p)}^N(df_pX, df_pY)$ for all $p \in M$ and $X,Y \in T_pM$, is called a *(Riemannian) isometry*. It is easy to show that Riemannian isometries map geodesics to geodesics, and hence preserve distance, that is, are also metric isometries. Conversely, metric isometries of a smooth Riemannian manifold are Riemannian isometries. Thus, we simply refer to these maps as *isometries*.

From the above, it also follows that isometries *commute* with the exponential map, in the sense that if $f: M \to M$ is an isometry, then:

$$f(\exp_p(v)) = \exp_{f(p)}(df_p v) \tag{2.2}$$

for all $p \in M$ and $v \in T_pM$ such that $\exp_p(v)$ is defined. This allows to prove that isometries of a connected manifold are determined by its value and derivative at a single point.

Exercise 2.11. Suppose $f_1: M \to M$ and $f_2: M \to M$ are isometries of a connected Riemannian manifold (M,g), such that there exists $p \in M$ with $f_1(p) = f_2(p)$ and $d(f_1)_p = d(f_2)_p$. Prove that $f_1(x) = f_2(x)$ for all $x \in M$.

Hint: Consider the closed subset $S = \{x \in M : f_1(x) = f_2(x) \text{ and } d(f_1)_x = d(f_2)_x\}$. Use (2.2) to prove that S is also open, and hence $S = M$, since M is connected.

Isometries of a Riemannian manifold (M,g) clearly form a group, denoted $\mathrm{Iso}(M,g)$, where the operation is given by composition. A fundamental result, due to Myers and Steenrod [173], is that this is a Lie group that acts smoothly on M.

Myers-Steenrod Theorem 2.12. *Let (M,g) be a Riemannian manifold. Any closed subgroup of $\mathrm{Iso}(M,g)$ in the compact-open topology is a Lie group. In particular, $\mathrm{Iso}(M,g)$ is a Lie group.*

Remark 2.13. A subset $G \subset \mathrm{Iso}(M,g)$ is closed in the compact-open topology if given any sequence $\{f_n\}$ of isometries in G that converges uniformly in compact subsets to a continuous map $f: M \to M$, then $f \in G$.

We now mention a special class of vector fields on M, closely related to $\mathrm{Iso}(M,g)$. A *Killing vector field* is a vector field whose local flow is a local isometry. Alternatively, Killing vector fields are characterized by the following property.

Proposition 2.14. *A vector field $X \in \mathfrak{X}(M)$ is a Killing vector field if and only if*

$$g(\nabla_Y X, Z) = -g(\nabla_Z X, Y), \quad \text{for all } Y, Z \in \mathfrak{X}(M).$$

Theorem 2.15. *The set $\mathfrak{iso}(M,g)$ of Killing fields on (M,g) is a Lie algebra. In addition, if (M,g) is complete, then $\mathfrak{iso}(M,g)$ is the Lie algebra of $\mathrm{Iso}(M,g)$.*

An essential concept in Riemannian geometry is that of *curvature*. The *curvature tensor* on (M,g) is defined as the $(1,3)$-tensor field given by the following expression,[1] for all $X,Y,Z \in \mathfrak{X}(M)$.

---

[1] We remark that some texts use the opposite sign convention for R, as there is no standard choice.

$$R(X,Y)Z := \nabla_{[X,Y]}Z - \nabla_X\nabla_Y Z + \nabla_Y\nabla_X Z$$

It is possible to use the metric to deal with this tensor as a $(0,4)$-tensor, given by

$$R(X,Y,Z,W) := g(R(X,Y)Z,W). \tag{2.3}$$

There are many important symmetries of this tensor that we state below. For each $X,Y,Z,W \in \mathfrak{X}(M)$,

(i) R is skew-symmetric in the first two and last two entries:

$$R(X,Y,Z,W) = -R(Y,X,Z,W) = R(Y,X,W,Z);$$

(ii) R is symmetric in the first and last pairs of entries:

$$R(X,Y,Z,W) = R(Z,W,X,Y);$$

(iii) R satisfies a cyclic permutation property, called the *first Bianchi identity*:

$$R(X,Y)Z + R(Z,X)Y + R(Y,Z)X = 0.$$

Using the curvature tensor, we can define the *sectional curvature* of the plane spanned by (linearly independent) vectors X and Y as

$$\sec(X,Y) := \frac{R(X,Y,X,Y)}{g(X,X)g(Y,Y) - g(X,Y)^2}. \tag{2.4}$$

It is possible to verify that $\sec(X,Y)$ depends only on the plane σ spanned by X and Y, and not specifically on the given basis $\{X,Y\}$ of σ. For this reason, we sometimes denote (2.4) by $\sec(\sigma)$, where $\sigma = \text{span}\{X,Y\} \subset T_pM$.

There are several interpretations of curvature, the most naive being that it measures the failure of second covariant derivatives to commute. The curvature tensor is also part of the *Jacobi equation* along a geodesic γ, given by

$$\frac{D}{dt}\frac{D}{dt}J + R(\gamma'(t),J(t))\gamma'(t) = 0. \tag{2.5}$$

This is an ODE whose solutions J are vector fields along γ, called *Jacobi fields*. A Jacobi field J along γ is the velocity (or variational field) of a variation of γ by geodesics. More precisely, $J(t) = \frac{\partial}{\partial s}\alpha(s,t)\big|_{s=0}$ where $\alpha: (-\varepsilon,\varepsilon) \times [a,b] \to M$ is a piecewise smooth map, such that $\alpha(0,t) = \gamma(t)$ and $\alpha(s,t)$ is a geodesic for each fixed $s \in (-\varepsilon,\varepsilon)$. Jacobi fields describe how quickly geodesics with the same starting point and different directions move away from each other.

Exercise 2.16 ($\star$). Let M be a surface with constant sectional curvature k, and let γ be a unit speed geodesic on M. Consider a variation $\alpha(s,t)$ by geodesics, such that

$\alpha(s,0) = \gamma(0)$ for all s. Consider the Jacobi field $J(t) := \frac{\partial}{\partial s}\alpha(s,t)\big|_{s=0}$, and assume that $\frac{D}{dt}J(0) = w$ is orthogonal to γ and $\|w\| = 1$. Let $w(t)$ be the parallel translate of w along $\gamma(t)$. Prove that $J(t) = \mathrm{sn}_k(t)w(t)$, where sn_k is the function given by

$$\mathrm{sn}_k(t) := \begin{cases} \frac{\sin(t\sqrt{k})}{\sqrt{k}} & \text{if } k > 0, \\ t & \text{if } k = 0, \\ \frac{\sinh(t\sqrt{-k})}{\sqrt{-k}} & \text{if } k < 0. \end{cases}$$

Hint: Write $J(t) = f_1(t)\gamma'(t) + f_2(t)w(t)$ and use the properties of the curvature tensor R described above, the hypothesis that $\langle \frac{D}{dt}J(0), \gamma'(0)\rangle = 0$, the fact that $\langle w(t), \gamma'(t)\rangle = 0$ (because parallel translation is an isometry), and unicity of solutions to ODEs, to conclude that $f_1 \equiv 0$ and $f_2(t) = \mathrm{sn}_k(t)$.

Remark 2.17. The above statement about Jacobi fields also holds on more general *space forms*, i.e., Riemannian manifolds $M(k)$ that have constant sectional curvature k, since their curvature tensor satisfies $R(X,Y,Z,W) = k(\langle X,Z\rangle\langle Y,W\rangle - \langle Y,Z\rangle\langle X,W\rangle)$. If a space form $M(k)$ is simply-connected and has dimension n, then it is isometric to the sphere $S^n(1/\sqrt{k})$, the Euclidean space $\mathbb{R}^n$ or the hyperbolic space $H^n(1/\sqrt{-k})$, according to the cases $k > 0$, $k = 0$ and $k < 0$ respectively.

As illustrated by the above exercise, on manifolds with nonnegative sectional curvature (sec ≥ 0), the distance between geodesics starting at the same point grows *slower* than in flat space; in other words, the exponential map $\exp_p$ is distance nonincreasing. Similarly, on manifolds with nonpositive sectional curvature (sec ≤ 0), distance between geodesics starting at the same point grows *faster* than in flat space; which means that $\exp_p$ is distance nondecreasing. Curvature also describes how parallel transport along short loops differs from the identity. More details on these interpretations can be found in Grove [106], Jost [135] and Petersen [183].

Besides analyzing sectional curvatures (which actually fully determine the curvature tensor R), it is convenient to summarize information contained in R by constructing simpler tensors. The *Ricci tensor* is a $(0,2)$-tensor field defined as a trace of the curvature tensor R. More precisely, if $\{e_1,\ldots,e_n\}$ is an orthonormal basis of T_pM,

$$\mathrm{Ric}(X,Y) := \mathrm{tr}\left(R(X,\cdot)Y\right) = \sum_{i=1}^n R(X,e_i,Y,e_i).$$

The *scalar curvature* is a function scal$\colon M \to \mathbb{R}$ given by the trace of Ric, i.e.,

$$\mathrm{scal} := \mathrm{tr}\left(\mathrm{Ric}\right) = \sum_{j=1}^n \mathrm{Ric}(e_j,e_j) = 2\sum_{1\leq i<j\leq n} R(e_i,e_j,e_i,e_j) = 2\sum_{1\leq i<j\leq n}\sec(e_i,e_j).$$

The curvatures Ric and scal respectively encode information on the volume distortion and volume defect of small balls in (M, g), compared with a space form.

A metric is called *Einstein* if it is proportional to its Ricci tensor:

$$\text{Ric}(X, Y) = \lambda \, g(X, Y). \tag{2.6}$$

In this case, the number $\lambda \in \mathbb{R}$ is called the *Einstein constant* of g. It is easy to see that if (M, g) has constant sectional curvature $\text{sec} = k$, then it is Einstein, with constant $\lambda = (n-1)k$, where $n = \dim M$. Moreover, any Einstein metric with constant λ has constant scalar curvature $\text{scal} = n\lambda$.

Remark 2.18. Einstein metrics originate from General Relativity, as solutions to the Einstein equations in vacuum, $\text{Ric} = \left(\frac{1}{2}\text{scal} - \Lambda\right)g$, where Λ is the cosmological constant. Details on this equation and its impact in Physics can be found in [33, 160]. As a curiosity, the reason why Riemannian metrics are denoted g also comes from General Relativity, since, in this context, g is interpreted as a gravitational field.

We now state a classical result that establishes a link between the curvature and topology of a manifold. Recall that the diameter $\text{diam}(X)$ of a metric space X is the smallest number d such that any two points in X are always at distance $\leq d$.

Bonnet-Myers Theorem 2.19. *Let (M, g) be a connected complete n-dimensional Riemannian manifold, with $n \geq 2$. Assume $\text{Ric} \geq (n-1)k\,g$ for some $k > 0$. Then:*

(i) $\text{diam}(M) \leq \frac{\pi}{\sqrt{k}}$; *in particular, M is compact;*
(ii) *The universal covering of M is compact, hence $\pi_1(M)$ is finite.*

We end this section with a quick discussion of Riemannian *immersions* and *submersions*, which are dual types of maps between Riemannian manifolds (of potentially different dimensions), that generalize the notion of isometries.

Let (M, g) and $(\overline{M}, \overline{g})$ be Riemannian manifolds and $i: M \to \overline{M}$ be an immersion (recall Definition A.4). Then i is a *Riemannian immersion* if for all $p \in M$, di_p is a linear isometry from T_pM onto its image, which is a subspace of $T_{i(p)}\overline{M}$. In other words, $g = i^*\overline{g}$. A vector field X on M can always be locally extended to a vector field $\overline{X}$ on $\overline{M}$. Let $\overline{\nabla}$ be the Levi-Civita connection of $(\overline{M}, \overline{g})$. It is easy to prove that the component of $\overline{\nabla}_{\overline{X}}\overline{Y}$ that is tangent to M, denoted $(\overline{\nabla}_{\overline{X}}\overline{Y})^\top$, gives the value of the Levi-Civita connection of (M, g) applied to vector fields X and Y along M, independent of the choice of local extensions $\overline{X}$ and $\overline{Y}$ (cf. Exercise 2.2). The difference between these connections is a bilinear symmetric map with values on the normal space to M, called the *second fundamental form* II of M:

$$\mathrm{II}_p: T_pM \times T_pM \to \nu_pM, \quad \mathrm{II}(X, Y)_p := (\overline{\nabla}_{\overline{X}}\overline{Y})_p - (\overline{\nabla}_{\overline{X}}\overline{Y})_p^\top,$$

where ν_pM denotes the *normal space* to M at p, i.e., such that $T_{i(p)}\overline{M} = T_pM \oplus \nu_pM$ is a $\overline{g}$-orthogonal direct sum. For each normal vector ξ to M at p, we can also define a bilinear symmetric form II_ξ by taking the inner product of ξ and II,

$$(\mathrm{II}_\xi)_p: T_pM \times T_pM \to \mathbb{R}, \quad \mathrm{II}_\xi(X, Y)_p := \overline{g}_p(\xi, \mathrm{II}(X, Y)).$$

If II vanishes identically, then M is called a *totally geodesic* submanifold. This is equivalent to each geodesic of M being a geodesic of $\overline{M}$, since the connections ∇ and $\overline{\nabla}$ agree and hence so do the corresponding geodesic equations. Being totally geodesic is a very strong property; however, this class of submanifolds appears many times along this text (see, e.g., Exercise 2.29 and Proposition 3.93).

The relation between curvatures on M and $\overline{M}$ is given by the *Gauss equation*

$$R(X,Y,X,Y)=\overline{R}(\overline{X},\overline{Y},\overline{X},\overline{Y})+\overline{g}(\mathrm{II}(X,X),\mathrm{II}(Y,Y))-\overline{g}(\mathrm{II}(X,Y),\mathrm{II}(X,Y)), \quad (2.7)$$

where R and $\overline{R}$ denote the curvature tensors of M and $\overline{M}$, respectively. Assuming X,Y and $\overline{X},\overline{Y}$ are pairs of orthonormal vectors, the quantities $\sec(X,Y)=R(X,Y,X,Y)$ and $\overline{\sec}(\overline{X},\overline{Y})=\overline{R}(\overline{X},\overline{Y},\overline{X},\overline{Y})$ are called *intrinsic* and *extrinsic* curvatures of the plane tangent to M spanned by these vectors. Note that both agree if M is totally geodesic.

Since II_ξ is symmetric, there exists a self-adjoint operator $\mathscr{S}_\xi$ with respect to g, called the *shape operator*, satisfying

$$g(\mathscr{S}_\xi X,Y)=\mathrm{II}_\xi(X,Y). \quad (2.8)$$

It is not difficult to prove that $\mathscr{S}_\xi(X)=(-\overline{\nabla}_X\overline{\xi})^\top$ where $\overline{\xi}$ is any normal field that extends ξ. Eigenvalues and eigenvectors of $\mathscr{S}_\xi(X)$ are respectively called *principal curvatures* and *principal directions*; see Remark 2.21 for a geometric interpretation.

Exercise 2.20. Compute the principal curvatures and directions of a round sphere and a round cylinder embedded in $\mathbb{R}^3$.

Hint: Use the explicit description of the normal vectors fields (e.g., the normal to the unit sphere is $\xi(x,y,z)=(x,y,z)$) and apply the chain rule, i.e., $\mathscr{S}_\xi(X)=(-\overline{\nabla}_X\overline{\xi})^\top=-\overline{\nabla}_X\overline{\xi}=\frac{d}{dt}(\xi\circ\alpha)$, where α is a curve in the surface.

The second fundamental form also allows us to define the *mean curvature vector* $\mathbf{H}$ of the submanifold M, which is a section of $v(M)$ given by the trace of the second fundamental form, i.e.,

$$\mathbf{H}(p):=\sum_i\mathrm{II}(e_i,e_i) \quad (2.9)$$

where $\{e_i\}$ is an orthonormal basis of T_pM. The length $\|\mathbf{H}\|$ of the mean curvature vector is called the *mean curvature* of M. A submanifold whose mean curvature is identically zero is called *minimal*. From the definitions, it is clear that totally geodesic submanifolds are automatically minimal, but the converse need not be true.

Remark 2.21. Let M be an embedded surface in $\overline{M}=\mathbb{R}^3$ with the induced metric. According to the Gauss equation (2.7), the sectional curvature of M coincides with *Gaussian curvature* of M, i.e., the product of eigenvalues $\lambda_1\lambda_2$ (principal curvatures) of the shape operator $\mathscr{S}_\xi(\cdot)=-\overline{\nabla}_{(\cdot)}\xi$, where ξ is a unitary normal vector to M. Notice that this is precisely the right-hand side of (2.7), since the determinant of II

is equal to the product of its eigenvalues. An important fact in differential geometry is that each embedded surface with nonzero Gaussian curvature is, up to rigid motions, locally given by the graph of $f(x_1, x_2) = \frac{1}{2}\left(\lambda_1 x_1^2 + \lambda_2 x_2^2\right) + O(\|x\|^3)$. In other words, if $\sec > 0$ (respectively $\sec < 0$) M is locally a small perturbation of elliptic (respectively hyperbolic) paraboloid.

The dual notion to an immersion is that of a submersion. Let $(\overline{M}, \overline{g})$ and (M, g) be Riemannian manifolds and $\pi \colon \overline{M} \to M$ be a submersion (recall Definition A.4). Then π is a *Riemannian submersion* if for all $p \in \overline{M}$, $d\pi_p$ is a linear isometry from $(\ker d\pi_p)^\perp$ onto $T_{\pi(p)}M$, where $(\ker d\pi_p)^\perp$ is the subspace of $T_p\overline{M}$ given by the $\overline{g}$-orthogonal complement to $\ker d\pi_p$. The subspace

$$\mathscr{V}_p := \ker d\pi_p \tag{2.10}$$

is called the *vertical space* at $p \in \overline{M}$, while

$$\mathscr{H}_p := (\ker d\pi_p)^\perp = \left\{ X \in T_p\overline{M} : \overline{g}(X, Y) = 0 \text{ for all } Y \in \mathscr{V}_p \right\} \tag{2.11}$$

is called the *horizontal space* at $p \in \overline{M}$. The collections of vertical and horizontal spaces on $\overline{M}$ form two complementary distributions $\mathscr{V}$ and $\mathscr{H}$ on $\overline{M}$, that we accordingly call *vertical* and *horizontal distributions*. The vertical distribution is clearly integrable, since it is tangent to the foliation $\mathscr{F}_\pi = \{\pi^{-1}(x) : x \in M\}$, but the horizontal distribution may be nonintegrable. Again, denote by $\overline{\nabla}$ and ∇ the Levi-Civita connections on $\overline{M}$ and M, respectively. The two fundamental tensors T and A of the Riemannian submersion $\pi \colon \overline{M} \to M$ are $(1,2)$-tensors on $\overline{M}$, given by

$$\begin{aligned}
T_X Y &= \left(\overline{\nabla}_{X^{\mathscr{V}}} Y^{\mathscr{V}}\right)^{\mathscr{H}} + \left(\overline{\nabla}_{X^{\mathscr{V}}} Y^{\mathscr{H}}\right)^{\mathscr{V}}, \\
A_X Y &= \left(\overline{\nabla}_{X^{\mathscr{H}}} Y^{\mathscr{H}}\right)^{\mathscr{V}} + \left(\overline{\nabla}_{X^{\mathscr{H}}} Y^{\mathscr{V}}\right)^{\mathscr{H}},
\end{aligned} \tag{2.12}$$

where $X^{\mathscr{V}}$ and $X^{\mathscr{H}}$ respectively denote vertical and horizontal components of a vector field X on $\overline{M}$. The restrictions of the tensors T and A to $\mathscr{V}$ and $\mathscr{H}$ can be interpreted respectively as the second fundamental form of the leaves of $\mathscr{F}_\pi$ and the integrability of the horizontal distribution $\mathscr{H}$. In particular, we say that a Riemannian submersion is *integrable* if the restriction of A to $\mathscr{H}$ vanishes identically.

Analogously to the local extension of vector fields for Riemannian immersions, there exists a horizontal lift property for Riemannian submersions. Namely, a vector field X on M can always be lifted to a horizontal vector field $\overline{X}$ on $\overline{M}$ that is π-related to X, i.e., such that $d\pi(\overline{X}) = X \circ \pi$. Using this, it is not hard to see that if X and Y are vector fields on M, then

$$\overline{\nabla}_{\overline{X}} \overline{Y} - \overline{\nabla_X Y} = A_{\overline{X}} \overline{Y} = \tfrac{1}{2}[\overline{X}, \overline{Y}]^{\mathscr{V}}, \tag{2.13}$$

see Petersen [183, p. 82]. The relation between curvatures on $\overline{M}$ and M was first studied by Gray [104] and O'Neill [177], who proved that

$$
\begin{aligned}
R(X,Y,X,Y) &= \overline{R}(\overline{X},\overline{Y},\overline{X},\overline{Y}) + 3\|A_{\overline{X}}\overline{Y}\|^2 \\
&= \overline{R}(\overline{X},\overline{Y},\overline{X},\overline{Y}) + \tfrac{3}{4}\|[\overline{X},\overline{Y}]^{\mathcal{V}}\|^2,
\end{aligned}
\tag{2.14}
$$

for all vectors X and Y on M. The above is called the *Gray-O'Neill formula*. In particular, it follows that if $\overline{M}$ satisfies a certain lower sectional curvature bound, then so does M. Formulas for the other sectional curvatures of $\overline{M}$ (containing vertical directions), as well as the implications for Ricci and scalar curvatures, can be found in Besse [33, Chap 9].

2.2   Bi-invariant Metrics

The main goal of this section is to study the special Riemannian structure on Lie groups given by *bi-invariant* metrics. We mostly use $\langle \cdot, \cdot \rangle$ to denote metrics on Lie groups, to avoid any confusion between a metric $g(\cdot,\cdot)$ and an element $g \in G$. Bi-invariant metrics on Lie groups are denoted $Q(\cdot,\cdot)$.

Definition 2.22. A Riemannian metric $\langle \cdot, \cdot \rangle$ on a Lie group G is *left-invariant* if L_g is an isometry for all $g \in G$, that is, if for all $g, h \in G$ and $X, Y \in T_h G$,

$$
\langle d(L_g)_h X, d(L_g)_h Y \rangle_{gh} = \langle X, Y \rangle_h.
$$

Similarly, *right-invariant* metrics are those for which the right translations R_g are isometries. Note that, given an inner product $\langle \cdot, \cdot \rangle_e$ in $T_e G$, it is possible to define a left-invariant metric on G by setting, for all $g \in G$ and $X, Y \in T_g G$,

$$
\langle X, Y \rangle_g := \langle d(L_{g^{-1}})_g X, d(L_{g^{-1}})_g Y \rangle_e,
$$

and the right-invariant case is analogous.

Definition 2.23. A *bi-invariant metric* Q on a Lie group G is a Riemannian metric that is simultaneously left- and right-invariant.

The natural extension of these concepts to k-forms is that a k-form $\omega \in \Omega^k(G)$ is *left-invariant* if it coincides with its pull-back by left translations, i.e., $L_g^* \omega = \omega$ for all $g \in G$. *Right-invariant* and *bi-invariant* forms are analogously defined. Once more, given any $\omega_e \in \Lambda^k(T_e G)$, it is possible to define a left-invariant k-form $\omega \in \Omega(G)$ by setting, for all $g \in G$ and $X_i \in T_g G$,

$$
\omega_g(X_1, \dots, X_k) := \omega_e(d(L_{g^{-1}})_g X_1, \dots, d(L_{g^{-1}})_g X_k).
$$

The right-invariant case is analogous.

Proposition 2.24. *Every compact Lie group admits a bi-invariant metric Q.*

Proof. Let ω be a right-invariant volume form[2] on G and $\langle \cdot, \cdot \rangle$ a right-invariant metric. Define for all $X, Y \in T_x G$,

$$Q(X,Y)_x := \int_G \langle dL_g X, dL_g Y \rangle_{gx} \, \omega.$$

First, we claim that Q is left-invariant. Fix $X, Y \in T_x G$ and consider the function $f \colon G \to \mathbb{R}$ given by $f(g) := \langle dL_g X, dL_g Y \rangle_{gx}$. Then,

$$
\begin{aligned}
Q(dL_h X, dL_h Y)_{hx} &= \int_G \langle dL_g (dL_h X), dL_g (dL_h Y) \rangle_{g(hx)} \, \omega \\
&= \int_G \langle dL_{gh} X, dL_{gh} Y \rangle_{(gh)x} \, \omega = \int_G f(gh) \, \omega \\
&= \int_G R_h^*(f \omega) = \int_G f \omega = \int_G \langle dL_g X, dL_g Y \rangle_{gx} \, \omega = Q(X,Y)_x,
\end{aligned}
$$

which proves that Q is left-invariant. We also have

$$
\begin{aligned}
Q(dR_h X, dR_h Y)_{xh} &= \int_G \langle dL_g(dR_h X), dL_g(dR_h Y) \rangle_{g(xh)} \, \omega \\
&= \int_G \langle dR_h dL_g X, dR_h dL_g Y \rangle_{(gx)h} \, \omega = \int_G \langle dL_g X, dL_g Y \rangle_{gx} \, \omega = Q(X,Y)_x,
\end{aligned}
$$

which proves that Q is right-invariant, concluding the proof. □

Exercise 2.25. Consider $\mathfrak{su}(n) = \{A \in \mathfrak{gl}(n, \mathbb{C}) : A^* + A = 0, \operatorname{tr} A = 0\}$, the Lie algebra of $\mathrm{SU}(n)$ (see Exercise 1.51). Verify that the inner product in $T_e \mathrm{SU}(n)$ defined by $Q(X,Y) = \frac{1}{2} \operatorname{Re} \operatorname{tr}(XY^*)$ can be extended to a bi-invariant metric.

Proposition 2.26. *Let G be a Lie group endowed with a bi-invariant metric Q, and $X, Y, Z \in \mathfrak{g}$. Then the following hold:*

(i) $Q([X,Y],Z) = -Q(Y,[X,Z])$;
(ii) $\nabla_X Y = \frac{1}{2}[X,Y]$;
(iii) $R(X,Y)Z = \frac{1}{4}[[X,Y],Z]$;
(iv) $R(X,Y,X,Y) = \frac{1}{4}\|[X,Y]\|^2$.

In particular, (G, Q) has nonnegative sectional curvature $\sec \geq 0$.

---

[2] i.e., $\omega \in \Omega^n(G)$ is a nonvanishing n-form, where $n = \dim G$, see Definition A.26.

Proof. Differentiating the formula $Q(\text{Ad}(\exp(tX))Y, \text{Ad}(\exp(tX))Z) = Q(Y,Z)$ it follows from Proposition 1.37 and (1.11) that

$$Q([X,Y],Z) + Q(Y,[X,Z]) = 0,$$

which proves (i). Furthermore, (ii) follows from the Koszul formula (2.1) using (i) and the fact that ∇ is symmetric.

To prove (iii), we use (ii) to compute $R(X,Y)Z$ as follows:

$$
\begin{aligned}
R(X,Y)Z &= \nabla_{[X,Y]}Z - \nabla_X\nabla_Y Z + \nabla_Y\nabla_X Z \\
&= \tfrac{1}{2}[[X,Y],Z] - \tfrac{1}{2}\nabla_X[Y,Z] + \tfrac{1}{2}\nabla_Y[X,Z] \\
&= \tfrac{1}{2}[[X,Y],Z] - \tfrac{1}{4}[X,[Y,Z]] + \tfrac{1}{4}[Y,[X,Z]] \\
&= \tfrac{1}{4}[[X,Y],Z] + \tfrac{1}{4}\left([[X,Y],Z] + [[Z,X],Y] + [[Y,Z],X]\right) \\
&= \tfrac{1}{4}[[X,Y],Z].
\end{aligned}
$$

Finally, to prove (iv), we use (i) to verify that

$$
\begin{aligned}
Q(R(X,Y)X,Y) &= \tfrac{1}{4}Q([[X,Y],X],Y) \\
&= -\tfrac{1}{4}Q([X,[X,Y]],Y) \\
&= \tfrac{1}{4}Q([X,Y],[X,Y]) \\
&= \tfrac{1}{4}\|[X,Y]\|^2.
\end{aligned}
$$

$\square$

Theorem 2.27. *The Lie exponential map and the Riemannian exponential map at the identity agree on Lie groups endowed with bi-invariant metrics. In particular, the Lie exponential map of a connected compact Lie group is surjective.*

Proof. Let G be a Lie group endowed with a bi-invariant metric and $X \in \mathfrak{g}$. To prove that the exponential maps coincide, it suffices to prove that the 1-parameter subgroup $\gamma: \mathbb{R} \to G$ given by $\gamma(t) = \exp(tX)$ is the geodesic with $\gamma(0) = e$ and $\gamma'(0) = X$. First, recall that γ is the integral curve of the left-invariant vector field X passing through $e \in G$ at $t = 0$, that is, $\gamma'(t) = X(\gamma(t))$ and $\gamma(0) = e$. Furthermore, from Proposition 2.26,

$$\frac{\mathrm{D}}{\mathrm{d}t}\gamma' = \frac{\mathrm{D}}{\mathrm{d}t}X(\gamma(t)) = \nabla_\gamma X = \nabla_X X = \tfrac{1}{2}[X,X] = 0.$$

Therefore, γ is a geodesic and the Lie exponential map coincides with the Riemannian exponential map.

From Proposition 2.24, if G is compact, it admits a bi-invariant metric Q. Using that exponential maps coincide, and that the Lie exponential map is defined for all $X \in \mathfrak{g}$, it follows from the Hopf-Rinow Theorem 2.9 that (G,Q) is a complete Riemannian manifold. Thus, $\exp = \exp_e : T_e G \to G$ is surjective. $\qquad\square$

Exercise 2.28. Use the fact that exp is not surjective in $\mathrm{SL}(2,\mathbb{R})$ to prove that $\mathrm{SL}(2,\mathbb{R})$ does not admit a metric such that the Lie exponential map and the Riemannian exponential map coincide in e.

Exercise 2.29. Let G be a compact Lie group endowed with a bi-invariant metric. Prove that each closed subgroup H is a totally geodesic submanifold.

In the next result, we prove that each Lie group G with bi-invariant metric is a *symmetric space*, i.e., for each $a \in G$ there exists an isometry I^a that reverses geodesics through a (see also Exercise 6.32).

Theorem 2.30. *Let G be a connected Lie group endowed with a bi-invariant metric. For each $g \in G$, set*

$$I^g : G \to G, \quad I^g(x) := gx^{-1}g.$$

Then I^g is an isometry that fixes g and reverses geodesics through g. In other words, $I^g \in \mathrm{Iso}(G)$ and if γ is a geodesic with $\gamma(0) = g$, then $I^g(\gamma(t)) = \gamma(-t)$.

Proof. Since $I^e(g) = g^{-1}$, the map $\mathrm{d}(I^e)_e : T_e G \to T_e G$ is the multiplication by -1, i.e., $\mathrm{d}(I^e)_e = -\mathrm{id}$. Hence it is an isometry of $T_e G$. Since $\mathrm{d}(I^e)_g = \mathrm{d}(R_{g^{-1}})_e \circ \mathrm{d}(I^e)_e \circ \mathrm{d}(L_{g^{-1}})_g$, for any $g \in G$, the map $\mathrm{d}(I^e)_g : T_g G \to T_{g^{-1}} G$ is also an isometry. Hence I^e is an isometry. It clearly reverses geodesics through e, and since $I^g = R_g I^e R_g^{-1}$, it follows that I^g is also an isometry that reverses geodesics through g. $\qquad\square$

2.3 Killing Form and Semisimple Lie Algebras

To continue our study of bi-invariant metrics, we introduce the Killing form, which allows to establish a classical algebraic condition for compactness of Lie groups.

Definition 2.31. Let G be a Lie group and $X,Y \in \mathfrak{g}$. The *Killing form* of $\mathfrak{g}$ (also said Killing form of G) is the symmetric bilinear form

$$B(X,Y) := \mathrm{tr}\left(\mathrm{ad}(X)\mathrm{ad}(Y)\right).$$

If B is nondegenerate, then $\mathfrak{g}$ is said to be *semisimple*.

We provide in Theorem 2.41 other equivalent definitions of semisimplicity for a Lie algebra. A Lie group G is said to be *semisimple* if its Lie algebra $\mathfrak{g}$ is semisimple.

Proposition 2.32. *The Killing form is Ad-invariant, that is, for all $X,Y \in \mathfrak{g}$ and $g \in G$, it holds that $B(\mathrm{Ad}(g)X, \mathrm{Ad}(g)Y) = B(X,Y)$.*

Proof. If $\varphi : \mathfrak{g} \to \mathfrak{g}$ is a Lie algebra automorphism, then $\mathrm{ad}(\varphi(X))\varphi(Y) = \varphi \circ \mathrm{ad}(X)Y$. Thus, $\mathrm{ad}(\varphi(X)) = \varphi \circ \mathrm{ad}(X) \circ \varphi^{-1}$ and hence

$$B(\varphi(X), \varphi(Y)) = \mathrm{tr}\left(\mathrm{ad}(\varphi(X))\mathrm{ad}(\varphi(Y))\right)$$
$$= \mathrm{tr}\left(\varphi \circ \mathrm{ad}(X)\mathrm{ad}(Y) \circ \varphi^{-1}\right)$$
$$= \mathrm{tr}\left(\mathrm{ad}(X)\mathrm{ad}(Y)\right)$$
$$= B(X,Y).$$

Since $\mathrm{Ad}(g)$ is a Lie algebra automorphism, the proof is complete.   □

Corollary 2.33. *Let G be a semisimple Lie group with negative-definite Killing form B. Then $-B$ is a bi-invariant metric.*

Remark 2.34. Let G be a Lie group endowed with a bi-invariant metric Q. From Proposition 2.26, it follows that

$$\mathrm{Ric}(X,Y) = \mathrm{tr}\, R(X,\cdot)Y = \mathrm{tr}\, \tfrac{1}{4}[[X,\cdot],Y] = -\tfrac{1}{4}\mathrm{tr}[Y,[X,\cdot]] = -\tfrac{1}{4}B(X,Y). \qquad (2.15)$$

Therefore, the Ricci tensor of (G,Q) is independent of the bi-invariant metric Q.

Theorem 2.35. *Let G be a n-dimensional semisimple connected Lie group. Then G is compact if and only if its Killing form B is negative-definite.*

Proof. First, suppose that B is negative-definite. From Corollary 2.33, $-B$ is a bi-invariant metric on G. Hence, the Hopf-Rinow Theorem 2.9 and Theorem 2.30 imply that $(G,-B)$ is a complete Riemannian manifold, whose Ricci curvature satisfies (2.15). It follows from the Bonnet-Myers Theorem 2.19 that G is compact.

Conversely, suppose G is compact. From Proposition 2.24, it admits a bi-invariant metric Q. Hence, using item (i) of Propositions 2.26 and 1.37, it follows that, if $(e_1,\ldots,e_n)$ is an orthonormal basis of $\mathfrak{g}$, then

$$B(X,X) = \mathrm{tr}\left(\mathrm{ad}(X)\mathrm{ad}(X)\right)$$
$$= \sum_{i=1}^{n} Q(\mathrm{ad}(X)\mathrm{ad}(X)e_i, e_i)$$
$$= -\sum_{i=1}^{n} Q(\mathrm{ad}(X)e_i, \mathrm{ad}(X)e_i)$$
$$= -\sum_{i=1}^{n} \|\mathrm{ad}(X)e_i\|^2 \leq 0.$$

Note that, if there exists $X \neq 0$ such that $\|\mathrm{ad}(X)e_i\|^2 = 0$ for all i, then by definition of the Killing form, $B(Y,X) = 0$ for each Y. This would imply that B is degenerate, contradicting the fact that $\mathfrak{g}$ is semisimple. Thus, for each $X \neq 0$, we have $B(X,X) < 0$. Therefore, B is negative-definite.   □

The next result follows directly from Corollary 2.33, Remark 2.34 and Theorem 2.35.

Corollary 2.36. *Let G be a semisimple compact connected Lie group with Killing form B. Then* $(G, -B)$ *is an Einstein manifold with positive Einstein constant* $\lambda = \frac{1}{4}$.

We conclude this section with a discussion on equivalent definitions of semisimple Lie algebras. The key concept in this discussion is that of ideal of a Lie algebra. A Lie subalgebra $\mathfrak{h}$ is an *ideal* of a Lie algebra $\mathfrak{g}$ if $[X, Y] \in \mathfrak{h}$, for all $X \in \mathfrak{h}$ and $Y \in \mathfrak{g}$. Using tools from the previous chapter, we now prove the following important relation between ideals of a Lie algebra and normal subgroups of a Lie group.

Proposition 2.37. *Let G be a Lie group with Lie algebra* $\mathfrak{g}$. *If* $\mathfrak{h}$ *is an ideal of* $\mathfrak{g}$, *then the connected Lie subgroup H with Lie algebra* $\mathfrak{h}$ *is a normal subgroup of G. Conversely, if H is a normal Lie subgroup of G, then its Lie algebra* $\mathfrak{h}$ *is an ideal of* $\mathfrak{g}$.

Proof. Suppose $\mathfrak{h}$ is an ideal of $\mathfrak{g}$, and let H be the connected Lie subgroup with Lie algebra $\mathfrak{h}$ given by Proposition 1.21. Then, by Proposition 1.37, we have that $\mathrm{ad}(X)Y \in \mathfrak{h}$ for all $X \in \mathfrak{g}$ and $Y \in \mathfrak{h}$. From (1.12), it follows that

$$\mathrm{Ad}(\exp(X))Y = \exp(\mathrm{ad}(X))Y = \sum_{k=0}^{\infty} \frac{\mathrm{ad}(X)^k}{k!} Y \in \mathfrak{h}.$$

On the other hand, from (1.10),

$$\exp(\mathrm{Ad}(\exp(X))Y) = \exp(X)\exp(Y)\exp(X)^{-1},$$

and hence $\exp(X)\exp(Y)\exp(X)^{-1} \in H$. From Proposition 1.20, it follows that H is a normal subgroup of G.

Conversely, suppose H is a normal subgroup of G, that is $gHg^{-1} = H$ for all $g \in G$. Then $\exp(t\mathrm{Ad}(g)X) = g\exp(tX)g^{-1} \in H$ for all $g \in G$ and $X \in \mathfrak{h}$. Differentiating at $t = 0$, we have that $\mathrm{Ad}(g)$ maps $\mathfrak{h}$ to itself. In particular, $\mathrm{Ad}(\exp(tX))Y \in \mathfrak{h}$ for all $X \in \mathfrak{g}$ and $Y \in \mathfrak{h}$. Differentiating again at $t = 0$ and using Proposition 1.37, we have that $[X, Y] \in \mathfrak{h}$ for all $X \in \mathfrak{g}$ and $Y \in \mathfrak{h}$, that is, $\mathfrak{h}$ is an ideal of $\mathfrak{g}$. $\square$

If an ideal $\mathfrak{h}$ has no ideals other than the trivial, $\{0\}$ and $\mathfrak{h}$, it is called *simple*. Following the usual convention, by *simple Lie algebras* we mean Lie subalgebras that are noncommutative simple ideals. We stress that simple ideals (which may be commutative) are not referred to as Lie algebras, but simply called *simple ideals*.

Given a Lie algebra $\mathfrak{g}$, consider the decreasing sequence of ideals

$$\mathfrak{g}^{(1)} := [\mathfrak{g}, \mathfrak{g}], \quad \mathfrak{g}^{(2)} := [\mathfrak{g}^{(1)}, \mathfrak{g}^{(1)}], \quad \cdots \quad \mathfrak{g}^{(k)} := [\mathfrak{g}^{(k-1)}, \mathfrak{g}^{(k-1)}], \quad \cdots$$

If there exists a positive integer m such that $\mathfrak{g}^{(m)} = \{0\}$, then $\mathfrak{g}$ is said to be *solvable*.

Example 2.38. Consider the ideal of all $n \times n$ matrices (a_{ij}) over $\mathbb{K} = \mathbb{R}$ or $\mathbb{K} = \mathbb{C}$, with $a_{ij} = 0$ if $i > j$. One can easily verify that this is a solvable Lie algebra. Other trivial examples are nilpotent Lie algebras.

It is possible to prove that every Lie algebra admits a maximal solvable ideal τ, called its *radical*. We recall some results whose proof can be found in Ise and Takeuchi [133].

Proposition 2.39. *The following hold:*

(i) *If $\mathfrak{h}$ is an ideal of $\mathfrak{g}$, then the Killing form $B_{\mathfrak{h}}$ of $\mathfrak{h}$ satisfies $B(X,Y) = B_{\mathfrak{h}}(X,Y)$, for all $X,Y \in \mathfrak{h}$;*

(ii) *If $\mathfrak{g} = \mathfrak{g}_1 \oplus \mathfrak{g}_2$ is direct sum of ideals, then $\mathfrak{g}_1$ is orthogonal to $\mathfrak{g}_2$ with respect to B. Thus B is the sum of the Killing forms B_1 and B_2 of $\mathfrak{g}_1$ and $\mathfrak{g}_2$, respectively.*

Cartan Theorem 2.40. *A Lie algebra $\mathfrak{g}$ is solvable if and only if $B(\mathfrak{g},\mathfrak{g}^{(1)}) = \{0\}$. In particular, if B vanishes identically, then $\mathfrak{g}$ is solvable.*

We are now ready to present a theorem that gives equivalent definitions of semisimple Lie algebras.

Theorem 2.41. *Let $\mathfrak{g}$ be a Lie algebra with Killing form B. Then the following are equivalent:*

(i) *$\mathfrak{g} = \mathfrak{g}_1 \oplus \cdots \oplus \mathfrak{g}_n$ is the direct sum of simple Lie algebras $\mathfrak{g}_i$ (i.e., noncommutative simple ideals);*

(ii) *$\mathfrak{g}$ has trivial radical $\tau = \{0\}$;*

(iii) *$\mathfrak{g}$ has no commutative ideal other than $\{0\}$;*

(iv) *B is nondegenerate, i.e., $\mathfrak{g}$ is semisimple.*

Before proving this theorem, we present two important properties of semisimple Lie algebras.

Remark 2.42. If the Lie algebra $\mathfrak{g}$ is the direct sum of noncommutative simple ideals $\mathfrak{g} = \mathfrak{g}_1 \oplus \cdots \oplus \mathfrak{g}_n$, then $[\mathfrak{g},\mathfrak{g}] = \mathfrak{g}$. Indeed, on the one hand, $[\mathfrak{g}_i,\mathfrak{g}_j] = \{0\}$ if $i \neq j$, since $\mathfrak{g}_i$ and $\mathfrak{g}_j$ are ideals. On the other hand, $[\mathfrak{g}_i,\mathfrak{g}_i] = \mathfrak{g}_i$, since $\mathfrak{g}_i^{(1)} = [\mathfrak{g}_i,\mathfrak{g}_i]$ is an ideal and $\mathfrak{g}_i$ is a noncommutative simple ideal.

Remark 2.43. If $\mathfrak{g} = \mathfrak{g}_1 \oplus \cdots \oplus \mathfrak{g}_n$ is the direct sum of simple Lie algebras, then this decomposition is unique up to permutations. In fact, consider another decomposition $\mathfrak{g} = \widetilde{\mathfrak{g}}_1 \oplus \cdots \oplus \widetilde{\mathfrak{g}}_m$ and let $\widetilde{X} \in \widetilde{\mathfrak{g}}_k$. Then $\widetilde{X} = \sum_i X_i$, where $X_i \in \mathfrak{g}_i$. Since $\mathfrak{g}_i$ is a noncommutative simple ideal, for each i such that $X_i \neq 0$, there exists $V_i \in \mathfrak{g}_i$ different from X_i, such that $[V_i,X_i] \neq 0$. Hence $[\widetilde{X},V_i] \neq 0$ is a vector that belongs to both $\mathfrak{g}_i$ and $\widetilde{\mathfrak{g}}_k$. Therefore, the ideal $\mathfrak{g}_i \cap \widetilde{\mathfrak{g}}_k$ is different from $\{0\}$. Since $\mathfrak{g}_i$ and $\widetilde{\mathfrak{g}}_k$ are simple ideals, it follows that $\mathfrak{g}_i = \mathfrak{g}_i \cap \widetilde{\mathfrak{g}}_k = \widetilde{\mathfrak{g}}_k$.

We now prove Theorem 2.41, following Ise and Takeuchi [133, pp. 71–72].

Proof. We proceed by proving the equivalences (i) $\Leftrightarrow$ (ii), (i) $\Leftrightarrow$ (iii) and (i) $\Leftrightarrow$ (iv).

(i) $\Leftrightarrow$ (ii). Assume that $\mathfrak{g} = \mathfrak{g}_1 \oplus \cdots \oplus \mathfrak{g}_n$ is the direct sum of noncommutative simple ideals. Let $\pi_i: \mathfrak{g} \to \mathfrak{g}_i$ denote the projection onto each factor and note that π_i is a Lie algebra homomorphism. Thus, the projection $\tau_i = \pi_i(\tau)$ is a solvable ideal of $\mathfrak{g}_i$. Since $\mathfrak{g}_i$ is simple, then either τ_i is equal to $\{0\}$ or to $\mathfrak{g}_i$. However, the solvable ideal τ_i cannot be equal to $\mathfrak{g}_i$, since $\mathfrak{g}_i = [\mathfrak{g}_i, \mathfrak{g}_i]$ (see Remark 2.42). Therefore, $\tau_i = \{0\}$ and hence $\tau = \{0\}$.

Conversely, assume that $\tau = \{0\}$. For any ideal $\mathfrak{h}$ of $\mathfrak{g}$, set $\mathfrak{h}^\perp = \{X \in \mathfrak{g} : B(X, \mathfrak{h}) = 0\}$. Note that $\mathfrak{h}^\perp$ and $\mathfrak{h}^\perp \cap \mathfrak{h}$ are ideals, and B restricted to $\mathfrak{h}^\perp \cap \mathfrak{h}$ vanishes identically. It then follows from Proposition 2.39 and the Cartan Theorem 2.40 that $\mathfrak{h}^\perp \cap \mathfrak{h}$ is solvable. Since the radical is trivial, we conclude that $\mathfrak{h}^\perp \cap \mathfrak{h} = \{0\}$. Therefore $\mathfrak{g} = \mathfrak{h} \oplus \mathfrak{h}^\perp$. Since $\mathfrak{g}$ is finite-dimensional, by induction, $\mathfrak{g}$ is the direct sum of simple ideals. Moreover, $\tau = \{0\}$ implies that each simple ideal is noncommutative.

(i) $\Leftrightarrow$ (iii). Suppose that there exists a nontrivial commutative ideal $\mathfrak{a}$. Then the radical must contain $\mathfrak{a}$, hence it is nontrivial. It then follows from (i) $\Leftrightarrow$ (ii) that $\mathfrak{g}$ is not a direct sum of noncommutative simple ideals.

Conversely, assume that $\mathfrak{g}$ is not a direct sum of noncommutative simple ideals. Then, from (i) $\Leftrightarrow$ (ii), the radical τ is nontrivial, and there exists a positive integer m such that $\tau^{(m-1)} \neq \{0\}$. Set $\mathfrak{a} = \tau^{(m-1)}$ and note that $\mathfrak{a}$ is a nontrivial commutative ideal.

(i) $\Leftrightarrow$ (iv). Assume that $\mathfrak{g} = \mathfrak{g}_1 \oplus \cdots \oplus \mathfrak{g}_n$ is the direct sum of noncommutative simple ideals. From Proposition 2.39, it suffices to prove that $B|_{\mathfrak{g}_i}$ is nondegenerate for each i. Consider a fixed i and let $\mathfrak{h} := \{X \in \mathfrak{g}_i : B(X, \mathfrak{g}_i) = 0\}$. Note that $\mathfrak{h}$ is an ideal of $\mathfrak{g}_i$. Since $\mathfrak{g}_i$ is a simple ideal, either $\mathfrak{h} = \mathfrak{g}_i$ or $\mathfrak{h} = \{0\}$. If $\mathfrak{h} = \mathfrak{g}_i$, then the Cartan Theorem 2.40 implies that $\mathfrak{g}_i$ is solvable, contradicting the fact that $[\mathfrak{g}_i, \mathfrak{g}_i] = \mathfrak{g}_i$. Therefore $\mathfrak{h} = \{0\}$, hence $B|_{\mathfrak{g}_i}$ is nondegenerate.

Conversely, assume that B is nondegenerate. From the last equivalence, it suffices to prove that $\mathfrak{g}$ has no commutative ideals other than $\{0\}$. Let $\mathfrak{a}$ be a commutative ideal of $\mathfrak{g}$. For each $X \in \mathfrak{a}$ and $Y \in \mathfrak{g}$ we have $\mathrm{ad}(X)\mathrm{ad}(Y)(\mathfrak{g}) \subset \mathfrak{a}$. Therefore $B(X, Y) = \mathrm{tr}\left(\mathrm{ad}(X)\mathrm{ad}(Y)\right)|_{\mathfrak{a}}$. On the other hand, since $\mathfrak{a}$ is commutative, $\mathrm{ad}(X)\mathrm{ad}(Y)|_{\mathfrak{a}} = 0$. Therefore $B(X, Y) = 0$ for each $X \in \mathfrak{a}$ and $Y \in \mathfrak{g}$. Since B is nondegenerate, it follows that $\mathfrak{a} = \{0\}$. $\qquad\square$

Exercise 2.44. Show that if a compact Lie group G is semisimple, then its center $Z(G)$ is finite. Conclude from Exercise 1.60 that $\mathrm{U}(n)$ is not semisimple.

2.4 Splitting Lie Groups with Bi-invariant Metrics

In Proposition 2.24, we proved that each compact Lie group admits a bi-invariant metric. In this section, we prove that the only simply-connected Lie groups that admit bi-invariant metrics are products of compact Lie groups with vector spaces. We also prove that, if the Lie algebra of a compact Lie group G is simple, then the bi-invariant metric on G is unique up to multiplication by constants.

Theorem 2.45. *Let $\mathfrak{g}$ be a Lie algebra endowed with a bi-invariant metric. Then $\mathfrak{g} = \mathfrak{g}_1 \oplus \ldots \oplus \mathfrak{g}_n$ is the direct orthogonal sum of simple ideals $\mathfrak{g}_i$. In addition, let $\widetilde{G}$ be the connected and simply-connected Lie group with Lie algebra isomorphic to $\mathfrak{g}$. Then $\widetilde{G}$ is isomorphic to the product of normal Lie subgroups $G_1 \times \ldots \times G_n$, such that $G_i = \mathbb{R}$ if $\mathfrak{g}_i$ is commutative, and G_i is compact if $\mathfrak{g}_i$ is noncommutative.*

Proof. In order to verify that $\mathfrak{g}$ is the direct orthogonal sum of simple ideals, it suffices to prove that, if $\mathfrak{h}$ is an ideal, then $\mathfrak{h}^\perp$ is also an ideal. Let $X \in \mathfrak{h}^\perp$, $Y \in \mathfrak{g}$ and $Z \in \mathfrak{h}$. Then, using Proposition 2.26, it follows that

$$Q([X,Y],Z) = -Q([Y,X],Z) = Q(X,[Y,Z]) = 0.$$

Hence $[X,Y] \in \mathfrak{h}^\perp$, and this proves the first assertion.

From Lie's Third Theorem 1.14, given $\mathfrak{g}_i$, there exists a unique connected and simply-connected Lie group G_i with Lie algebra isomorphic to $\mathfrak{g}_i$. In particular, $G_1 \times \ldots \times G_n$ is a connected and simply-connected Lie group, with Lie algebra $\mathfrak{g} = \mathfrak{g}_1 \oplus \ldots \oplus \mathfrak{g}_n$. From uniqueness in Lie's Third Theorem, it follows that $\widetilde{G} = G_1 \times \ldots \times G_n$. If $\mathfrak{g}_i$ is commutative and simple, then $\mathfrak{g}_i = \mathbb{R}$. Hence, since G_i is connected and simply-connected, $G_i \cong \mathbb{R}$. Else, if $\mathfrak{g}_i$ is noncommutative, observe that there does not exist $X \in \mathfrak{g}_i$, $X \neq 0$, such that $[X,Y] = 0$ for all $Y \in \mathfrak{g}_i$. Indeed, if there existed such X, then $\mathrm{span}\{X\} \subset \mathfrak{g}_i$ would be a nontrivial ideal. From the proof of Theorem 2.35, the Killing form of $\mathfrak{g}_i$ is negative-definite; and from the same theorem, G_i is compact. Finally, since $\mathfrak{g}_i$ is an ideal of $\mathfrak{g}$, it follows from Proposition 2.37 that G_i is a normal subgroup of $\widetilde{G}$.   □

The above theorem and Remark 2.42 imply the following corollary.

Corollary 2.46. *Let $\mathfrak{g}$ be a Lie algebra with a bi-invariant metric. Then $\mathfrak{g} = \widetilde{\mathfrak{g}} \oplus Z(\mathfrak{g})$ is a direct sum of ideals, where $\widetilde{\mathfrak{g}}$ is semisimple. In particular, $[\widetilde{\mathfrak{g}},\widetilde{\mathfrak{g}}] = \widetilde{\mathfrak{g}}$.*

Remark 2.47. Let G be a compact connected Lie group with a nontrivial connected and simply-connected normal Lie subgroup H. From the proof of Theorem 2.45, it follows that there exists a connected normal Lie subgroup L of G such that $G \cong (H \times L)/\Gamma$, where Γ is a finite normal subgroup of $H \times L$. In other words, if G admits a nontrivial normal subgroup, then *up to a finite covering* it splits as a product of Lie groups. For instance, $SU(n)$ is a normal subgroup of $U(n)$, and $U(n) \cong (SU(n) \times S^1)/\mathbb{Z}_n$, see Exercise 1.60. Furthermore, the Lie algebra of $U(n)$ splits as a direct sum of ideals $\mathfrak{u}(n) = \mathfrak{su}(n) \oplus Z(\mathfrak{u}(n))$, see Corollary 2.46. From Remark 1.56, another example is $SO(4) \cong (SU(2) \times SU(2))/\mathbb{Z}_2$, that has two normal subgroups isomorphic to $SU(2)$.

Proposition 2.48. *Let G be a compact simple Lie group with Killing form B and bi-invariant metric Q. Then the bi-invariant metric is unique up to rescaling. In addition, (G,Q) is an Einstein manifold, that is, it satisfies (2.6).*

Proof. Let $\widetilde{Q}$ be another bi-invariant metric on G, and let $P : \mathfrak{g} \to \mathfrak{g}$ be the positive-definite self-adjoint operator such that $\widetilde{Q}(X,Y) = Q(PX,Y)$. We claim that P and ad commute, that is, $P\,\mathrm{ad}(X) = \mathrm{ad}(X)P$ for all $X \in \mathfrak{g}$. Indeed,

$$Q(P\,\mathrm{ad}(X)Y,Z) = \widetilde{Q}(\mathrm{ad}(X)Y,Z)$$

$$= -\widetilde{Q}(Y,\mathrm{ad}(X)Z)$$

$$= -Q(PY,\mathrm{ad}(X)Z)$$

$$= Q(\mathrm{ad}(X)PY,Z).$$

Furthermore, eigenspaces of P are $\mathrm{ad}(X)$-invariant, that is, they are ideals. In fact, let $Y \in \mathfrak{g}$ be an eigenvector of P associated to an eigenvalue μ. Then $P\mathrm{ad}(X)Y = \mathrm{ad}(X)PY = \mu\,\mathrm{ad}(X)Y$. Since $\mathfrak{g}$ is simple, it follows that $P = \mu\,\mathrm{id}$, hence $\widetilde{Q} = \mu\,Q$.

Since G is compact, it follows from Theorem 2.35 and Corollary 2.33 that $-B$ is a bi-invariant metric. Hence, there exists λ such that $-B(X,Y) = 4\lambda Q(X,Y)$. Therefore, from Remark 2.34, $\mathrm{Ric}(X,Y) = -\frac{1}{4}B(X,Y) = \lambda Q(X,Y)$, so G is Einstein. $\square$

Exercise 2.49. Let G be a compact semisimple Lie group with Lie algebra $\mathfrak{g} = \mathfrak{g}_1 \oplus \cdots \oplus \mathfrak{g}_n$ given by the direct sum of noncommutative simple ideals. Consider a bi-invariant metric Q on G. Prove that there exist positive numbers λ_j such that

$$Q = \sum_j -\lambda_j B_j,$$

where $B_j = B|_{\mathfrak{g}_j}$ is the restriction of the Killing form to $\mathfrak{g}_j$.

Exercise 2.50. In order to compute the Killing form B of $\mathrm{SU}(n)$, recall that its Lie algebra is

$$\mathfrak{su}(n) = \big\{A \in \mathfrak{gl}(n,\mathbb{C}) : A^* + A = 0,\ \mathrm{tr}\,A = 0\big\},$$

see Exercise 1.51. Consider the following special diagonal matrices in $\mathfrak{su}(n)$,

$$X = \begin{pmatrix} i\theta_1 & & \\ & \ddots & \\ & & i\theta_n \end{pmatrix} \quad \text{and} \quad Y = \begin{pmatrix} i\zeta_1 & & \\ & \ddots & \\ & & i\zeta_n \end{pmatrix}.$$

Use the fact that $\mathrm{tr}\,X = \mathrm{tr}\,Y = 0$, hence $\sum_{i=1}^n \theta_i = \sum_{i=1}^n \zeta_i = 0$, and Exercise 1.38 to verify that computing B in the above X and Y gives:

$$B(X,Y) = \mathrm{tr}\,(\mathrm{ad}(X)\mathrm{ad}(Y)) = -2n\sum_{i=1}^n \theta_i\,\zeta_i.$$

From Exercise 2.25, the inner product $Q(Z,W) = \mathrm{Re}\,\mathrm{tr}(ZW^*)$ in $T_e\mathrm{SU}(n)$ can be extended to a bi-invariant metric Q. Since $\mathrm{SU}(n)$ is simple, from Proposition 2.48, there exists a constant $c \in \mathbb{R}$ such that $B = cQ$. Conclude that $c = -4n$ and that the Killing form of $\mathrm{SU}(n)$ is

$$B(Z,W) = -2n\,\mathrm{Re}\,\mathrm{tr}(ZW^*), \quad \text{for all } Z,W \in \mathfrak{su}(n).$$

Part II
Isometric Actions

Chapter 3
Proper and Isometric Actions

In this chapter, we present a concise introduction to the theory of proper and isometric actions. We begin with basic definitions and a quick primer on fiber bundles, leading to the Slice Theorem 3.49 and the Tubular Neighborhood Theorem 3.57. These fundamental results are used throughout the rest of the book; in particular, to establish a strong correspondence between proper and isometric actions in Sect. 3.3. Finally, the stratification of a manifold by orbit types of a proper action is studied in Sect. 3.5.

The following references were a source of inspiration for this chapter, and could serve as further reading material: Bredon [53], Duistermaat and Kolk [79], Kawakubo [137], Onishchik [179], Palais and Terng [182], Pedrosa [11, Part I], Spindola [197] and Walschap [222].

3.1 Proper Actions and Fiber Bundles

In this section, proper actions are introduced together with a preliminary study of fiber bundles, and accompanied by several examples.

Definition 3.1. Let G be a Lie group and M a smooth manifold. A smooth map $\mu: G \times M \to M$ is called a *(left) action* of G on M, or a *(left) G-action* on M, if

(i) $\mu(e,x) = x$, for all $x \in M$;
(ii) $\mu(g_1, \mu(g_2,x)) = \mu(g_1 g_2, x)$, for all $g_1, g_2 \in G$ and $x \in M$.

Whenever μ is implicit, the manifold M is called a *G-manifold* or *G-space*, and it is common to denote $\mu(g,x)$ by $g \cdot x$, or even gx, but we avoid this slight abuse of notation until later in the book.

Similarly, one can define a *right action* $\mu: M \times G \to M$ of G on M to be a smooth map satisfying properties analogous to (i) and (ii) above, see Example 3.8. For a right action, the short notation for $\mu(x,g)$ is $x \cdot g$, or xg.

© Springer International Publishing Switzerland 2015
M.M. Alexandrino, R.G. Bettiol, *Lie Groups and Geometric Aspects of Isometric Actions*, DOI 10.1007/978-3-319-16613-1_3

For example, evaluating matrices on vectors defines a left action of $G = \mathrm{GL}(n,\mathbb{R})$ on $M = \mathbb{R}^n$, given by $\mu(A,x) = Ax$. In Chap. 2, we have seen that every Lie group G acts on its Lie algebra $\mathfrak{g}$ via the adjoint action $\mu \colon G \times \mathfrak{g} \to \mathfrak{g}$, $\mu(g,X) = \mathrm{Ad}(g)X$, and this action is studied in detail in Chap. 4. Any Lie subgroup $H \subset G$ of a Lie group G determines a left (and a right) H-action on G, given by multiplication on the left (and, respectively, on the right). In addition, conjugation by elements in H is also an H-action on G.

Definition 3.2. Let $\mu \colon G \times M \to M$ be a left action and $x \in M$. The subgroup

$$G_x := \{g \in G : \mu(g,x) = x\} \subset G$$

is called *isotropy group* or *stabilizer* of $x \in M$, and

$$G(x) := \{\mu(g,x) : g \in G\} \subset M$$

is called the *orbit* of $x \in M$. If $G(x) = \{x\}$, then x is called a *fixed point* of the action.

The subgroup $\bigcap_{x \in M} G_x$ is called the *ineffective kernel* of the action; if it is the trivial group $\{e\}$, then the action is said to be *effective*. Moreover, if $G_x = \{e\}$, for all $x \in M$, the action is said to be *free*. If for all $x, y \in M$, there exists $g \in G$ with $\mu(g,x) = y$, then the action is said to be *transitive*. Notice that the isotropy group G_x of a fixed point is the entire group G, and an action with fixed points cannot be free.

Remark 3.3. Every G-action can be reduced to an effective action of the quotient of G by the ineffective kernel $\bigcap_{x \in M} G_x$, which is a normal subgroup of G.

Given a (left) G-action on M, let us also define two auxiliary maps[1]:

$$\begin{aligned} \mu^g \colon M &\longrightarrow M & \mu_x \colon G &\longrightarrow M \\ x &\longmapsto \mu(g,x) & g &\longmapsto \mu(g,x). \end{aligned} \tag{3.1}$$

The key idea of an action is that each $g \in G$ determines a *transformation* of M, namely $\mu^g \colon M \to M$. Since our actions are assumed smooth, μ^g are diffeomorphisms, and hence $\mu^G := \{\mu^g : g \in G\}$ can be identified with a subgroup[2] of the diffeomorphism group $\mathrm{Diff}(M)$. An orbit $G(x)$ consists of all possible images $\mu^g(x)$ for $g \in G$, and the isotropy group G_x consist of all $g \in G$ that fix $x = \mu^g(x)$. If M is endowed with a Riemannian metric g, an action on (M,g) is said to be *isometric*, or *by isometries*, if μ^g is an isometry of (M,g) for all $g \in G$. In this case, the metric g is said to be G-*invariant*, and μ^G can be identified with a subgroup of $\mathrm{Iso}(M,\mathrm{g})$.

---

[1] Analogous maps can be defined for a right G-action, simply replacing $\mu(g,x)$ by $\mu(x,g)$.

[2] Notice that μ^G is isomorphic to G only if the action μ is effective.

Example 3.4. The following are three basic examples of actions on $M = \mathbb{R}^3$.

(i) Let $G = SO(3)$ and set $\mu(A,x) = Ax$. The orbit $G(x)$ of a nonzero vector $x \in \mathbb{R}^3$ is the sphere of radius $\|x\|$ centered at the origin, and the isotropy group G_x is isomorphic to the group $SO(2)$ of rotations around the line spanned by x. The origin $x = 0$ is a fixed point of this action.

(ii) Let $G = SO(2)$ and set $\mu(A,x) = (A(x_1,x_2),x_3)$. If $x \in \mathbb{R}^3$ is such that $r^2 = x_1^2 + x_2^2 > 0$, then the orbit $G(x)$ is a circle of radius r contained in the plane perpendicular to the z-axis at $(0,0,x_3)$, and the isotropy group is $G_x = \{e\}$. If $r^2 = x_1^2 + x_2^2 = 0$, then x is a fixed point of the action.

(iii) Let $G = SO(2) \times \mathbb{R}$ and set $\mu((A,b),x) = (A(x_1,x_2),b+x_3)$. The orbit $G(x)$ of a vector $x \in \mathbb{R}^3$ with $r^2 = x_1^2 + x_2^2 > 0$ is a round cylinder of radius r whose rotation axis is the z-axis, and the isotropy group is $G_x = \{e\}$. The orbit $G(x)$ of a vector $x = (0,0,x_3)$ is the z-axis, and the isotropy group $G_x = SO(2)$ is the group of rotations around the z-axis.

Action (ii) is called a *subaction* of action (i), since it is the restriction of the action (i) to the subgroup of $SO(3)$ consisting of block diagonal matrices whose first 2×2 block is an element of $SO(2)$. Action (ii) is also clearly a subaction of action (iii). Both (i) and (ii) are called *linear actions* (or *(linear) representations*, see Definition 1.35) since μ^g is a linear transformation for all $g \in G$. Action (iii) is not linear.

Example 3.5. Let us describe a few operations to construct new actions using other actions. As mentioned in Example 3.4, restricting a G-action on M to a subgroup H defines an H-action on M called a *subaction*. Given (left) actions $\mu_1 : G_1 \times M_1 \to M_1$ and $\mu_2 : G_2 \times M_2 \to M_2$, we may define a *product action* of $G_1 \times G_2$ on $M_1 \times M_2$ by

$$\mu : (G_1 \times G_2) \times (M_1 \times M_2) \to M_1 \times M_2,$$
$$\mu((g_1,g_2),(x_1,x_2)) := (\mu_1(g_1,x_1),\mu_2(g_2,x_2)). \tag{3.2}$$

If $G_1 = G_2 = G$, then restricting the product action (3.2) to the diagonal subgroup $\Delta G := \{(g,g) \in G \times G\}$ defines a subaction called *diagonal action* of G on $M_1 \times M_2$.

Definition 3.6. Consider (left) G-actions $\mu_1 : G \times M_1 \to M_1$ and $\mu_2 : G \times M_2 \to M_2$. A map $f : M_1 \to M_2$ is called *G-equivariant* if $\mu_2(g,f(x)) = f(\mu_1(g,x))$ for all $x \in M_1$ and $g \in G$. Equivariant diffeomorphisms provide a natural notion of equivalence among G-manifolds.

Exercise 3.7. Let $\mu : G \times M \to M$ be a G-action. Verify that $\overline{\mu} : G \times TM \to TM$ defined by $\overline{\mu}(g,(p,v)) := (\mu(g,p),d(\mu^g)_p v)$ is a G-action on TM, and that the projection $\pi : TM \to M$, $\pi(p,v) = p$, is a G-equivariant map. Notice that $\overline{\mu}$ induces a subaction by the isotropy group G_p on $T_p M$, which is called *isotropy representation*.

Example 3.8. Let $\mu_L: G \times M \to M$ be a left G-action, and define $\mu_R: M \times G \to M$ by setting $\mu_R(x,g) := \mu_L(g^{-1},x)$. Then μ_R is a right G-action, and the identity map on M is a G-equivariant diffeomorphism with respect to these actions. Right actions can be analogously transformed into left actions.

Exercise 3.9. Let $\mu: G \times M \to M$ be a left action. Prove that the isotropy group G_x changes by conjugation as x moves along its orbit $G(x)$. More precisely, show that $G_{\mu(g,x)} = g\, G_x\, g^{-1}$.

If two orbits $G(x)$ and $G(y)$ have nontrivial intersection, then they coincide. This means that orbits of a G-action on M form a partition of M; and hence we can consider the quotient

$$M/G := \big\{G(x) : x \in M\big\},$$

called the *orbit space* or *quotient space* of the G-action on M. The natural projection $\pi: M \to M/G$, $\pi(x) := G(x)$, is called the *quotient map*, or *projection map*, and the topology on M/G is determined by declaring that $U \subset M/G$ is open if its preimage $\pi^{-1}(U) \subset M$ is open. This implies that π is continuous, and it is possible to prove that π is also an open map, i.e., maps open subsets of M to open sets of M/G.

Exercise 3.10. Find the orbit spaces M/G of each G-action on $M = \mathbb{R}^3$ from Example 3.4, by identifying a subset of $\mathbb{R}^3$ that contains a point representing each G-orbit.

Example 3.11. Consider the S^1-action on $\mathbb{C}$ by complex multiplication $e^{i\theta} \cdot z = e^{i\theta}z$, and the resulting product action[3] of the torus $S^1 \times S^1$ on $\mathbb{C} \times \mathbb{C}$. The diagonal action

$$\mu: S^1 \times (\mathbb{C} \times \mathbb{C}) \to (\mathbb{C} \times \mathbb{C}), \quad \mu\big(e^{i\theta},(z_1,z_2)\big) = \big(e^{i\theta}z_1, e^{i\theta}z_2\big) \tag{3.3}$$

is called the *Hopf action*. This S^1-action on $\mathbb{C}^2$ can be restricted to an S^1-action on the unit sphere $S^3 = \{(z_1,z_2) \in \mathbb{C}^2 : |z_1|^2 + |z_2|^2 = 1\}$, using the same formula (3.3), which is also called Hopf action.

In order to describe the orbit space S^3/S^1 of the Hopf action, consider the Northern hemisphere $\Sigma := \{(z_1,x_3) \in \mathbb{C} \times \mathbb{R} : |z_1|^2 + x_3^2 = 1, x_3 > 0\}$ of a 2-dimensional subsphere in S^3. If $z = (z_1,z_2) \in S^3$ with $z_2 \neq 0$, then there exists a unique $e^{i\theta} \in S^1$ such that $e^{i\theta} z_2$ is a (real) positive number, i.e., such that $\mu(e^{i\theta},z) \in \Sigma$. On the other hand, the orbit of $z = (z_1,0) \in S^3$ is the equator $\partial\Sigma$. Therefore, *topologically*, $S^3/S^1 = \Sigma \cup [(z_1,0)] = S^2$ is the compactification of the 2-disk, hence a 2-sphere. More details on its geometry are discussed later, see Exercise 3.31.

Exercise 3.12. Let $\theta: [0,+\infty) \to \mathbb{R}$ be a smooth function. Verify that $\mu(t,z) := e^{it\theta(|z|)}z$ defines an $\mathbb{R}$-action on $\mathbb{C}$, and that an orbit of this action is either a circle

---

[3]For more details on this action, see Example 6.48.

centered at the origin or a fixed point. Describe the corresponding orbit space, in terms of the set $\theta^{-1}(0) = \{r \in [0, +\infty) : \theta(r) = 0\}$ where θ vanishes.

In order to study the geometry and topology of orbit spaces, we first need to develop more tools. The next result asserts that, given a smooth action, we can associate to each element of $\mathfrak{g}$ a vector field on M.

Proposition 3.13. *Consider a smooth action* $\mu : G \times M \to M$.

(i) *Each* $X \in \mathfrak{g}$ *induces a smooth vector field* X^* *on* M, *called* action field, *by*

$$X^*(p) := \tfrac{\mathrm{d}}{\mathrm{d}t}\mu(\exp(tX), p)\big|_{t=0}.$$

(ii) *The flow of* X^* *is given by* $\varphi_t^{X^*} = \mu^{\exp(tX)}$.

Proof. Define $\sigma^X : M \to TG \times TM$ by $\sigma^X(p) := (X_e, 0_p)$, and note that $\mathrm{d}\mu \circ \sigma^X$ is smooth. Item (i) follows from the fact that $X^*(p)$ is given by

$$\mathrm{d}\mu \circ \sigma^X(p) = \mathrm{d}\mu \left(\tfrac{\mathrm{d}}{\mathrm{d}t}(\exp(tX), p)\big|_{t=0}\right) = \tfrac{\mathrm{d}}{\mathrm{d}t}\mu(\exp(tX), p)\big|_{t=0}.$$

For each $X \in \mathfrak{g}$ and $p \in M$, notice that $\mu^{\exp(tX)}$ satisfies $\mu^{\exp(tX)} = \mathrm{id}$ if $t = 0$, and

$$
\begin{aligned}
\tfrac{\mathrm{d}}{\mathrm{d}t}\mu^{\exp(tX)}(p)\big|_{t=t_0} &= \tfrac{\mathrm{d}}{\mathrm{d}s}\mu^{\exp((s+t_0)X)}(p)\big|_{s=0} \\
&= \tfrac{\mathrm{d}}{\mathrm{d}s}\mu\left(\exp(sX) \cdot \exp(t_0X), p\right)\big|_{s=0} \\
&= X^*(\mu(\exp(t_0X), p)) \\
&= X^*\left(\mu^{\exp(t_0X)}(p)\right),
\end{aligned}
$$

proving that $\varphi_t^{X^*} = \mu^{\exp(tX)}$ is the flow of the action field X^*. $\qquad\square$

Remark 3.14. It is easy to verify that the association $\mathfrak{g} \ni X \mapsto X^* \in \mathfrak{X}(M)$ defined in (i) is a *Lie anti-homomorphism*, i.e., $[X, Y]^* = -[X^*, Y^*]$.

A consequence of Proposition 3.13 is that, if all orbits of a G-action on M have the same dimension, then the partition

$$\mathscr{F} = \{G(p)\}_{p \in M} \tag{3.4}$$

is a *foliation* of M, see Definition A.16. The fact that orbits of a G-action are indeed smooth submanifolds is proved in Proposition 3.41, and the fact that for each $v \in T_pG(p)$ there exists $X \in \mathfrak{g}$ such that $X_p^* = v$ follows from (3.11). If the orbits of a G-action have varying dimension, then $\mathscr{F}$ is a *singular* foliation (see Remark 3.42 and Definition 5.1). Foliations of the form (3.4) are called *homogeneous foliations*.

Exercise 3.15. Consider the action $\mu: SO(3) \times \mathbb{R}^3 \to \mathbb{R}^3$ in Example 3.4 (i).

(i) Verify that if $X \in \mathfrak{so}(3) \cong \mathbb{R}^3$, then $X^*(p) = X \times p$;
(ii) Consider A_X defined in Exercise 1.9. Verify that e^{tA_X} is a rotation in $\mathbb{R}^3$ about the axis spanned by X with angular speed $\|X\|$.

For future reference, let us state the result of a simple computation regarding the derivative of the map μ_x in (3.1), for both left and right actions.

Proposition 3.16. *If μ is a left action, then the derivative of $\mu_x: G \to M$ in (3.1) satisfies $\ker d(\mu_x)_{g_0} = T_{g_0}(g_0 G_x)$. If μ is a right action, then it satisfies $\ker d(\mu_x)_{g_0} = T_{g_0}(G_x g_0)$.*

We now introduce the concept of *proper actions*, originally due to Palais [180]. Properness is a key hypothesis to extend the theory of actions of compact groups to the noncompact case.

Definition 3.17. An action $\mu: G \times M \to M$ is *proper* if the map

$$G \times M \ni (g,x) \longmapsto (\mu(g,x),x) \in M \times M \tag{3.5}$$

is proper, i.e., if the preimage of any compact subset of $M \times M$ under (3.5) is a compact subset of $G \times M$.

It follows directly from the definition that each isotropy group of a proper action is compact, since the preimage of $(x,x) \in M \times M$ under the above map is $G_x \times \{x\}$.

Exercise 3.18. Prove that only proper actions on a compact manifold M are those by compact groups.

The following characterization of proper actions is easy to prove, and implies that all actions by compact groups are proper.

Proposition 3.19. *An action $\mu: G \times M \to M$ is proper if and only if for any sequence $\{g_n\}$ in G and any convergent sequence $\{x_n\}$ in M, such that $\{\mu(g_n, x_n)\}$ converges, the sequence $\{g_n\}$ admits a convergent subsequence.*

Exercise 3.20. Let H be a closed subgroup of the Lie group G. Prove that right multiplication $G \times H \ni (g,h) \mapsto gh \in G$ is a free proper right H-action.

Exercise 3.21. An action $G \times M \to M$ is called *properly discontinuous* if for all $x \in M$, there exists a neighborhood $U \ni x$ such that $gU \cap U = \emptyset$ for all $g \in G \setminus \{e\}$. Let G be a discrete group that acts on a manifold M. Prove that this action is properly discontinuous if and only if it is free and proper.

Exercise 3.22. Prove that the $\mathbb{R}$-action on $\mathbb{C}$ defined in Exercise 3.12 is not proper if $\theta^{-1}(0)$ is nonempty.

Proper actions are closely related to principal bundles. Before describing this relation, we need some more definitions. We start with that of a *fiber bundle*.

Intuitively, a fiber bundle is a space that *locally* looks like a product space, but *globally* may fail to be a product. In the local scale, such objects are products of a *base* and a *fiber*, where we think of the fiber as a space that is "attached" to each point on the base space to form the fiber bundle. The way in which this attachment is made is locally coherent to make it look like a product at this level, however, it allows certain twists in a global scale. Information about these "twists" is encoded in a Lie group, called the *structure group*. Let us now give the rigorous definition:

Definition 3.23. Let E, B and F be manifolds and G a Lie group. Assume that $\pi: E \to B$ a smooth submersion and there is an effective left G-action on F. Furthermore, suppose that B admits an open covering $\{U_\alpha\}$ and that there exist diffeomorphisms $\psi_\alpha: U_\alpha \times F \to \pi^{-1}(U_\alpha)$ satisfying:

(i) $(\pi \circ \psi_\alpha)(b, f) = b$, for all $(b, f) \in U_\alpha \times F$;
(ii) If $U_\alpha \cap U_\beta \neq \emptyset$, then $(\psi_\beta^{-1} \circ \psi_\alpha)(b, f) = (b, \theta_{\alpha,\beta}(b)f)$, where $\theta_{\alpha,\beta}(b) \in G$ and $\theta_{\alpha,\beta}: U_\alpha \cap U_\beta \to G$ is smooth.

Then $(E, \pi, B, F, G, \{U_\alpha\}, \{\psi_\alpha\})$ is called a *coordinate bundle*. Moreover, $(E, \pi, B, F, G, \{U_\alpha\}, \{\psi_\alpha\})$ and $(E, \pi, B, F, G, \{V_\beta\}, \{\varphi_\beta\})$ are called *equivalent* if $(\varphi_\beta^{-1} \circ \psi_\alpha)(b, f) = (b, \widetilde{\theta}_{\alpha,\beta}(b)f)$, where $\widetilde{\theta}_{\alpha,\beta}: U_\alpha \cap V_\beta \to G$ is smooth. An equivalence class of coordinate bundles, denoted (E, π, B, F, G), is called a *fiber bundle*. In this case, E is called the *total space*, π the *projection*, B the *base*, F the *fiber* and G the *structure group*. For each $b \in B$, the preimage $\pi^{-1}(b)$ is called the *fiber over b* and is often denoted E_b. Furthermore, ψ_α are called *bundle charts* and $\theta_{\alpha,\beta}$ *transition functions*. A fiber bundle is usually denoted by $F \to E \to B$, or simply by its total space E, if the underlying structure is implicitly understood. We also say that E is an F-bundle over B.

Example 3.24. The product manifold $E = B \times F$ is the total space of the *trivial bundle*, for which $\pi: E \to B$ is the projection onto the first factor, $G = \{e\}$ and the bundle charts are the identity.

Remark 3.25. The total space E of a fiber bundle can be reconstructed by *gluing* trivial products $U_\alpha \times F$ according to transition functions $\theta_{\alpha,\beta}$. More precisely, consider the disjoint union $\bigsqcup_\alpha (U_\alpha \times F)$. Whenever $x \in U_\alpha \cap U_\beta$, identify $(x, f) \in U_\alpha \times F$ with $(x, \theta_{\alpha,\beta}(x)(f)) \in U_\beta \times F$. Then the quotient space $E = \bigsqcup_\alpha (U_\alpha \times F)/\sim$ and the projection onto the first factor is the bundle projection.

Examples of fiber bundles were already encountered in the previous chapters, such as the tangent bundle TM of a manifold M. A point in TM consists of a pair (p, v), where $v \in T_p M$. The projection $\pi: TM \to M$ is the map $(p, v) \mapsto p$, and the fiber over $p \in M$ is $\pi^{-1}(p) = T_p M$. The tangent bundle of a manifold is a particular example of a special class of fiber bundles called *vector bundles*. Vector bundles are, by definition, fiber bundles whose fiber is a vector space V and whose structure group is the group $GL(V)$ of automorphisms of V. The dimension of V is called the *rank* of the vector bundle. Another example of vector bundle is the *cotangent bundle* TM^* of M, whose fiber at each $p \in M$ is the dual space $T_p M^*$. Verifying that these are indeed vector bundles of rank $\dim M$ is an easy exercise.

Another important example of vector bundle is the *normal bundle* of a submanifold. Let N be a submanifold of M, and consider their tangent bundles TN and TM, respectively. There is a natural inclusion of T_pN in T_pM for all $p \in N$, and hence one can consider the quotient space $v_p(N) := T_pM/T_pN$. This is the fiber of the normal bundle of N over p. In other words, the normal bundle of N is defined as

$$v(N) := \bigsqcup_{p \in N} v_p(N) = \bigsqcup_{p \in N} T_pM/T_pN, \tag{3.6}$$

and is a vector bundle over N with rank equal to the codimension of N in M. Notice that the above definition does not require a Riemannian metric, i.e., in this context, there is no canonical way of measuring the *angle* between the normal space v_pN and the tangent space T_pN. However, if we consider M endowed with a Riemannian metric g, then the above definition is equivalent to

$$v(N) := \bigsqcup_{p \in N} \{v \in T_pM : g_p(v,w) = 0 \text{ for all } w \in T_pN\},$$

which explains why it is called the *normal* bundle. For the same reason, the normal bundle $v(N)$ is sometimes denoted $TN^\perp$.

Remark 3.26. Certain geometric structures on a fiber bundle can be encoded as a *reduction* of its structural group G to a Lie subgroup $H \subset G$. This corresponds to the bundle admitting an atlas with H-valued transition functions. For instance, if M is a manifold with $\dim M = n$, the presence of a Riemannian metric on M allows to reduce the structural group $GL(n, \mathbb{R})$ of TM to the subgroup $O(n)$, and furthermore to $SO(n)$, provided M is orientable.

Definition 3.27. A *smooth section* of a fiber bundle $F \to E \to B$ is a smooth map $\sigma: B \to E$ with $\pi \circ \sigma = \mathrm{id}_B$, that is, σ is a smooth choice of $\sigma(b) \in E_b$ in each fiber.

Example 3.28. Smooth functions on M are sections of the trivial bundle $M \times \mathbb{R}$. Vector fields are sections of the tangent bundle TM, and differential 1-forms are sections of the cotangent bundle TM^*. If N is a submanifold of M, then sections of the normal bundle $v(N)$ are normal vector fields along N.

A fiber bundle (P, ρ, B, F, G) is called a *principal G-bundle*, or *principal bundle*, if $F = G$ and the action of G on itself is by left translations. As explained in Proposition 3.33 and Theorem 3.34 below, a fiber bundle $G \to P \to B$ is principal if and only if its fibers are orbits of a free proper G-action (whose orbit space is hence B).

Example 3.29. There are two very familiar examples of principal G-bundles, related to basic structures of a smooth manifold M. First, the universal covering $\rho: \widetilde{M} \to M$ is a principal G-bundle over M, where $G = \pi_1(M)$ is its fundamental group. Notice that $\pi_1(M)$ determines a properly discontinuous action on $\widetilde{M}$ by deck transformations, see Exercise 3.21. Second, consider the *frame bundle* of M, defined by

$$B(TM) := \bigsqcup_{x \in M} B(T_x M),$$

where $B(T_x M)$ is the set of all frames (i.e., ordered bases) on the vector space $T_x M$. Then $B(TM)$ is a principal G-bundle over M, with $G = \mathrm{GL}(n, \mathbb{R})$, where $n = \dim M$. Indeed, note that $B(T_x M)$ is diffeomorphic to $\mathrm{GL}(n, \mathbb{R})$, since each frame $\xi = \{\xi_i\}$ in $B(T_x M)$ determines a linear isomorphism $\xi \colon \mathbb{R}^n \to T_x M$ given by $\xi(e_i) = \xi_i$, where $\{e_i\}$ is the canonical basis of $\mathbb{R}^n$. Note also that composition of isomorphisms $\mu(\xi, A) := \xi \circ A$ determines a transitive right action of $\mathrm{GL}(n, \mathbb{R})$ on $B(T_x M)$.

Example 3.30. The *Hopf bundles* $S^1 \to S^{2n+1} \to \mathbb{C}P^n$ and $S^3 \to S^{4n+3} \to \mathbb{H}P^n$ are respectively principal S^1-bundles and S^3-bundles. Recall that the complex projective space $\mathbb{C}P^n := S^{2n+1}/S^1$ and the quaternionic projective space $\mathbb{H}P^n := S^{4n+3}/S^3$ are, respectively, the spaces of complex lines in $\mathbb{C}^{n+1}$ and quaternionic lines in $\mathbb{H}^{n+1}$. These are parametrized by equivalence classes $[v]$ of unit vectors $v \in S^{2n+1} \subset \mathbb{C}^{n+1}$, or $v \in S^{4n+3} \subset \mathbb{H}^{n+1}$, that span the same complex or quaternionic line, respectively. Such an equivalence class $[v]$ is precisely the orbit of the unit vector v under the multiplication action of the group S^1 of unit complex numbers or the group $S^3 \cong \mathrm{Sp}(1)$ of unit quaternions. For more details, see Example 6.10.

Exercise 3.31 ($\star$). In this exercise, we study the Hopf bundle $S^1 \to S^3 \to \mathbb{C}P^1$, see also Example 3.11. Recall that $\mathbb{C}P^1$ is isometric to the sphere $S^2\left(\frac{1}{2}\right) \subset \mathbb{R}^3$ of radius $\frac{1}{2}$. Write $S^3 = \{(z_1, z_2) \in \mathbb{C}^2 : |z_1|^2 + |z_2|^2 = 1\}$ as a subset of $\mathbb{C}^2$ and $S^2\left(\frac{1}{2}\right) = \{(z, x) \in \mathbb{C} \times \mathbb{R} : |z|^2 + x^2 = \frac{1}{4}\}$ as a subset of $\mathbb{C} \times \mathbb{R}$. Consider the map

$$\pi \colon S^3 \to S^2\left(\tfrac{1}{2}\right), \quad \pi(z_1, z_2) := \left(z_1 \overline{z_2}, \frac{|z_1|^2 - |z_2|^2}{2}\right). \tag{3.7}$$

(i) Prove that π is well-defined, i.e., it maps S^3 to $S^2\left(\frac{1}{2}\right)$;

(ii) Verify that $\pi(z_1, z_2) = \pi(w_1, w_2)$ if and only if there exists $e^{i\theta} \in S^1$ such that $(w_1, w_2) = \left(e^{i\theta} z_1, e^{i\theta} z_2\right)$;

(iii) Let $N = \left(0, \frac{1}{2}\right) \in S^2\left(\frac{1}{2}\right)$ be the North pole, and consider the open subsets $U_\pm = S^2\left(\frac{1}{2}\right) \setminus \{\pm N\}$. Verify that $\pi^{-1}(U_\pm) = S^3 \setminus \pi^{-1}(\pm N)$;

(iv) Describe bundle charts $\psi_\pm \colon U_\pm \times S^1 \to \pi^{-1}(U_\pm)$, to verify that $\pi \colon S^3 \to S^2\left(\frac{1}{2}\right)$ is a fiber bundle, and moreover, a principal S^1-bundle. Observe that π is also a Riemannian submersion, see also Exercise 3.81.

Remark 3.32. It is possible to prove that a principal bundle admits a smooth section if and only if it is trivial, see Walschap [222, p. 68]. This follows from the correspondence between sections and equivariant bundle charts on a principal bundle. In particular, the Hopf bundles in Example 3.30 do not admit any smooth sections.

Proposition 3.33. *Principal G-bundles $G \to P \to B$ have an underlying free proper right G-action $\mu \colon P \times G \to P$ whose orbits are the fibers of the bundle.*

Proof. Denote by $\rho: P \to B$ the projection map. Given $x \in M$, let $\psi_\alpha: U_\alpha \times G \to \rho^{-1}(U_\alpha)$ be a bundle chart with $x = \psi_\alpha(b, f) \in \rho^{-1}(U_\alpha)$, and set

$$\mu(x, g) := \psi_\alpha\big(\psi_\alpha^{-1}(x) \cdot g\big), \tag{3.8}$$

where $(b, f) \cdot g := (b, f \cdot g)$, for all $b \in B$ and $f, g \in G$. Let us verify that this action is well-defined, i.e., that the above definition does not depend on the choice of ψ_α. This follows from the fact that the structure group acts on the *left*, while (3.8) is a *right* action (see Exercise 3.40). More precisely, for any $x \in \rho^{-1}(U_\alpha) \cap \rho^{-1}(U_\beta)$,

$$\begin{aligned}
\psi_\alpha\big(\psi_\alpha^{-1}(x) \cdot g\big) &= \psi_\alpha(b, fg) \\
&= \psi_\beta(b, \theta_{\alpha,\beta}(b)fg) \\
&= \psi_\beta\big((b, \theta_{\alpha,\beta}(b)f) \cdot g\big) \\
&= \psi_\beta\big(\psi_\beta^{-1}(x) \cdot g\big).
\end{aligned}$$

The definition of μ and Exercise 3.20 imply that μ is a free proper right action. The fact that its orbits coincide with the fibers is immediate from (3.8). $\square$

The next theorem provides a converse to the above result, and a method to build principal bundles. It also shows that, in special cases, M/G is a smooth manifold.

Theorem 3.34. *Let* $\mu: M \times G \to M$ *be a free proper right action. Then the orbit space* M/G *admits a smooth structure such that* $G \to M \to M/G$ *is a principal G-bundle, where the bundle projection map* $\rho: M \to M/G$ *is the quotient map.*

Remark 3.35. The smooth structure on M/G is such that $\rho: M \to M/G$ is smooth, and a map $h: M/G \to N$ is smooth if and only if $h \circ \rho$ is smooth. These properties uniquely characterize the smooth structure of M/G.

Proof (Sketch). Let S be a submanifold containing x, with[4] $T_x M = T_x S \oplus \mathrm{d}(\mu_x)_e(\mathfrak{g})$, and let $\varphi: S \times G \to M$ be the restriction of the G-action to $S \times G$, that is, $\varphi := \mu|_{S \times G}$.

Claim 3.36. Up to shrinking S, there exists a G-invariant neighborhood U of $G(x)$ such that $\varphi: S \times G \to U$ is a G-equivariant diffeomorphism, where the G-action on $S \times G$ is by right multiplication on the second factor.

In order to prove this claim, we first verify that $\mathrm{d}\varphi_{(s,e)}$ is a linear isomorphism and hence, by the Inverse Function Theorem, φ is a local diffeomorphism on a neighborhood of $S \times \{e\}$, up to shrinking S. Indeed, note that $\varphi(\cdot, e)$ is the identity map on S, and the linearization of $\varphi(s, \cdot)$ at e also has trivial kernel, since the action is free, see Proposition 3.16. It then follows from

$$\mathrm{d}\varphi_{(s,g)}(X, \mathrm{d}R_g Y) = \big(\mathrm{d}(\mu^g)_s \circ \mathrm{d}\varphi_{(s,e)}\big)(X, Y)$$

---

[4] A posteriori, this condition means that the submanifold S is transverse to the orbit $G(x)$.

that φ is a local diffeomorphism near each point of $S \times G$. Using that the action is proper and Proposition 3.19, a standard argument with sequences implies that φ must be injective, which completes the proof of Claim 3.36. Note that φ is not a chart for M/G, since S is contained in M and not in M/G.

The next part of the proof is to identify S with an open subset of M/G. To this aim, we recall that the quotient map $\rho: M \to M/G$ is continuous and open. Moreover, using that the action is proper, it is possible to prove that M/G is Hausdorff.[5] The above considerations and Claim 3.36 imply that $\rho(S)$ is an open subset of M/G and $\rho|_S: S \to \rho(S)$ is a homeomorphism. This allows to define charts on M/G, and bundle charts on the bundle $G \to M \to M/G$. The compatibility of these charts and bundle charts follows from the following.

Claim 3.37. Let S_1 and S_2 be submanifolds transverse to two different orbits, with $W := \rho(S_1) \cap \rho(S_2) \neq \emptyset$. Set $\rho_i := \rho|_{S_i}: S_i \to \rho(S_i)$, $V_i := \rho_i^{-1}(W)$ and $\varphi_i := \mu|_{S_i \times G}$.

(i) $(\rho_1^{-1} \circ \rho_2): V_2 \to V_1$ is a diffeomorphism;

(ii) $(\psi_2^{-1} \circ \psi_1)(b, g) = (b, \theta(b)g)$, where $\psi_i: \rho(S_i) \times G \to \rho^{-1}(\rho(S_i))$ is given by $\psi_i(b, g) = \varphi_i(\rho_i^{-1}(b), g)$ and $b \in W$.

In order to prove (i), note that $\varphi_2^{-1}(V_1)$ is a graph, i.e., $\varphi_2^{-1}(V_1) = \{(s, \widetilde{\theta}(s)) : s \in V_2\}$, which implies that $\widetilde{\theta}: S \to G$ is smooth. Thus, $(\rho_1^{-1} \circ \rho_2)(s) = \varphi(s, \widetilde{\theta}(s))$ is smooth, and also a diffeomorphism. As for (ii), defining the transition function as $\theta(b) := \widetilde{\theta}(\rho_2^{-1}(b))$ guarantees that ψ_1 and ψ_2 satisfy (ii) in Definition 3.23. $\qquad\square$

Corollary 3.38. *Let G be a Lie group and H be a closed subgroup. Consider the right H-action on G by multiplication on the right. The corresponding orbit space G/H is a manifold, and the quotient map $\rho: G \to G/H$, $\rho(g) = gH$, determines a principal H-bundle $H \to G \to G/H$. In addition, if H is a normal subgroup of G, then G/H is a Lie group and ρ is a Lie group homomorphism.*

Proof. The first part of the above statement follows directly from Theorem 3.34 and Exercise 3.20. To show that G/H is a Lie group if H is a normal subgroup, define

$$\alpha: G \times G \ni (g_1, g_2) \longmapsto g_1 g_2^{-1} \in G$$
$$\widetilde{\alpha}: G/H \times G/H \ni (g_1 H, g_2 H) \longmapsto g_1 g_2^{-1} H \in G/H. \tag{3.9}$$

Note that, since H is normal, $\widetilde{\alpha}$ is well-defined. The fact that $\rho: G \to G/H$ is a projection of a bundle and $\rho \circ \alpha = \widetilde{\alpha} \circ (\rho \times \rho)$ imply that $\widetilde{\alpha}$ is smooth. Therefore G/H is a Lie group, see Remark 1.2. $\qquad\square$

---

[5]In fact, since (3.5) is a proper map between locally compact Hausdorff spaces, it is also *closed*, i.e., maps closed subsets to closed subsets. Thus, its image $\mathscr{R} = \{(x, y) \in M \times M : G(x) = G(y)\}$ is closed. As $\rho: M \to M/G$ is an open map, $\rho(M \times M \setminus \mathscr{R})$ is open. This is easily seen to be the complement of the diagonal in $M/G \times M/G$, which is hence closed, proving that M/G is Hausdorff.

Example 3.39. Consider the free right action of a Lie group G on itself by multiplication on the right (see Exercise 3.20) and the induced diagonal action on $G \times G$,

$$\mu: (G \times G) \times \Delta G \to G \times G, \quad \mu\big((g_1, g_2), (g, g)\big) := (g_1 g, g_2 g), \qquad (3.10)$$

see Example 3.5. This action is clearly free, hence by Theorem 3.34, there is a principal bundle $\Delta G \to G \times G \to (G \times G)/\Delta G$. The map $\alpha: G \times G \to G$ defined in (3.9) descends to a map $\overline{\alpha}: (G \times G)/\Delta G \to G$, $\overline{\alpha}([g_1, g_2]) = g_1 g_2^{-1}$, which is well-defined since $\alpha(\Delta G) = \{e\}$. It is easy to verify that $\overline{\alpha}$ is determines a Lie group isomorphism $(G \times G)/\Delta G \cong G$. This provides a convenient way of rewriting G as an orbit space, which is useful in several applications (see Remark 4.12).

Exercise 3.40. Suppose that M is a manifold with a free proper right G_1-action $\mu_1: M \times G_1 \to M$ and a left G_2-action $\mu_2: G_2 \times M \to M$. Suppose that μ_1 and μ_2 are actions that *commute*,[6] that is, $\mu_1(\mu_2(g_2, x), g_1) = \mu_2(g_2, \mu_1(x, g_1))$ for all $x \in M$ and $g_i \in G_i$. Show that μ_2 descends to a left G_2-action $\widetilde{\mu}_2: G_2 \times M/G_1 \to M/G_1$, given by $\widetilde{\mu}_2(g_2, G_1(x)) := G_1(\mu_2(g_2, x))$, and verify that $\pi: M \to M/G_1$ is G_2-equivariant.

Each isotropy group $H = G_x$ of a G-action on M is such that G/H is a smooth manifold, by Corollary 3.38. We now prove that the orbit $G(x)$ is the image of an immersion of G/H into M.

Proposition 3.41. *Let* $\mu: G \times M \to M$ *be a left action and define* $\widetilde{\mu}_x: G/H \to M$ *by* $\widetilde{\mu}_x \circ \rho = \mu_x$, *where* $H = G_x$ *is the isotropy at* $x \in M$ *and* $\rho: G \to G/H$ *is the quotient map. Then* $\widetilde{\mu}_x$ *is a G-equivariant[7] injective immersion, with image $G(x)$. In particular, $G(x)$ is an immersed submanifold of M whose tangent space at $x \in M$ is $T_x G(x) = \mathrm{d}(\mu_x)_e(\mathfrak{g})$. In addition, if the action is proper, then $\widetilde{\mu}_x$ is an embedding and $G(x)$ is an embedded submanifold of M.*

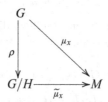

Proof. According to Corollary 3.38, $H \to G \to G/H$ is a principal H-bundle. This implies that $\widetilde{\mu}_x$ is smooth. It follows from Proposition 3.16 that the derivative of $\widetilde{\mu}_x$ at every point is injective, and hence $\widetilde{\mu}_x$ is an injective immersion. The fact that $\widetilde{\mu}_x$ is an embedding when the action is proper can be proved using Proposition 3.19. $\square$

---

[6] More generally, two (either left or right) actions μ_1 and μ_2 on M are said to *commute* if the induced transformations (3.1) on M satisfy $(\mu_1)^{g_1} \circ (\mu_2)^{g_2} = (\mu_2)^{g_2} \circ (\mu_1)^{g_1}$ for all $g_i \in G_i$.

[7] The G-action on G/H is by left translations, that is, $\overline{g} \cdot gH := \overline{g}gH$, see Exercise 3.40 and (6.13).

As observed above, the tangent space $T_x G(x)$ to an orbit is the image of the linear map $d(\mu_x)_e \colon \mathfrak{g} \to T_x M$. Denote by $\mathfrak{g}_x$ the Lie algebra of the isotropy group G_x, and let $\mathfrak{m}_x$ be a complement, i.e., $\mathfrak{g} = \mathfrak{g}_x \oplus \mathfrak{m}_x$ as vector spaces. Then, since $\ker d(\mu_x)_e = \mathfrak{g}_x$, we have that $\mathfrak{m}_x$ can be identified with $T_x G(x)$. More precisely, there is a natural identification

$$\mathfrak{m}_x \ni X \longmapsto X^*(x) \in T_x G(x), \tag{3.11}$$

where X^* is the action field on M induced by $X \in \mathfrak{m}_x$, see Proposition 3.13.

Remark 3.42. From Proposition 3.41, orbits $G(x)$ of a left action $\mu \colon G \times M \to M$ are immersed submanifolds of M. The dimension of these submanifolds $G(x)$ is *lower semi-continuous* on x, i.e., for each $x_0 \in M$, the dimension of orbits $G(x)$ for x near x_0 is greater than or equal to $\dim G(x_0)$. This can be proved using that $\dim G(x)$ is the rank of the linear map $d(\mu_x)_e \colon \mathfrak{g} \to T_x M$, which is in fact constant along $G(x)$. From continuity of $x \mapsto d(\mu_x)_e$ and lower semi-continuity of the rank of a continuous family of linear maps, it follows that $\dim G(x)$ is lower semi-continuous.

Example 3.43. Consider the 2-torus $T^2 = \mathbb{R}^2 / \mathbb{Z}^2$ as the orbit space of the right $\mathbb{Z}^2$-action by translations, $\mu_1 \colon \mathbb{Z}^2 \times \mathbb{R}^2 \to \mathbb{R}^2$, $\mu_1\big((n,m),(x,y)\big) = (x+n, y+m)$, recall 1.14. Choose a real number $\alpha \in \mathbb{R}$ and define a left $\mathbb{R}$-action on $\mathbb{R}^2$ by $\mu_2(t,(x,y)) := (x+t, y+\alpha t)$. By Exercise 3.40, μ_2 descends to an $\mathbb{R}$-action $\tilde{\mu}_2$ on T^2. If $\alpha \in \mathbb{Q}$ is rational, then the orbits of $\tilde{\mu}_2$ are embedded circles in T^2, which correspond to straight line segments in $\mathbb{R}^2$ joining two points on the lattice $\mathbb{Z}^2$. On the other hand, if $\alpha \notin \mathbb{Q}$ is irrational, then the orbits of $\tilde{\mu}_2$ are *dense*[8] (and uniformly distributed) in T^2; in particular, they are not embedded submanifolds. Recall that, by Exercise 3.18, the $\mathbb{R}$-action $\tilde{\mu}_2$ on T^2 cannot be proper, despite μ_2 being proper. Analogous statements can be proved for higher dimensional tori.

Exercise 3.44. Verify that the following are equivariantly diffeomorphic:

(i) $S^n = \mathrm{SO}(n+1)/\mathrm{SO}(n)$;
(ii) $\mathbb{R}P^n = \mathrm{SO}(n+1)/S(\mathrm{O}(n) \times \mathrm{O}(1))$;
(iii) $\mathbb{C}P^n = \mathrm{SU}(n+1)/S(\mathrm{U}(n) \times \mathrm{U}(1))$;
(iv) $\mathbb{H}P^n = \mathrm{Sp}(n+1)/\mathrm{Sp}(n) \times \mathrm{Sp}(1)$.

Here, $S(\mathrm{O}(n) \times \mathrm{O}(1))$ is the subgroup of $\mathrm{SO}(n+1)$ formed by the matrices

$$A = \begin{pmatrix} B & 0 \\ 0 & \pm 1 \end{pmatrix},$$

where $B \in \mathrm{O}(n)$ and $\det A = 1$, and analogously for $S(\mathrm{U}(1) \times \mathrm{U}(n))$.

---

[8]This is a consequence of the so-called Kronecker Approximation Theorem.

Hint: Use Example 3.30 and Proposition 3.41. To prove (ii), note that the $SO(n+1)$-action on S^n commutes with the antipodal $\mathbb{Z}_2$-action and hence descends to an action on $\mathbb{R}P^n$, see Exercise 3.40. More details on the solution to this exercise are given in Example 6.10.

Remark 3.45. The same idea in Exercise 3.44 can be used to find equivariant diffeomorphisms between the k-Grassmannians on $\mathbb{R}^n$, $\mathbb{C}^n$ and $\mathbb{H}^n$ and quotients of classical Lie groups. Recall that if V is a finite-dimensional vector space, the *k-Grassmannian* on V is defined as

$$\mathrm{Gr}_k(V) := \{W \text{ linear subspace of } V : \dim W = k\}.$$

In particular, $\mathrm{Gr}_1(\mathbb{R}^{n+1}) = \mathbb{R}P^n$, $\mathrm{Gr}_1(\mathbb{C}^{n+1}) = \mathbb{C}P^n$ and $\mathrm{Gr}_1(\mathbb{H}^{n+1}) = \mathbb{H}P^n$. Observe that $SO(n)$, $SU(n)$ and $Sp(n)$ act on, respectively, $\mathrm{Gr}_k(\mathbb{R}^n)$, $\mathrm{Gr}_k(\mathbb{C}^n)$ and $\mathrm{Gr}_k(\mathbb{H}^n)$. These transitive actions provide the equivariant diffeomorphisms:

(ii') $\mathrm{Gr}_k(\mathbb{R}^n) = SO(n)/S(O(n-k) \times O(k))$;
(iii') $\mathrm{Gr}_k(\mathbb{C}^n) = SU(n)/S(U(n-k) \times U(k))$;
(iv') $\mathrm{Gr}_k(\mathbb{H}^n) = Sp(n)/Sp(n-k) \times Sp(k)$.

Remark 3.46. The above also be used to prove that $SU(n)/T$ is a *complex flag manifold*, where T is the subgroup of $SU(n)$ formed by diagonal matrices. Indeed, let $F_{1,\ldots,n-1}(\mathbb{C}^n)$ be the set of *complex flags* on $\mathbb{C}^n$, i.e.,

$$F_{1,\ldots,n-1}(\mathbb{C}^n) := \{\{0\} \subset E_1 \subset \ldots \subset E_{n-1} : E_i \in \mathrm{Gr}_i(\mathbb{C}^n)\}.$$

Set $\widetilde{F} := \{(l_1,\ldots,l_n) : l_1 \ldots l_n \text{ orthogonal lines of } \mathbb{C}^n\}$. On the one hand, there is a natural bijection between $\widetilde{F}$ and $F_{1,\ldots,n-1}(\mathbb{C}^n)$ given by $(l_1,\ldots,l_n) \mapsto \{0\} \subset E_1 \subset \ldots \subset E_{n-1}$, where $E_i = l_1 \oplus \cdots \oplus l_i$. On the other hand, $SU(n)$ acts naturally on $\widetilde{F}$ and this action is transitive. Note that, for an orthogonal basis $\{e_1,\ldots,e_n\}$ of $\mathbb{C}^n$, the isotropy group of $(\mathbb{C}e_1,\ldots,\mathbb{C}e_n) \in \widetilde{F}$ is T. Therefore, $SU(n)/T = \widetilde{F} = F_{1,\ldots,n-1}(\mathbb{C}^n)$.

3.2 Slices and Tubular Neighborhoods

In this section, we discuss the construction of associated bundles to a principal bundle, and prove two foundational results for the theory discussed in the remainder of the book: the Slice Theorem 3.49 and the Tubular Neighborhood Theorem 3.57.

Definition 3.47. Let $\mu: G \times M \to M$ be an action. A *slice* at $x_0 \in M$ is an embedded submanifold S_{x_0} containing x_0 and satisfying the following properties:

(i) $T_{x_0}M = d\mu_{x_0}\mathfrak{g} \oplus T_{x_0}S_{x_0}$ and $T_xM = d\mu_x\mathfrak{g} + T_xS_{x_0}$, for all $x \in S_{x_0}$;
(ii) S_{x_0} is invariant under G_{x_0}, i.e., if $x \in S_{x_0}$ and $g \in G_{x_0}$, then $\mu(g,x) \in S_{x_0}$;
(iii) If $x \in S_{x_0}$ and $g \in G$ are such that $\mu(g,x) \in S_{x_0}$, then $g \in G_{x_0}$.

Example 3.48. Consider the action of $SO(2) \times \mathbb{R}$ on $\mathbb{R}^3$ defined in Example 3.4 (iii). A slice for this action at the point $x = (x_1,x_2,x_3) \in \mathbb{R}^3$ is either a straight line segment $S_x = \{(tx_1, tx_2, x_3) : |t - 1| < \varepsilon\}$ if $r^2 = x_1^2 + x_2^2 > 0$, or else, a disk $S_x = \{(y_1, y_2, x_3), \in \mathbb{R}^3 : y_1^2 + y_2^2 < \varepsilon\}$ if $r^2 = x_1^2 + x_2^2 = 0$. Notice that these slices are invariant under the corresponding isotropies $G_x = \{e\}$ and $G_x = SO(2)$, respectively.

Slice Theorem 3.49. *Let $\mu \colon G \times M \to M$ be a proper action and $x_0 \in M$. Then there exists a slice S_{x_0} at x_0.*

Proof. We start by constructing a Riemannian metric on M for which the subaction of the isotropy group $H = G_{x_0}$ is by isometries. This metric is defined by averaging the H-action, that is,

$$\langle X, Y \rangle_p := \int_H \langle\langle d\mu^g X, d\mu^g Y \rangle\rangle_{\mu(g,p)} \, \omega, \tag{3.12}$$

where ω is a right-invariant volume form on H and $\langle\langle \cdot, \cdot \rangle\rangle$ is an arbitrary Riemannian metric on M. Arguments similar to those in the proof of Proposition 2.24 imply that H acts by isometries on M equipped with the metric $\langle \cdot, \cdot \rangle$.

It is easy to see that, if $g \in H$, then $d\mu^g \colon T_{x_0}M \to T_{x_0}M$ leaves invariant the tangent space $T_{x_0}G(x_0)$. Since H acts by isometries, we have that $d\mu^g$ also leaves invariant the orthogonal complement to $T_{x_0}G(x_0)$ in $T_{x_0}M$, which is the normal space $v_{x_0}G(x_0)$. Define a candidate to slice at x_0 by setting

$$S_{x_0} := \exp_{x_0}\left(B_\varepsilon(0)\right), \tag{3.13}$$

where $B_\varepsilon(0)$ is an open ball of radius $\varepsilon > 0$ around the origin in the normal space $v_{x_0}G(x_0)$. Since $d\mu^g$ leaves invariant $v_{x_0}G(x_0)$ and isometries map geodesics to geodesics, S_{x_0} is invariant under the H-action. Therefore, item (ii) of Definition 3.47 is satisfied. Item (i) is satisfied by the construction of S_{x_0} and continuity of $d\mu$.

It only remains to verify item (iii), which we prove by contradiction. Suppose (iii) is not satisfied for any $\varepsilon > 0$ in (3.13). Then there exists a sequence $\{x_n\}$ in S_{x_0} and a sequence $\{g_n\}$ in G such that $\lim x_n = x_0$, $\lim \mu(g_n, x_n) = x_0$, $\mu(g_n, x_n) \in S_{x_0}$ and $g_n \notin H$. Since the action is proper, there exists a subsequence, which we also denote $\{g_n\}$, that converges to $g \in H$. Set $\widetilde{g}_n := g^{-1}g_n$. Then $\lim \widetilde{g}_n = e$, $\lim \mu(\widetilde{g}_n, x_n) = x_0$, $\mu(\widetilde{g}_n, x_n) \in S_{x_0}$ and $\widetilde{g}_n \notin H$. Using Proposition 3.16 and the Inverse Function Theorem, one can prove the next claim, up to possibly shrinking S_{x_0}.

Claim 3.50. There exists a submanifold $C \subset G$ containing e, such that $\mathfrak{g} = \mathfrak{h} \oplus T_e C$, where $\mathfrak{h}$ is the Lie algebra of H. Furthermore, $\varphi \colon C \times S_{x_0} \ni (c, s) \mapsto \mu(c, s) \in M$ is a diffeomorphism onto its image.

It follows from the Inverse Function Theorem that for each $\widetilde{g}_n$, there exists a unique $c_n \in C$ and $h_n \in H$, such that $\widetilde{g}_n = c_n h_n$. Since $\widetilde{g}_n \notin H$, we conclude that $c_n \neq e$. On the other hand, $\mu(h_n, x_n) \in S_{x_0}$ because $h_n \in H$. The fact that $c_n \neq e$ and Claim 3.50 imply that $\mu(c_n, \mu(h_n, x_n)) \notin S_{x_0}$. This is a contradicts the fact that $\mu(\widetilde{g}_n, x_n) \in S_{x_0}$, proving that (iii) must be satisfied. $\square$

We now describe the construction of so-called *associated bundles* to a principal bundle, which plays a fundamental role in the description of tubular neighborhoods of orbits of proper actions.

Let $H \to P \to B$ be a principal H-bundle and denote by $\mu_1 \colon P \times H \to P$ the underlying free proper right proper H-action, see Proposition 3.33. Suppose that F is a manifold that admits a left H-action $\mu_2 \colon H \times F \to F$. Then the diagonal action[9]

$$\mu \colon H \times (P \times F) \to (P \times F), \quad \mu\big(h, (x, f)\big) := \big(\mu_1(x, h^{-1}), \mu_2(h, f)\big) \qquad (3.14)$$

is a proper left H-action on $P \times F$, whose orbit space we denote by $P \times_H F$. Furthermore, given $(x, f) \in P \times F$, denote its projection to the orbit space by $[x, f] \in P \times_H F$.

Theorem 3.51. *The above orbit space $P \times_H F$ is a manifold, called* twisted space, *and it is the total space of a fiber bundle*

$$F \to P \times_H F \to B,$$

called associated bundle with fiber F, *whose projection $\pi \colon P \times_H F \to B$ is given by $\pi\big([x, f]\big) = \rho(x)$, where $\rho \colon P \to B$ is the projection of the principal H-bundle P, and whose structural group is H.*

Proof (Sketch). Given an H-orbit $H(x)$ in P, let S be a submanifold transverse to $H(x)$ and let U be an H-invariant neighborhood of $H(x)$ in P, as in Claim 3.36.

Claim 3.52. *The map $\varphi \colon S \times F \to U \times_H F$, $\varphi(s, f) := [s, f]$, is a diffeomorphism.*

Let $\widetilde{\varphi} \colon H \times (S \times F) \to U \times F$ be the restriction of the H-action (3.14) to $S \times F$, that is, $\widetilde{\varphi} := \mu|_{H \times (S \times F)}$. Similarly to Claim 3.36, it is possible to prove that $\widetilde{\varphi}$ is an H-equivariant diffeomorphism, where the H-action on $H \times (S \times F)$ is by left multiplication on the first factor, and the H-action on $U \times F$ is given by (3.14).

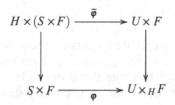

Thus, dividing by the H-action, $\widetilde{\varphi}$ descends to the desired diffeomorphism φ. This construction provides charts for the smooth manifold $P \times_H F$.

Let us now define bundle charts for the fiber bundle $F \to P \times_H F \to B$ and verify their compatibility. Let S_1 and S_2 be submanifolds transverse to two different orbits, with $\rho(S_1) \cap \rho(S_2) \neq \emptyset$, and set $\rho_i := \rho|_{S_i} \colon S_i \to \rho(S_i)$. From Claim 3.52,

$$\psi_i \colon \rho(S_i) \times F \to U \times_H F, \quad \psi_i(b, f) := \big[\rho_i^{-1}(b), f\big]$$

---

[9]Recall Examples 3.5 and 3.8.

are diffeomorphisms. Furthermore, we claim that $(\psi_2^{-1} \circ \psi_1)(b, f) = (b, \theta(b)f)$, where θ is as in the proof of Claim 3.37. In order to verify this, it suffices to show that if $[s_1, f_1] = [s_2, f_2]$, then $f_2 = \mu_2(\widetilde{\theta}(s_2), f_1)$, where $\widetilde{\theta}$ is also as in the proof of Claim 3.37. The fact that $[s_1, f_1] = [s_2, f_2]$ implies that there exists h such that

$$\mu_2(h, f_1) = f_2, \quad \text{and} \quad \mu_1(s_2, h) = s_1. \tag{3.15}$$

From the proof of Theorem 3.34, $s_1 = \mu_1(s_2, \widetilde{\theta}(s_2))$. Thus, since the action on P is free, it follows that $\widetilde{\theta}(s_2) = h$, which proves the above claim. □

A familiar example of fiber bundle associated to a principal G-bundle is given by the tangent bundle TM of a manifold M. This is the associated bundle with fiber $\mathbb{R}^n$ to the principal G-bundle $B(TM)$, defined in Example 3.29, where $n = \dim M$ and $G = GL(n, \mathbb{R})$ acts on $\mathbb{R}^n$ by evaluating matrices on vectors.

Exercise 3.53. Define $\psi \colon B(TM) \times_G \mathbb{R}^n \to TM$ by $\psi([\xi, v]) = \xi(v)$, where $G = GL(n, \mathbb{R})$ and a frame $\xi \in B(T_xM)$ is interpreted as a linear isomorphism $\xi \colon \mathbb{R}^n \to T_xM$. Prove that ψ is an identification between these two vector bundles.

Hint: To show that ψ is well-defined, note that $\psi([\xi \cdot g^{-1}, g(v)]) = \xi(v)$ for all $g \in G$. To prove that ψ is injective, verify that $\psi([\xi_1, v_1]) = \psi([\xi_2, v_2])$ if and only if $\xi_2 = \xi_1 \cdot g^{-1}$ and $v_2 = g(v_1)$ for some $g \in G$.

Remark 3.54. More generally, if E is a vector bundle over M of rank n and $B(E)$ is the principal $GL(n, \mathbb{R})$-bundle of frames of E, then the associated bundle to $B(E)$ with fiber $\mathbb{R}^n$ is canonically identified with the original vector bundle E.

Example 3.55. Let us describe a family of associated bundles with fiber S^2 to the principal S^1-bundle $S^1 \to S^3 \to S^2$, discussed in detail in Exercise 3.31. Writing $S^2 = \{(z, x) \in \mathbb{C} \times \mathbb{R} : |z|^2 + x^2 = 1\}$, define a family μ_k, $k \in \mathbb{N}$, of S^1-actions on S^2,

$$\mu_k \colon S^1 \times S^2 \to S^2, \quad \mu_k(e^{i\theta}, (z, x)) := (e^{ki\theta}z, x),$$

that is, μ_k is a rotation of *speed* k on S^2. By Theorem 3.51, the corresponding twisted spaces $M_k = S^3 \times_{S^1} S^2$ are manifolds which are the total space of an S^2-bundle over S^2, that is, $S^2 \to M_k \to S^2$. It is not difficult to show that M_k is diffeomorphic to $S^2 \times S^2$ if k is even, and to $\mathbb{C}P^2 \# \overline{\mathbb{C}P}^2$ if k is odd.[10] In particular, we stress that the notation $P \times_G F$ hides the fact that there may be different G-actions on F which give rise to different twisted spaces.

Exercise 3.56. Reinterpret the isomorphism $(G \times G)/\Delta G \cong G$ described in Example 3.39 in terms of the associated bundle $G \times_G G$ to the trivial principal G-bundle $G \to G \to \{e\}$, where the left G-action on G considered is multiplication on the left.

---

[10]Recall that $\overline{\mathbb{C}P}^2$ denotes the complex projective plane $\mathbb{C}P^2$ endowed with the orientation opposite to the standard. It is well-known that the connected sum $\mathbb{C}P^2 \# \overline{\mathbb{C}P}^2$ is the *only* nontrivial S^2-bundle over S^2, since such bundles are classified by $\pi_1(\mathrm{Diff}(S^2)) \cong \pi_1(SO(3)) \cong \mathbb{Z}_2$.

Let μ be a proper G-action on M. Given $x_0 \in M$, let S_{x_0} be a slice at x_0. We define a *tubular neighborhood* of the orbit $G(x_0)$ as the image of S_{x_0} under the G-action, that is,

$$\mathrm{Tub}\big(G(x_0)\big) := \mu\big(G, S_{x_0}\big).$$

The next theorem gives $\mathrm{Tub}\big(G(x_0)\big)$ the structure of an associated fiber bundle; in particular, it shows that manifolds with proper G-actions are *locally* G-equivariant to associated bundles.

Tubular Neighborhood Theorem 3.57. *Let $\mu \colon G \times M \to M$ be a proper action. For every $x_0 \in M$, there exists a G-equivariant diffeomorphism between $\mathrm{Tub}(G(x_0))$ and the total space of the associated bundle with fiber S_{x_0},*

$$S_{x_0} \to G \times_H S_{x_0} \to G/H,$$

to the principal H-bundle $H \to G \to G/H$, where $H = G_{x_0}$ is the isotropy at x_0.

Remark 3.58. The G-action considered on $G \times_H S_{x_0}$ is given by $\overline{g} \cdot [g, s] := [\overline{g}g, s]$. Details on the principal H-bundle $H \to G \to G/H$ are found in Corollary 3.38.

Proof. Let $\varphi \colon G \times S_{x_0} \to \mathrm{Tub}\big(G(x_0)\big)$ be the restriction of the G-action to $G \times S_{x_0}$, that is, $\varphi := \mu|_{G \times S_{x_0}}$. Similarly to the proof of Claim 3.36, note that $\mathrm{d}\varphi_{(e,x)}$ is surjective and $\mathrm{d}\varphi_{(g,x)}(\mathrm{d}L_g X, Y) = \mathrm{d}(\mu^g)_x \circ \mathrm{d}\varphi_{(e,x)}(X, Y)$, hence $\mathrm{d}\varphi_{(g,x)}$ is surjective for all $g \in G$ and $x \in S_{x_0}$. In particular, $\mathrm{Tub}(G(x_0))$ is an open neighborhood of $G(x_0)$, which is clearly G-invariant.

Claim 3.59. $\varphi(g, x) = \varphi(h, y)$ *if and only if* $h = gk^{-1}$ *and* $y = \mu(k, x)$, *where* $k \in H$.

Assume that $\varphi(g, x) = \varphi(h, y)$. Then $\mu(g, x) = \mu(h, y)$ and $y = \mu(k, x)$, where $k = h^{-1}g$. Since $x, y \in S_{x_0}$, it follows from Definition 3.47 that $k \in H$. The converse statement is clear. We now define the candidate to G-equivariant diffeomorphism,

$$\psi \colon G \times_H S_{x_0} \to \mathrm{Tub}\big(G(x_0)\big), \quad \psi\big([g, s]\big) := \mu(g, s). \tag{3.16}$$

The fact that $\pi \colon G \times S_{x_0} \to G \times_H S_{x_0}$, $\pi(g, s) = [g, s]$, is the projection of a principal bundle, and Claim 3.59, imply that ψ is well-defined, smooth, bijective and G-equivariant. It only remains to prove that ψ is a diffeomorphism.

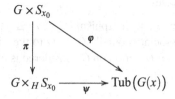

Note that $d\pi$ and $d\varphi$ are surjective, and $\varphi = \psi \circ \pi$, hence $d\psi$ is surjective. As

$$\dim G \times_H S_{x_0} = \dim G/H + \dim S_{x_0} = \dim M = \dim \text{Tub}(G(x_0)), \qquad (3.17)$$

it follows that $d\psi$ is an isomorphism. Thus, by the Inverse Function Theorem, ψ is a local diffeomorphism, and hence a diffeomorphism since it is bijective. $\qquad \square$

Remark 3.60. By the above result, the image of a slice at x_0 under the transformation $\mu^g \colon M \to M$ is a slice at $\mu(g,x_0)$, that is, $S_{\mu(g,x_0)} = \mu^g(S_{x_0})$. It also follows from the Tubular Neighborhood Theorem 3.57 that there is a unique G-equivariant retraction $r \colon \text{Tub}(G(x_0)) \to G(x_0)$ such that $r \circ \psi = \tilde{\mu}_{x_0} \circ \pi$, where $\pi \colon G \times_H S_{x_0} \to G/H$ is the projection map of the associated bundle, $\tilde{\mu}_{x_0}$ is defined as in Proposition 3.41, and ψ is the G-equivariant diffeomorphism (3.16).

Remark 3.61. Let $\mu \colon G \times M \to M$ be an action whose orbits have constant dimension. The Tubular Neighborhood Theorem 3.57 implies that the holonomy group $\text{Hol}(G(x_0),x_0)$ of the leaf $G(x_0)$ of the foliation $\{G(x)\}_{x \in M}$ coincides with the image of the *slice representation* of $H = G_{x_0}$, see Definition 3.72 and Remarks 5.13 and 5.14. In the general case of foliations with compact leaves and finite holonomy, there is an analogous result to the Tubular Neighborhood Theorem 3.57, known as the *Reeb Local Stability Theorem* (see Moerdijk and Mrčun [163]).

3.3 Isometric Actions

In this section, we study the relation between *proper* and *isometric* actions. In Proposition 3.62, we show that actions by closed subgroups of isometries are proper, and conversely, in Theorem 3.65, we show that every proper action can be made isometric with respect to a certain Riemannian metric (see also Remark 3.67).

Proposition 3.62. *Let (M,g) be a Riemannian manifold and G a closed subgroup of the isometry group $\text{Iso}(M,g)$. The action $\mu \colon G \times M \ni (g,x) \mapsto g(x) \in M$ is proper.*

Proof. For the sake of brevity, we only prove the result for the case in which (M,g) is complete. Consider sequences $\{g_n\}$ in G and $\{x_n\}$ in M, such that $\lim \mu(g_n,x_n) = y$ and $\lim x_n = x$. We have to prove that there exists a convergent subsequence of $\{g_n\}$ in G. Choose n_0 such that $\text{dist}(y,\mu(g_n,x_n)) < \frac{\varepsilon}{2}$ and $\text{dist}(x,x_n) < \frac{\varepsilon}{2}$ for $n > n_0$.

Claim 3.63. For fixed $R_1 > 0$, we have $\mu(g_n, B_{R_1}(x)) \subset B_{R_1+\varepsilon}(y)$ for all $n > n_0$.

Indeed, for $z \in B_{R_1}(x)$, we have

$$\begin{aligned}
\text{dist}(\mu(g_n,z),y) &\leq \text{dist}(y,\mu(g_n,x_n)) + \text{dist}(\mu(g_n,x_n),\mu(g_n,z)) \\
&< \tfrac{\varepsilon}{2} + \text{dist}(x_n,z) \\
&\leq \tfrac{\varepsilon}{2} + \text{dist}(z,x) + \text{dist}(x,x_n) \\
&< \varepsilon + R_1.
\end{aligned}$$

The fact that each g_n is an isometry of (M,g) implies that $\{g_n\}$ is an equicontinuous family. Thus, by the Arzelà-Ascoli Theorem,[11] Claim 3.63, and the fact that closed balls in M are compact (since M was assumed complete), there exists a subsequence $\{g_n^{(1)}\}$ of $\{g_n\}$ that converges uniformly on $\overline{B_{R_1}(x)}$ to a continuous map from $\overline{B_{R_1}(x)}$ to M. Using this argument inductively, define a subsequence $\{g_n^{(i)}\}$ of $\{g_n^{(i-1)}\}$ that converges uniformly on $\overline{B_{R_i}(x)}$ to a continuous map from $\overline{B_{R_i}(x)}$ to M, where $R_i > R_{i-1}$. Note that the *diagonal* subsequence $\{g_i^{(i)}\}$ converges uniformly on each $\overline{B_{R_i}(x)}$ to a continuous map $g\colon M \to M$. It follows from the Myers-Steenrod Theorem 2.12 that g is an isometry of (M,g) that belongs to G.   □

Remark 3.64. The hypothesis that G is a *closed* subgroup of $\mathrm{Iso}(M,g)$ is crucial for Proposition 3.62 to hold. For instance, the $\mathbb{R}$-action μ_2 on the Euclidean space $\mathbb{R}^2$ defined in Example 3.43 is clearly isometric, and hence so is the induced $\mathbb{R}$-action $\tilde{\mu}_2$ on the flat torus T^2. However, $\tilde{\mu}_2$ is not a proper action, see Exercise 3.18.

Using the Slice Theorem 3.49, we now prove a converse result, following Palais and Terng [182, Chapter 5].

Theorem 3.65. *Let $\mu\colon G \times M \to M$ be a proper action. There exists a G-invariant metric g on M such that $\mu^G = \{\mu^g : g \in G\}$ is a closed subgroup of $\mathrm{Iso}(M,g)$.*

Proof. The strategy to construct g is to define it locally, using an averaging procedure similar to (3.12), and then glue these together using G-invariant partitions of unity. Note that M/G is paracompact,[12] since it is a Hausdorff locally compact space and the union of countably many compact spaces. Thus, there exists a locally finite open covering $\{\pi(S_{x_\alpha})\}$ of M/G, where S_{x_α} is a slice at a point $x_\alpha \in M$. Set $U_\alpha := \pi^{-1}(\pi(S_{x_\alpha}))$ and let $\{f_\alpha\}$ be a G-invariant partition of unity subordinate to $\{U_\alpha\}$, which exists by the following result of Palais [180].

Claim 3.66. *If $\{U_\alpha\}$ is a locally finite open cover of M by G-invariant open sets, then there exists a smooth partition of unity $\{f_\alpha\}$ subordinate to $\{U_\alpha\}$, such that each $f_\alpha\colon U_\alpha \to [0,1]$ is G-invariant.*

Define a Riemannian metric $\langle \cdot, \cdot \rangle^\alpha$ on M along the slice S_{x_α}, i.e., a section of $(TM^* \otimes TM^*)|_{S_{x_\alpha}}$, by setting

$$\langle X, Y \rangle_p^\alpha := \int_{H_\alpha} \langle\!\langle \mathrm{d}\mu^g X, \mathrm{d}\mu^g Y \rangle\!\rangle_{\mu(g,p)} \, \omega,$$

---

[11]This theorem gives a criterion for convergence of continuous maps in the compact-open topology. More precisely, a sequence of continuous functions $\{g_n\colon K \to B\}$ between compact metric spaces K and B admits a uniformly convergent subsequence if $\{g_n\}$ is *equicontinuous*, i.e., for every $\varepsilon > 0$ there exists $\delta > 0$ such that $\mathrm{dist}(g_n(x), g_n(y)) < \varepsilon$ for all $\mathrm{dist}(x,y) < \delta$, $x,y \in K$, $n \in \mathbb{N}$.

[12]This means that any open cover of M/G admits a locally finite refinement.

where ω is a right-invariant volume form on the isotropy group $H_\alpha = G_{x_\alpha}$ and $\langle\langle\cdot,\cdot\rangle\rangle$ is an arbitrary Riemannian metric on S_{x_α}. It is easy to see that $\langle\cdot,\cdot\rangle^\alpha$ is H_α-invariant. Define a G-invariant Riemannian metric g_α on U_α by setting

$$g_\alpha\big(d\mu^g X, d\mu^g Y\big)_{\mu(g,p)} := \langle X, Y\rangle_p^\alpha. \tag{3.18}$$

The above metric is well-defined, since $\langle\cdot,\cdot\rangle^\alpha$ is H_α-invariant on S_{x_α}. Finally, define the desired metric g on M as $g := \sum_\alpha f_\alpha g_\alpha$.

It remains only to verify that μ^G is a closed subgroup of $\mathrm{Iso}(M,g)$. Assume that μ^{g_n} converges uniformly in each compact of M to an isometry $f: M \to M$. Thus, for each $x \in M$ we have that $\lim \mu(g_n, x) = f(x)$. Properness of the action implies that a subsequence $\{g_{n_i}\}$ converges to $g \in G$. It is not difficult to check that if $\lim g_{n_i} = g$, then $\mu^{g_{n_i}}$ converges uniformly in each compact of M to μ^g, and hence $\mu^g = f$. □

Remark 3.67. Theorem 3.65 clearly provides a type of converse statement to Proposition 3.62. Nevertheless, there are two subtle issues that might prevent one from exchanging isometric actions by proper actions when proving results about these. First, and most important, the G-invariant metric g provided by Theorem 3.65 is not necessarily complete.[13] We stress that completeness is a necessary hypothesis in many results for isometric actions. Second, the closed subgroup μ^G of $\mathrm{Iso}(M,g)$ is only isomorphic to G if $\mu: G \times M \to M$ is effective, see Remark 3.3.

Remark 3.68. Theorem 3.65 can be generalized to the context of *proper groupoids*, see Pflaum, Posthuma, and Tang [185]. These objects admit a Riemannian metric such that the induced foliation is a singular Riemannian foliation. Similarly to Remark 3.67, the ambient space with this invariant metric may fail to be complete, and many results on Riemannian foliations that assume completeness do not hold. For example, orbits of proper groupoids with trivial holonomy may not be diffeomorphic to each other. Nevertheless, when M is compact, foliations induced by a proper groupoid seem to be a particular case of *orbit-like foliations* on compact manifolds, i.e., singular Riemannian foliations whose restriction to slices are homogenous foliations, see [16] and Chap. 5.

Exercise 3.69 ($\star$). Let $\mu: G \times M \to M$ be an effective isometric proper action. Prove that if $\dim M = n$, then $\dim G \leq \frac{n(n+1)}{2}$.

Hint: Use that $\dim G - \dim G_x = \dim G(x) \leq \dim M = n$.

Given an isometric G-action on a Riemannian manifold (M,g), we usually refer to the directions tangent to the orbits as *vertical directions* and directions normal to the orbits as *horizontal directions*. In this way, we define the *vertical space* at $x \in M$ as

$$\mathcal{V}_x := T_x G(x) = \{X^*(x) : X \in \mathfrak{g}\}, \tag{3.19}$$

---

[13] Note that completeness can be guaranteed if, e.g., M is compact.

see (3.11), and the *horizontal space* as

$$\mathcal{H}_x := \{v \in T_x M : g(v, X^*(x)) = 0 \text{ for all } X \in \mathfrak{g}\}, \tag{3.20}$$

cf. (2.10) and (2.11). In this way, we get an orthogonal splitting $T_x M = \mathcal{V}_x \oplus \mathcal{H}_x$. Note that the dimensions of $\mathcal{V}_x$ and $\mathcal{H}_x$ may vary with $x \in M$, so these are *not* distributions on M. However, they become distributions when restricted to each *stratum* of M, see Theorem 3.102. Accordingly, we say that a curve $\gamma \colon [a, b] \to M$ is *vertical* or *horizontal* if its tangent vector γ' is always vertical or horizontal, respectively.

The following result for isometric actions was observed by Kleiner in his PhD thesis [141], and is usually referred to as *Kleiner's lemma*.

Lemma 3.70. *Let* $\mu \colon G \times M \to M$ *be an isometric action and let* $\gamma \colon [0, 1] \to M$ *be a geodesic segment between* $G(\gamma(0))$ *and* $G(\gamma(1))$ *that realizes the distance[14] between these orbits. There exists a subgroup H of G such that $G_{\gamma(t)} = H$ for $t \in (0, 1)$ and H is a subgroup of $G_{\gamma(0)}$ and $G_{\gamma(1)}$.*

Proof. Let $H := \{g \in G : \mu(g, \gamma(t)) = \gamma(t) \text{ for all } t \in [0, 1]\}$. Suppose there exists some $t_0 \in (0, 1)$ and $g \in G_{\gamma(t_0)}$ such that $g \notin H$. In particular, $d(\mu^g)_{\gamma(t_0)}(\dot\gamma(t_0)) \neq \dot\gamma(t_0)$. Define a piecewise smooth path $\widetilde\gamma \colon [0, 1] \to M$ by setting $\widetilde\gamma|_{[0, t_0]} := \gamma_{[0, t_0]}$ and $\widetilde\gamma|_{[t_0, 1]} := \mu^g(\gamma_{[t_0, 1]})$. Then $\widetilde\gamma$ joins $G(\gamma(0))$ to $G(\gamma(1))$ and has the same length as γ, contradicting the fact that minimizing geodesic segments are smooth. $\square$

Exercise 3.71. Use the isometric action in $\mathbb{R}^3$ described in Example 3.4 (iii) to show that the conclusion of Lemma 3.70 fails if γ is only a minimal geodesic segment.

We conclude this section with the important definition of *slice representation*, which is the restriction of the isotropy representation of an isometric action to the normal space to the orbit, see Exercise 3.7. More precisely, if $\mu \colon G \times M \to M$ is a proper isometric action, note that $S_x = \exp_x (B_\varepsilon(0))$ is a slice at x, called *normal slice*, where $B_\varepsilon(0) \subset \nu_x G(x)$ is an open ball in the normal space to $G(x)$, cf. (3.13).

Definition 3.72. Let $\mu \colon G \times M \to M$ be an isometric proper action and S_x a normal slice at x. The *slice representation* of G_x is the linear (orthogonal) representation

$$G_x \ni g \longmapsto d(\mu^g)_x \in O(\nu_x G(x)) \subset GL(\nu_x G(x)).$$

---

[14]Note that the assumption that γ realizes the distance between the orbits $G(\gamma(0))$ and $G(\gamma(1))$ is stronger than γ being a minimal geodesic segment between its endpoints $\gamma(0)$ and $\gamma(1)$.

3.4 Principal Orbits

In this section, we study some geometric properties of principal orbits. In particular, we prove the Principal Orbit Theorem 3.82, that implies that the set of points in principal orbits is open and dense.

Definition 3.73. Let $\mu \colon G \times M \to M$ be a proper action. Then $G(x)$ is a *principal orbit* if there exists a neighborhood V of x in M such that for each $y \in V$, $G_x \subset G_{\mu(g,y)}$ for some $g \in G$.

In rough terms, principal orbits are the ones that have the *smallest* isotropy group (i.e., the largest *type*, see Definition 3.84) among the nearby orbits.

Proposition 3.74. *Let $\mu \colon G \times M \to M$ be a proper left action. Then the following are equivalent:*

(i) $G(x)$ *is a principal orbit;*
(ii) *If S_x is a slice at x, then $G_x = G_y$ for all $y \in S_x$.*

Proof. First, let us prove that (i) implies (ii). From the definition of slice, for each $y \in S_x$ we have $G_y \subset G_x$. On the other hand, since $G(x)$ is a principal orbit, there exists g such that $g G_x g^{-1} \subset G_y$. Thus, $g G_x g^{-1} \subset G_y \subset G_x$. Since G_x is compact, we conclude that the above inclusions are equalities. In particular, $G_y = G_x$.

To prove the converse, we have to prove that for z in a tubular neighborhood of $G(x)$, there exists g such that $G_x \subset g G_z g^{-1}$. The orbit $G(z)$ intersects the slice S_x at least at one point, say y. Since y and z are in the same orbit, there exists g such that $G_y = g G_z g^{-1}$, recall Exercise 3.9. Therefore $G_x = G_y = g G_z g^{-1}$. $\qquad\square$

Remark 3.75. As we will see in Chap. 5, the above proposition implies that each principal orbit is a leaf with trivial holonomy of the foliation $\mathscr{F} = \{G(x)\}_{x \in M}$.

Exercise 3.76. Let $\mu \colon G \times M \to M$ be a proper action, and let $K := \bigcap_{x \in M} G_x$ be the ineffective kernel of the action. Consider the induced effective action $\widetilde{\mu} \colon G/K \times M \to M$, see Remark 3.3. Prove that $\widetilde{\mu}$ is smooth and proper. Check that if an orbit is principal with respect to $\widetilde{\mu}$, then it is also principal with respect to μ.

Exercise 3.77. Let $\mu \colon G \times M \to M$ be an isometric proper action and S_x a normal slice at x. Prove that $G(x)$ is a principal orbit if and only if the slice representation of G_x is trivial.

Let us explore some geometric properties of orbits of isometric actions.

Proposition 3.78. *Let $\mu \colon G \times M \to M$ be an isometric proper action on a complete Riemannian manifold M. Then the following hold:*

(i) *A geodesic γ orthogonal to an orbit $G(\gamma(0))$ remains orthogonal to all orbits it intersects, i.e., γ is a horizontal geodesic.*

In the following, suppose $G(x)$ is a principal orbit:

(ii) *Given $\xi \in \nu_x G(x)$, $\widehat{\xi}_{\mu(g,x)} := d(\mu^g)_x \xi$ is a well-defined normal vector field along $G(x)$, called* equivariant normal field;

(iii) $\mathscr{S}_{\widehat{\xi}_{\mu(g,x)}} = d\mu^g \mathscr{S}_{\widehat{\xi}_x} d\mu^{g^{-1}}$, *where $\mathscr{S}_{\widehat{\xi}}$ is the shape operator of $G(x)$;*

(iv) *Principal curvatures of $G(x)$ with respect to an equivariant normal field $\widehat{\xi}$ are constant along $G(x)$;*

(v) $\{ \exp_y (\widehat{\xi}_y) : y \in G(x) \}$ *is an orbit of μ.*

Proof. In order to prove item (i), by Proposition 3.13 and (3.11), it suffices to prove that if a Killing vector field X is orthogonal to $\gamma'(0)$, then X is orthogonal to $\gamma'(t)$ for all t. Since $\langle \nabla_{\gamma'(t)} X, \gamma'(t) \rangle = 0$ (see Proposition 2.14) and γ is a geodesic, we have that $\frac{d}{dt}\langle X, \gamma'(t)\rangle = 0$. Thus, X is always orthogonal to $\gamma'(t)$.

Item (ii) follows from Exercise 3.77. Item (iii) follows from:

$$
\begin{aligned}
\left\langle d\mu^{g^{-1}} \mathscr{S}_{\widehat{\xi}_{\mu(g,x)}} d\mu^g W, Z \right\rangle_x &= \left\langle \mathscr{S}_{\widehat{\xi}_{\mu(g,x)}} d\mu^g W, d\mu^g Z \right\rangle_{\mu(g,x)} \\
&= \left\langle -\nabla_{d\mu^g W} d(\mu^g)_x \xi, d\mu^g Z \right\rangle_{\mu(g,x)} \\
&= \left\langle -\nabla_W \widehat{\xi}, Z \right\rangle_x \\
&= \left\langle \mathscr{S}_{\widehat{\xi}_x} W, Z \right\rangle_x.
\end{aligned}
$$

As for (iv), note that if $\mathscr{S}_{\widehat{\xi}} X = \lambda X$, then $d\mu^{g^{-1}} \mathscr{S}_{\widehat{\xi}_{\mu(g,x)}} d\mu^g X = \lambda X$. Hence $\mathscr{S}_{\widehat{\xi}_{\mu(g,x)}} d\mu^g X = \lambda d\mu^g X$. Finally, item (v) follows from (2.2), because $\exp_{\mu(g,x)}(\widehat{\xi}_{\mu(g,x)}) = \exp_{\mu(g,x)}(d\mu^g \xi_x) = \mu^g(\exp_x(\xi))$. $\square$

Remark 3.79. The above proposition illustrates a few concepts and results of Chaps. 4 and 5. Item (i) implies that the homogeneous foliation (3.4) by orbits of a proper isometric action is a *singular Riemannian foliation* (see Definition 5.2). Item (v) implies that one can reconstruct this foliation taking all parallel submanifolds to a principal orbit. This is a consequence of *equifocality*, which is valid for every singular Riemannian foliation. Item (iv) and the fact that equivariant normal fields are parallel normal fields when the action is *polar* imply that principal orbits of a polar action on Euclidean space are *isoparametric* (see Definitions 4.8 and 4.14).

Let us explore a little further item (i) of Proposition 3.78 to obtain some metric information about the orbit space M/G of an isometric proper action. If a geodesic segment $\gamma: [0,1] \to M$ minimizes distance between the orbit $G(\gamma(0))$ and $\gamma(1)$, then it must be orthogonal to the submanifold $G(\gamma(0))$. Hence, by the above, γ must be horizontal. Now, for any $g \in G$, since $\mu^g: M \to M$ is an isometry, the curve $(\mu^g \circ \gamma): [0,1] \to M$ is a horizontal minimal geodesic segment joining $G(\gamma(0))$ and $\mu(g, \gamma(1)) \in G(\gamma(1))$, with the same length as γ. This means that for any $g \in G$, the distances from $G(\gamma(0))$ to $\gamma(1)$ and $\mu(g, \gamma(1))$ are the same. Thus, orbits of an isometric group action are *equidistant*, and hence there is a well-defined *orbital distance*, i.e., a distance on M/G. The orbit space M/G equipped with this distance becomes a metric space, more precisely, a *length metric space*.[15]

Remark 3.80. There are many important aspects of the geometry of the metric space M/G that reflect the geometry of an isometric G-action on M, especially if M has a lower curvature bound, making M/G an *Alexandrov space*.[16] Much recent progress on areas involving the interface of transformation groups and Riemannian geometry was achieved through the use of Alexandrov geometry of orbit spaces; however, we do not go in this direction in this book. For a brief glance at some of these ideas, see Example 3.107, where the appropriate generalization of tangent space at a nonregular point in M/G, called *tangent cone*, is mentioned in a specific example. Details on how Alexandrov techniques apply to study isometric actions can be found in the survey of Grove [107].

Exercise 3.81. Let $\mu \colon G \times M \to M$ be a free isometric action and recall Theorem 3.34. Show that the quotient map $\pi \colon M \to M/G$ is Riemannian submersion, where M/G is equipped with the natural orbit metric, whose associated distance is the above orbital distance.[17]

We conclude this section proving the so-called Principal Orbit Theorem, which guarantees that the subset of points in M on principal orbits is open and dense, and that M/G has an open and dense subset which is a smooth manifold (see Theorem 3.95 and Remark 3.106).

Principal Orbit Theorem 3.82. *Let* $\mu \colon G \times M \to M$ *be a proper action, where* M *is connected, and denote by* M_{princ} *the set of points of* M *contained in principal orbits. Then the following hold:*

(i) M_{princ} *is open and dense in* M;
(ii) *The subset* M_{princ}/G *of* M/G *is a connected manifold;*
(iii) *If* $G(x)$ *and* $G(y)$ *are principal orbits, there exists* $g \in G$ *such that* $G_x = g G_y g^{-1}$.

Proof. We first prove the existence of a principal orbit. Since G has finite dimension and the isotropy groups are compact, we can choose $x \in M$ such that G_x has the lowest dimension among isotropy groups and, for that dimension, the smallest number of connected components. If S_x is a slice at x, by definition, $G_y \subset G_x$ for every $y \in S_x$. By construction, we conclude that $G_y = G_x$. Hence, from Proposition 3.74, $G(x)$ is a principal orbit.

In order to prove that M_{princ} is open, let $x \in M_{\mathrm{princ}}$ and let S_x be a slice at x. Proposition 3.74 implies that $G_y = G_x$ for every $y \in S_x$. We claim that each $y \in S_x$ belongs to a principal orbit. If a point z is close to y, then z is in a tubular

---

[15] A metric space is a length metric space if the distance between any two points is realized by a shortest curve, called a geodesic.

[16] An *Alexandrov space* is a finite-dimensional length metric space with a lower curvature bound, in a comparison geometry sense. For details, see [59, 107].

[17] More generally, if μ is not free, then $\pi \colon M \to M/G$ is a *submetry*, which is a generalization of submersions to metric spaces.

neighborhood of $G(x)$. Note that $G(z)$ intersects S_x in at least one point, so there exists g such that $\mu(g,z) = w \in S_x$. Thus, $G_y = G_x = G_w = g G_z g^{-1}$ and hence each $y \in S_x$ belongs to a principal orbit, as claimed. Therefore, every point in the tubular neighborhood $\mathrm{Tub}(G(x)) = \mu(G, S_x)$ belongs to a principal orbit, proving M_{princ} is open.

If μ is an *isometric* action on a *complete* Riemannian manifold, then the rest of the proof follows easily from Proposition 3.74 and Kleiner's Lemma 3.70, since given any two orbits, there exists a geodesic segment that realizes the distance between these orbits. In this case, one can also prove that the space M_{princ}/G is convex. We encourage the reader to verify these claims, and, in what follows, we finish the proof in the general case.

To prove that M_{princ} is dense, consider $p \notin M_{\mathrm{princ}}$ and U a neighborhood of p. Choose $x \in U$ such that G_x has the lowest dimension among isotropy groups and, for that dimension, the smallest number of connected components. Then, from the argument above, we conclude that $x \in M_{\mathrm{princ}}$.

Proposition 3.74 and some arguments from the proof of Theorem 3.34 can be used to prove that M_{princ}/G is a manifold. In order to prove that M_{princ}/G is connected, assume that the action is proper and isometric. A set $A \subset M/G$ *does not locally disconnect* M/G if each $p \in M/G$ has a neighborhood U such that $U \setminus A$ is path-connected. Using (i), it is easy to verify that if $(M \setminus M_{\mathrm{princ}})/G$ does not locally disconnect M/G, then M_{princ}/G is path-connected. Thus, it suffices to prove that $(M \setminus M_{\mathrm{princ}})/G$ does not locally disconnect M/G. This can be done using the fact that $S_x/G_x = \mathrm{Tub}(G(x))/G$, the slice representation, Exercise 3.76 and the next:

Claim 3.83. Let K be a closed subgroup of $O(n)$, acting on $\mathbb{R}^n$ by multiplication. Then $\mathbb{R}^n_{\mathrm{princ}}/K$ is path-connected.

We prove Claim 3.83 by induction. If $n = 1$, then $K = \mathbb{Z}_2$ or $K = \{1\}$. In both cases, $\mathbb{R}_{\mathrm{princ}}/K$ is path-connected. For each sphere S^{n-1} centered at the origin, we can apply the induction hypothesis and the slice representation to conclude that $(S^{n-1} \setminus S^{n-1}_{\mathrm{princ}})/K$ does not locally disconnect S^{n-1}/K. Therefore, $S^{n-1}_{\mathrm{princ}}/K$ is path-connected. Consider $x, y \in \mathbb{R}^n_{\mathrm{princ}}$ and let $\widetilde{x}$ be the projection of x on the sphere S^{n-1} that contains y. The points along the straight line segment that joins x to $\widetilde{x}$ belong to principal orbits. Thus, $K(x)$ and $K(\widetilde{x})$ are connected by a path in $\mathbb{R}^n_{\mathrm{princ}}/K$. As we have already proved, $K(y)$ and $K(\widetilde{x})$ are connected by a path in $S^{n-1}_{\mathrm{princ}}/K$ and hence in $\mathbb{R}^n_{\mathrm{princ}}/K$. Therefore $K(x)$ and $K(y)$ are connected by a path in $\mathbb{R}^n_{\mathrm{princ}}/K$.

Finally, item (iii) is a direct consequence of item (ii). $\square$

3.5 Orbit Types

The *principal orbits* discussed in the last section are one of many possible *orbit types*. We now discuss some properties of orbit types of proper actions, e.g., we prove that on compact manifolds there is only a finite number of these, see

Theorem 3.91. In addition, it is also proved that each connected component of a set of orbits of the same type is a connected component of the total space of a certain fiber bundle, see Theorem 3.95. Finally, it is also proved that the connected components of sets of orbits of the same type give a *stratification* of M, see Theorem 3.102.

Definition 3.84. Let $\mu : G \times M \to M$ be a proper action.

(i) The orbit $G(x)$ has a *larger orbit type* than $G(y)$ if there is a G-equivariant map $\varphi : G(x) \to G(y)$, or, equivalently, if there is $g \in G$ such that $G_x \subset G_{\mu(g,y)}$;

(ii) The orbits $G(x)$ and $G(y)$ have the *same orbit type* if there is a G-equivariant diffeomorphism $\varphi : G(x) \to G(y)$, or, equivalently, if there is $g \in G$ such that $G_x = G_{\mu(g,y)}$;

(iii) An orbit $G(x)$ is said to be *regular* if the dimension of $G(x)$ coincides with the dimension of principal orbits;

(iv) A nonprincipal regular orbit is called *exceptional*;

(v) A nonregular orbit is called *singular*.[18]

These definitions are also used referring to points, e.g., x and y have the *same orbit type* if $G(x)$ and $G(y)$ have the same type, and x is *singular* if $G(x)$ is singular.

It is not difficult to see that $G(x)$ and $G(y)$ have the same orbit type if and only if $G(x)$ has a larger orbit type than $G(y)$ and $G(y)$ has a larger orbit type than $G(x)$, see Exercise 3.9. It is also clear that having the same orbit type is an equivalence relation (on both M/G and M). Note that Theorem 3.82 asserts not only that M_{princ} is open and dense in M, but also that there exists a unique type of principal orbit.

Exercise 3.85. Let $\mu : G \times M \to M$ be a proper action, and let $K := \bigcap_{x \in M} G_x$ be the ineffective kernel of this action. Consider the induced effective action $\widetilde{\mu} : G/K \times M \to M$, see Remark 3.3 and Exercise 3.76. Prove that orbits of the same type with respect to $\widetilde{\mu}$ are also of the same type with respect to μ.

Exercise 3.86. Let $\mu : G \times M \to M$ be a proper action and $G(p)$ a principal orbit. Prove that $G(x)$ is an exceptional orbit if and only if $\dim G(x) = \dim G(p)$ and the number of connected components of G_x is greater than the number of connected components of G_p.

Exercise 3.87. Consider the isometric action $\widetilde{\mu} : S^1 \times S^2 \to S^2$ of the circle S^1 on the round sphere S^2 by rotations. Use Exercise 3.40 to verify that $\widetilde{\mu}$ induces an isometric action $\mu : S^1 \times \mathbb{R}P^2 \to \mathbb{R}P^2$, which has an exceptional orbit.

Exercise 3.88 ($\star$). Consider the action of SU(3) on itself by conjugation. Prove that the orbits of this action are diffeomorphic to one of the following:

---

[18]It is important to notice that *singular* orbits are smooth submanifolds, whose *singular* nature is simply to have a lower dimension.

(i) $\{\lambda I\}$, where $I \in SU(3)$ is the identity and $\lambda \in \mathbb{C}$ satisfies $\lambda^3 = 1$;
(ii) $SU(3)/S(U(2) \times U(1))$;
(iii) $SU(3)/T$, where T is the subgroup of diagonal matrices in $SU(3)$.

Conclude, using Exercise 3.44 and Remark 3.46, that an orbit is diffeomorphic to either a point, or complex projective plane $\mathbb{C}P^2$, or the complex flag manifold $SU(3)/T$.

Hint: Use the fact that each matrix of $SU(3)$ is conjugate to a matrix of T, by diagonalization.

Remark 3.89. As explained in the next chapter, Exercises 3.87 and 3.88 give examples of polar actions. An isometric action is a *polar action* if for each regular point x, the set $\exp_x(v_x G(x))$ is a (totally geodesic) submanifold that intersects every orbit orthogonally, see Definition 4.8 for details. In Exercise 3.87, the polar action admits an exceptional orbit, and in Exercise 3.88, the polar action admits only principal and singular orbits. Note that $\mathbb{R}P^2$ is not simply-connected and $SU(3)$ is simply-connected. As explained in Chap. 5, polar actions do not admit exceptional orbits if the ambient space is simply-connected (Corollary 5.35).

Remark 3.90. For an example of nonpolar isometric action, see Example 3.11.

Theorem 3.91. *Let* $\mu : G \times M \to M$ *be a proper action. For each* $x \in M$, *there exists a slice* S_x *such that the tubular neighborhood* $\mathrm{Tub}(G(x)) = \mu(G, S_x)$ *contains only finitely many different orbit types. In particular, if M is compact, there is only a finite number of different orbit types in M.*

Proof. Using Theorem 3.65, let $\mathbf{g}$ be a Riemannian metric on M such that $\mu^G \subset \mathrm{Iso}(M, \mathbf{g})$. It is not difficult to verify the following:

Claim 3.92. *If* $y, z \in S_x$ *have the same* G_x-*orbit type, then y and z have the same G-orbit type.*

Therefore, it suffices to prove the result for the G_x-action on S_x. By the slice representation (see Definition 3.72) and Exercise 3.85, we can further reduce the problem to proving the result for a linear (orthogonal) action of $K \subset O(n)$ on $\mathbb{R}^n$ by multiplication. With this goal, we proceed by induction on n. If $n = 1$, then $K = \mathbb{Z}_2$ or $K = \{1\}$. In both cases, there exists only a finite number of different orbit types (at most 2). For a sphere S^{n-1}, we can apply the induction hypothesis, Claim 3.92 and the slice representation, to conclude that there exists a finite number of K-orbit types. For each $p \in S^{n-1}$, the points along the straight line segment joining the origin to p (except for the origin itself) have the same isotropy type (cf. Kleiner's Lemma 3.70). Thus, there exists a finite number of K-orbit types in $\mathbb{R}^n$. $\qquad\square$

In order to study how different orbit types of a proper action *stratify* a manifold, we first need to develop structural results for sets of points that have the same orbit type. The first step in this direction is the following result on fixed point sets.

Proposition 3.93. *Let* $\mu: G \times M \to M$ *be a proper action and* $H \subset G$ *a subgroup. The connected components of the* fixed *point set*

$$M^H := \{x \in M : \mu(h,x) = x, \text{ for all } h \in H\}$$

are embedded submanifolds of M *(of possibly different dimensions), whose tangent space is given by*

$$T_x M^H = \{v \in T_x M : d(\mu^h)_p v = v, \text{ for all } h \in H\}.$$

In addition, if the action μ *is isometric, then these submanifolds are totally geodesic.*

Proof. The fixed point set of H coincides with the fixed point set of its closure $\overline{H}$, and hence we may assume H is a closed subgroup of G. From Theorem 3.65, there exists an H-invariant Riemannian metric g on M. Given $x \in M^H$, let $\varepsilon > 0$ be such that $\exp_x : T_x M \to M$ is a diffeomorphism from the open ball $B_\varepsilon(0) \subset T_x M$ onto its image. Notice that, from (2.2), this diffeomorphism is H-equivariant, where the H-action on $B_\varepsilon(0)$ is the isotropy representation defined in Exercise 3.7. In particular, the exponential map provides a submanifold chart for M^H near x, which is mapped to the linear subspace $T_x M^H$ of $T_x M$. This proves that each connected component of M^H is an embedded submanifold. Finally, it is easy to deduce from (2.2) that each of these submanifolds is totally geodesic if the G-action is isometric. $\qquad\Box$

Besides the notion of orbits (or points) that have the *same orbit type*, it is convenient to define the following refinement, containing more geometric information.[19]

Definition 3.94. Two orbits $G(x)$ and $G(y)$ of a proper action have the *same local orbit type* if there is a G-equivariant diffeomorphism $\varphi: \text{Tub}(G(x)) \to \text{Tub}(G(y))$.

Thus, we have 2 equivalence relations: *to have the same orbit type* and *to have the same local orbit type*, denoted respectively by $\sim$ and $\approx$, following [79]. Clearly, having the same local orbit type is a stronger condition than having the same orbit type (cf. Definition 3.84), so equivalence classes with respect to $\sim$ are partitioned into equivalence classes with respect to $\approx$. These equivalence classes are denoted

$$M_x^\sim := \{y \in M : x \sim y\} \quad \text{and} \quad M_x^\approx := \{y \in M : x \approx y\},$$

and respectively called *orbit type* and *local orbit type* of $x \in M$. In this way, an orbit type is the union of local orbit types. Using the Tubular Neighborhood Theorem 3.57, one can prove that $x \approx y$ if and only if $x \sim y$ and the isotropy representations of G_x on $T_x M$ and G_y on $T_y M$ descend to equivalent representations

---

[19]To our knowledge, the notion of *local orbit types* was introduced in Duistermaat and Kolk [79].

on $T_x M/T_x G(x)$ and on $T_y M/T_y G(y)$, that is, there exists an equivariant linear isomorphism between these spaces,[20] recall Exercise 3.7 and Proposition 3.41.

In what follows, we show that $M_x^{\approx}$ is a (possibly disconnected) embedded submanifold of M. As proved in [79, Theorem 2.6.7], $M_x^{\approx}$ is an open and closed subset of $M_x^{\sim}$. In particular, an orbit type $M_x^{\sim}$ is the disjoint union of submanifolds of possibly different dimensions, and connected components of the same dimension are arranged into the local orbit types $M_y^{\approx}$, for $y \in M_x^{\sim}$.

Orbit types $M_x^{\sim}$ are commonly denoted $M_{(H)}$ in the literature, in reference to the set of points with isotropy group conjugate to $H = G_x$. We prefer to use the notation $M_x^{\sim}$ (in association with $M_x^{\approx}$), in the same way as [79], to stress the dependence on the point $x \in M$ rather than on the conjugacy class of its isotropy group H. Indeed, conjugates of H may have different isotropy representations at different points. Furthermore, the notations $M_x^{\sim}$ and $M_x^{\approx}$ are also suggestive of leaves of a foliation.

Theorem 3.95. *Let* $\mu : G \times M \to M$ *be a proper action. Each local orbit type* $M_x^{\approx}$ *is a union of G-invariant embedded submanifolds (of the same dimension). Moreover, the restriction of the quotient map* $\pi : M \to M/G$ *to* $M_x^{\approx}$ *determines a fiber bundle*

$$G/H \to M_x^{\approx} \to M_x^{\approx}/G,$$

whose structure group is $N(H)/H$, *where* $H = G_x$ *is the isotropy at x and $N(H)$ is the normalizer of H in G. In particular, subsets* $M_x^{\approx}/G$ *of orbits that have the same local orbit type are smooth manifolds inside* M/G.

Proof (Sketch). The proof of this result is analogous to the proof of the Tubular Neighborhood Theorem 3.57, and details can be found in [79, pp. 109–111]. The main idea is to prove that $M_x^{\approx}$ is the total space of the associated bundle with fiber G/H,

$$G/H \to G/H \times_K P \to P/K, \tag{3.21}$$

to the principal K-bundle $K \to P \to P/K$, where $P := M_x^{\approx} \cap M^H$ and $K := N(H)/H$.

Using the properties in Definition 3.47 and the Tubular Neighborhood Theorem 3.57, one can prove the following, where $\text{Tub}(G(x)) = \mu(G, S_x)$.

Claim 3.96. $M_x^{\sim} \cap \text{Tub}(G(x))$ *is a submanifold of* M. *Moreover, there is a G-equivariant diffeomorphism between* $M_x^{\sim} \cap \text{Tub}(G(x))$ *and* $G/H \times (S_x)^H$.

Note that if $y \in M_x^{\sim} \cap M^H$, then $G_y = gHg^{-1}$ for some $g \in G$ and $H \subset G_y$, and hence $H = G_y$ and $g \in N(H)$. Conversely, if $G_y = H$, then $y \in M_x^{\sim} \cap M^H$. Thus, we proved the first item in the following claim, whose other items are proved similarly.

---

[20]If the action is isometric, this simply means that the slice representations of G_x and G_y are equivalent, i.e., related by an equivariant linear map $\phi : v_x G(x) \to v_y G(y)$.

Claim 3.97. The following hold:

(i) $y \in M_x^{\sim} \cap M^H$ if and only if $G_y = H$.
(ii) Let $y \in M_x^{\sim} \cap M^H$. Then $\mu(g, y) \in M_x^{\sim} \cap M^H$ if and only if $g \in N(H)$;
(iii) $P = M_x^{\approx} \cap M^H$ is invariant under the subaction of $N(H)$, and the ineffective kernel of this action is H;
(iv) Each $y \in P$ has a neighborhood diffeomorphic to $(N(H)/H) \times (S_x)^H$.

Altogether, we have that P is a manifold with a free proper (left) K-action, so that $K \to P \to P/K$ is a principal K-bundle, and $P/K = M_x^{\approx}/G$. Moreover, it is not difficult to see that there is also a free (right) K-action on G/H (see Exercise 6.9), which allows us to construct the associated bundle (3.21) with fiber G/H.[21]

In order to prove that there exists a G-equivariant diffeomorphism between $M_x^{\approx}$ and the total space of this associated bundle, note that

$$\varphi: G/H \times P \to M_x^{\approx}, \quad \varphi(gH, s) := \mu(g, s),$$

is a well-defined map, by Claim 3.97. Given $y \in M_x^{\approx}$, we have $G_{\mu(g^{-1}, y)} = H$ for some $g \in G$. From Claim 3.97, $\mu(g^{-1}, y) \in M_x^{\sim} \cap M^H$. Since $M_x^{\approx}$ is G-invariant, $\mu(g^{-1}, y) \in P$ and hence $y = \varphi(g, \mu(g^{-1}, y))$. Thus, φ is surjective. Furthermore, the restriction $\varphi: G/H \times (S_x)^H \to M_x^{\sim} \cap \text{Tub}(G(x))$ is a diffeomorphism; and, since

$$T_{(gG_x, y)}(G/H \times (S_x)^H) \subset T_{(gH, y)}(G/H \times P), \tag{3.22}$$

we conclude that $d\varphi$ is surjective. We now observe the following (cf. Claim 3.59):

Claim 3.98. $\varphi(gH, y) = \varphi(hH, z)$ if and only if $h = gn^{-1}$ and $z = \mu(n, y)$ for some $n \in N(H)$.

Assume that $\varphi(gH, y) = \varphi(hH, z)$. Then $\mu(g, y) = \mu(h, z)$ and $z = \mu(n, y)$, where $n = h^{-1}g$. Since $z, y \in P$, it follows from Claim 3.97 that $n \in N(H)$. The converse statement is clear. We now define the candidate to G-equivariant diffeomorphism,

$$\psi: G/H \times_K P \to M_x^{\approx}, \quad \psi([gH, z]) := \mu(g, z),$$

cf. (3.16). The fact that $\pi: G/H \times P \to G/H \times_K P$, $\pi(gH, z) = [g, z]$, is the projection of a principal bundle, and Claim 3.98, imply that ψ is well-defined, smooth, bijective and G-equivariant. In only remains to prove that ψ is a diffeomorphism.

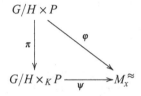

---

[21] Here, it is convenient to use a free *left* action on the principal bundle instead of a *right* action, cf. Proposition 3.33. Notice that, accordingly, the notation (3.21) for the twisted space is also reversed, cf. Theorem 3.51.

Note that $d\pi$ and $d\varphi$ are surjective, and $\varphi = \psi \circ \pi$, hence $d\psi$ is surjective. By a dimension argument analogous to (3.17), it follows that $d\psi$ is an isomorphism. Thus, by the Inverse Function Theorem, ψ is a local diffeomorphism, and hence a diffeomorphism, since it is bijective. $\qquad\square$

Remark 3.99. By Theorem 3.95, the result in Exercise 3.81 can be extended to the case of isometric actions that are not necessarily free. Namely, the projection map $\pi \colon M_x^{\approx} \to M_x^{\approx}/G$ is a Riemannian submersion, where $M_x^{\approx}/G$ is equipped with the natural orbit metric. Note that $M_x^{\approx}/G$ is *locally totally geodesic* in M/G, since by Claim 3.96, it is locally isometric to the fixed point set $(S_x)^H$, where S_x is a slice at $x \in M$, which is totally geodesic by Proposition 3.93.

We conclude this section with a discussion of the *orbit type stratification*.

Definition 3.100. A *stratification* of a manifold M is a partition of M by embedded submanifolds $\{M_i\}_{i \in I}$ of M, called *strata*, such that:

(i) The partition is *locally finite*, i.e., each compact subset of M only intersects a finite number of strata;
(ii) For each $i \in I$, there exists a subset $I_i \subset I \setminus \{i\}$ such that the closure of M_i is $\overline{M_i} = M_i \cup \bigcup_{j \in I_i} M_j$;
(iii) $\dim M_j < \dim M_i$, for all $j \in I_i$.

Example 3.101. Consider the action of $SO(2)$ on $\mathbb{R}^3$ described in Example 3.4 (ii). Then $\{M_i\}_{i \in I}$, given by

$$M_1 := \{(0,0,x_3) \in \mathbb{R}^3 : x_3 \in \mathbb{R}\}, \quad M_2 := \mathbb{R}^3 \setminus M_1, \qquad (3.23)$$

is a stratification of $\mathbb{R}^3$, where $I = \{1,2\}$, $I_1 = \emptyset$ and $I_2 = \{1\}$. The stratum M_1 consists of fixed points (which are singular orbits), and the stratum M_2 consists of regular points (which lie in principal orbits). Note that $M_2 = M_{\mathrm{princ}}$ is open an dense and M_{princ}/G is connected, as asserted by the Principal Orbit Theorem 3.82.

Theorem 3.102. *Let* $\mu \colon G \times M \to M$ *be a proper action. The partition of M into connected components of the orbit types $M_x^{\sim}$, $x \in M$, determines a stratification.*

Proof. The partition of M into connected components of orbit types is locally finite by the same arguments from the proof of Theorem 3.91. Given $y \in M$, we denote by $(M_y^{\sim})^0$ the connected component of $M_y^{\sim}$ that contains y.

Claim 3.103. *Let* $x \in \overline{(M_y^{\sim})^0}$ *be such that* $x \notin (M_y^{\sim})^0$. *Then:*

(i) $(M_x^{\sim})^0$ *is contained in* $\overline{(M_y^{\sim})^0}$;
(ii) $\dim(M_x^{\sim})^0 < \dim(M_y^{\sim})^0$.

In order to prove (i), we may assume that $y \in S_x$. From Claim 3.96, it suffices to prove that if $\tilde{x} \in (S_x)^H$, where $H = G_x$, then $\tilde{x} \in \overline{(M_y^{\sim})^0}$. Using the slice representation, we may also assume that $(S_x)^H$ is a linear subspace fixed by the linear action of $K = \mu^H$ on a vector space, and $x = 0$. Since the K-action fixes each point of $(S_x)^H$, it leaves invariant the normal space $\nu_p(S_x)^H$ that contains y. Consider the projection $\widetilde{y} = y - p + \widetilde{x}$ of y onto the normal space $\nu_{\widetilde{x}}(S_x)^H$. As $K_{\widetilde{x}} = K = K_p$, we have that $K_y = K_{\widetilde{y}}$. Moreover, as the K-action is linear, $K_z = K_{\widetilde{y}}$ for each z different from $\widetilde{x}$ along the straight line segment joining $\widetilde{x}$ to $\widetilde{y}$. Thus, $\widetilde{x} \in \overline{(M_{\widetilde{y}})^0}$, concluding the proof of (i).

As for (ii), let $(S_x)_{\widetilde{y}}$ be the set of points in S_x that have the same H-orbit type of $y \in S_x$. From the above discussion, we infer that $\dim((S_x)_{\widetilde{y}})^0 \geq \dim(S_x)^H + \dim \mu^H(y) + 1$. In particular, $\dim((S_x)_{\widetilde{y}})^0 > \dim(S_x)^H$. Thus, using Claim 3.96 and the fact that each point in $((S_x)_{\widetilde{y}})^0$ is also in $(M_{\widetilde{y}})^0$, we have that

$$\dim(M_{\widetilde{x}})^0 = \dim G(x) + \dim(S_x)^H < \dim G(x) + \dim((S_x)_{\widetilde{y}})^0 \leq \dim(M_{\widetilde{y}})^0,$$

which concludes the proof of (ii). □

Remark 3.104. Furthermore, it is possible to prove that an orbit type stratification is a *Whitney stratification*, see Duistermaat and Kolk [79].

Remark 3.105. Each stratum of the orbit type stratification of an isometric action is a *minimal submanifold*. In fact, the mean curvature vector of a stratum (see (2.9)) is tangent to the stratum by equivariance, but is also normal to the stratum by definition, and hence vanishes.

Remark 3.106. The orbit type stratification $\{M_i\}_{i \in I}$ of a manifold M induces a stratification $\{M_i/G\}_{i \in I}$ of the corresponding orbit space M/G, i.e., a partition into manifolds satisfying (i)–(iii) in Definition 3.100. From Remark 3.99, the orbit strata are locally totally geodesic. This plays an important role when M/G is studied from the viewpoint of Alexandrov geometry, see Remark 3.80.

Example 3.107. Consider the conjugation action of SU(3) on itself. From Exercise 3.88, there are 3 orbit types, since an orbit is either a fixed point, the complex projective plane $\mathbb{C}P^2$, or the complex flag manifold SU(3)/T. Clearly, the latter are the principal orbits, and it is not difficult to show that each of the 2 nonprincipal orbit types have 3 connected components (of the same dimension). This describes the orbit type stratification of SU(3), see Theorem 3.102. The orbit space of this conjugation action is a flat equilateral triangle, whose induced stratification is the obvious partition of a closed triangle into the union of its vertices (corresponding to fixed points in SU(3)), edges (corresponding to the orbits diffeomorphic to $\mathbb{C}P^2$), and interior (corresponding to principal orbits).

The linearization of the conjugation action of SU(3) at the identity $I \in$ SU(3) is its adjoint representation, see (1.8). Notice that this is precisely the isotropy representation (and the slice representation) at $I \in$ SU(3), since I is a fixed point of conjugation. The orbit structure of this adjoint action is studied in Example 4.42, where it is proved that its orbit space is a (closed) wedge of angle $\pi/3$ in the plane. We remark that this is precisely the so-called *tangent cone* to the orbit space of the conjugation action (the equilateral triangle) at the identity orbit (one of the vertices).

This is an instance of a general relation between tangent cones at points of the orbit space of an isometric action and the orbit space of the corresponding slice representation.

Remark 3.108. As demonstrated many times along this chapter (e.g., Theorems 3.82, 3.91 and 3.102), the local study of proper (or isometric) actions can be reduced to the local study of an isometric action in Euclidean space, by using the slice representation. Molino [165] pointed out that the same idea can be used in the local study of singular Riemannian foliations (see Definition 5.2). More precisely, after a suitable change of metrics, the local study of any singular Riemannian foliation is reduced to the study of a singular Riemannian foliation on the Euclidean space (with the standard metric). This idea is used, for example, in [8, 20], see also the survey [12].

Chapter 4
Adjoint and Conjugation Actions

Two actions play a central role in the theory of compact Lie groups G: the G-action on itself by *conjugation* and the *adjoint action* of G on its Lie algebra $\mathfrak{g}$, which is the linearization of the latter (see Definition 1.36 and Exercise 3.7). The goal of this chapter is to describe classical results on such actions from a geometric viewpoint.

The adjoint action is an example of *polar action* (see Remark 3.89 and Definition 4.8) and each of its regular orbits is an *isoparametric submanifold* (see Definition 4.14). Remarkably, several results on adjoint actions can be extended, not only to the theory of isoparametric submanifolds and polar actions, but also in a general context of *polar foliations* (see Definition 5.2). This parallel is made in Chap. 5.

The following references were a source of inspiration for this chapter, and could serve as further reading material: Duistermaat and Kolk [79], Fegan [85], Fulton and Harris [90], Hall [121], Helgason [126], Hsiang [129], Hunt [132], Onishchik [179], Palais and Terng [182], Serre [195] and Thorbergsson [208, 209].

4.1 Maximal Tori and Polar Actions

The goal of this section is twofold. First, we prove the Maximal Torus Theorem 4.1, that generalizes basic diagonalization results from linear algebra. Second, we present the definitions of polar actions and isoparametric submanifolds, briefly discussing some results from these theories, and relating them to the adjoint and conjugation actions of compact Lie groups.

Recall that a Lie group T is a *torus* if it is isomorphic to the product $S^1 \times \cdots \times S^1$. Alternatively, these are the compact connected abelian Lie groups, see Theorem 1.41. We say that a vector $X \in \mathfrak{t}$ is an *infinitesimal generator* if the corresponding 1-parameter subgroup $\{\exp(tX) : t \in \mathbb{R}\}$ is dense in T. It is not difficult to prove that every torus admits infinitesimal generators, see Example 3.43.

© Springer International Publishing Switzerland 2015
M.M. Alexandrino, R.G. Bettiol, *Lie Groups and Geometric Aspects of Isometric Actions*, DOI 10.1007/978-3-319-16613-1_4

Let T be the subgroup of diagonal matrices of $SU(n)$. It follows from Theorem 1.41 that T is a torus, and we know from linear algebra that every matrix $g \in SU(n)$ is conjugate to an element of T. This is a particular case of the next classical result, see Example 4.7.

Maximal Torus Theorem 4.1. *Let G be a connected, compact Lie group.*

(i) *There exists a maximal[1] torus T in G;*
(ii) *Any two maximal tori in G are conjugate;*
(iii) *Every element of G is contained in a maximal torus;*
(iv) *For each bi-invariant metric Q on G, the orbits of the conjugation action intersect every maximal torus Q-orthogonally.*

Proof. Our proof is based on Hunt [132] and Thorbergsson [207].

To prove (i), let $\mathfrak{t} \subset \mathfrak{g}$ be a maximal abelian subalgebra of the Lie algebra of G. By Theorem 1.21, there exists a unique connected subgroup $T \subset G$ with Lie algebra $\mathfrak{t}$. Note that its closure $\overline{T}$ is an abelian, connected, compact Lie group, see Proposition 1.39. Since $\mathfrak{t}$ is maximal, $\overline{T} = T$. From Theorem 1.41, T is a torus. If $T' \subset G$ is a torus with $T \subset T'$, then $[X, Z] = 0$ for all $X \in \mathfrak{t}'$ and $Z \in \mathfrak{t}$. By maximality of $\mathfrak{t}$, we have $X \in \mathfrak{t}$ and hence $\mathfrak{t}' = \mathfrak{t}$. Thus, from Theorem 1.21, $T' = T$.

Claim 4.2. *If $X \in \mathfrak{t}$ is an infinitesimal generator of T, $\mathfrak{t} = \{Y \in \mathfrak{g} : [X, Y] = 0\}$.*

Note that $\mathfrak{t} \subset \{Y \in \mathfrak{g} : [X, Y] = 0\}$ by Proposition 1.39. Conversely, let $Y \in \mathfrak{g}$ be such that $[X, Y] = 0$. From commutativity of X and Y, by Remark 1.40, we have

$$\exp(sX)\exp(tY) = \exp(tY)\exp(sX), \quad \text{for all } s, t \in \mathbb{R}.$$

Thus, $\exp(tY)$ commutes with all elements of $T = \overline{\{\exp(sX) : s \in \mathbb{R}\}}$ and hence Y commutes with all vectors of $\mathfrak{t}$. From Theorem 1.21, there is a unique connected Lie subgroup T' of G with Lie algebra $\mathfrak{t}' = \mathfrak{t} \oplus \operatorname{span}\{Y\}$. Note that $\overline{T'}$ is an abelian, connected, compact Lie group, and hence a torus, by Theorem 1.41. Since T is a maximal torus contained in $\overline{T'}$, we have that $T = \overline{T'}$ and hence $Y \in \mathfrak{t}$, proving Claim 4.2.

In order to prove (ii), let X_1 and X_2 be infinitesimal generators of maximal tori T_1 and T_2 respectively, and set

$$f: G \to \mathbb{R}, \quad f(g) := Q(\operatorname{Ad}(g)X_1, X_2),$$

where Q is a bi-invariant metric on G (see Proposition 2.24). Since G is compact, f attains a minimum, say at the point $g_0 \in G$. Therefore, for all $Y \in \mathfrak{g}$,

$$0 = \frac{d}{dt} f\left(\exp(tY)g_0\right)\Big|_{t=0}$$

$$= \frac{d}{dt} Q\left(\operatorname{Ad}(\exp(tY))\operatorname{Ad}(g_0)X_1, X_2\right)\Big|_{t=0}$$

---

[1]i.e., such that if $T' \subset G$ is a torus and $T \subset T'$, then $T = T'$.

$$=Q([Y,\mathrm{Ad}(g_0)X_1],X_2)$$

$$=Q(Y,[\mathrm{Ad}(g_0)X_1,X_2]),$$

and hence $[\mathrm{Ad}(g_0)X_1,X_2]=0$. From Claim 4.2, $\mathrm{Ad}(g_0)X_1 \in \mathfrak{t}_2$, which means that $g_0\exp(tX_1)g_0^{-1} \in T_2$. Thus, $g_0 T_1 g_0^{-1} \subset T_2$ and, by maximality of T_1, we conclude that equality holds, proving (ii).

As for (iii), let $g \in G$. Since G is compact, by Theorem 2.27, there exists $Y \in \mathfrak{g}$ such that $\exp(Y)=g$. Thus, g is contained in the maximal torus that contains the corresponding 1-parameter subgroup $\{\exp(tY) : t \in \mathbb{R}\}$. Item (iii) now follows from item (ii). In order to prove (iv), let $p \in T$ be a point in a maximal torus, and consider its orbit $G(p)=\{g\,p\,g^{-1} : g \in G\}$ by the conjugation action. It is easy to see that

$$T_pG(p)=\left\{\mathrm{d}R_pX-\mathrm{d}L_pX : X \in \mathfrak{g}\right\}, \quad \text{and} \quad T_pT=\left\{\mathrm{d}R_pZ : Z \in \mathfrak{t}\right\},$$

cf. (1.13). If Q is a bi-invariant metric on G, then L_g and R_g are isometries for all $g \in G$. Thus, for all $X \in \mathfrak{g}$ and $Z \in \mathfrak{t}$, we have

$$\begin{aligned}
Q\big(\mathrm{d}R_pX-\mathrm{d}L_pX,\mathrm{d}R_pZ\big) &= Q(X,Z)-Q(\mathrm{d}L_pX,\mathrm{d}R_pZ) \\
&= Q(X,Z)-Q\big(X,\mathrm{d}(L_{p^{-1}})_p\,\mathrm{d}R_pZ\big) \\
&= Q\big(X,Z-\mathrm{Ad}\big(p^{-1}\big)Z\big) \\
&= 0,
\end{aligned}$$

since T is abelian, and hence $Z \in \mathfrak{t}$ satisfies $\mathrm{Ad}(g)Z=Z$ for all $g \in T$, see (1.8). □

Remark 4.3. A maximal torus T in G is its own centralizer, that is, $Z_G(T)=T$, see Exercise 1.59. In fact, it is clear that $T \subset Z_G(T)$, and the existence of $g \in Z_G(T)$ with $g \notin T$ would contradict the maximality of T as an abelian subgroup of G.

Definition 4.4. The *rank* of a connected, compact Lie group G is defined to be the dimension of a maximal torus[2] in G, and is denoted by $\mathrm{rank}\,G$.

Example 4.5. Consider the Lie group $S^3=\{(z_1,z_2) \in \mathbb{C}^2 : |z_1|^2+|z_2|^2=1\}$, with the product defined in Exercise 1.52, and recall that $S^3 \cong \mathrm{SU}(2)$. We claim that the great circle

$$T=\left\{\big(e^{i\theta},0\big) \in S^3 : e^{i\theta} \in S^1\right\} \tag{4.1}$$

is a maximal torus in S^3. In fact, if $(w_1,w_2) \in S^3$ commutes with every element of T, then $e^{i\theta}w_2=e^{-i\theta}w_2$ for all $\theta \in \mathbb{R}$, so $w_2=0$, and hence $(w_1,w_2) \in T$. Thus, the Lie group $S^3 \cong \mathrm{SU}(2) \cong \mathrm{Sp}(1)$ has rank one. In particular, the maximal tori in S^3

---

[2] By the Maximal Torus Theorem 4.1 (ii), all maximal tori in G are conjugate and hence have the same dimension.

are its 1-parameter subgroups, which are the great circles through the origin $(1,0)$ and its antipodal point $(-1,0)$. These are all conjugate to (4.1), cf. Maximal Torus Theorem 4.1 (ii) and (iii).

In fact, the above is essentially the only rank one Lie group, due to the following result, see Duistermaat and Kolk [79].

Theorem 4.6. *If G is a connected compact Lie group of rank one, then G is isomorphic to either* $\mathrm{SU}(2)$, $\mathrm{SO}(3)$, *or* S^1.

Example 4.7. The subgroup $T \subset \mathrm{U}(n)$ of diagonal matrices given by

$$
T = \left\{ \begin{pmatrix} e^{i\theta_1} & & \\ & \ddots & \\ & & e^{i\theta_n} \end{pmatrix} : \theta_j \in \mathbb{R} \right\}, \tag{4.2}
$$

is clearly an n-dimensional torus. It can be shown that T is a maximal torus in $\mathrm{U}(n)$, and also in $\mathrm{Sp}(n) \supset \mathrm{U}(n)$. In particular, $\operatorname{rank}\mathrm{U}(n) = \operatorname{rank}\mathrm{Sp}(n) = n$. Using Exercise 1.60, it is easy to see that the group $T/Z(\mathrm{U}(n))$, which can be identified with the subgroup of T formed by matrices with $\theta_1 + \cdots + \theta_n = 0$, is a maximal torus[3] in $\mathrm{SU}(n)$, and hence $\operatorname{rank}\mathrm{SU}(n) = n-1$. The image of T under the standard embeddings $\mathrm{U}(n) \subset \mathrm{SO}(2n) \subset \mathrm{SO}(2n+1)$ is also a maximal torus in both $\mathrm{SO}(2n)$ and $\mathrm{SO}(2n+1)$, which hence also have rank n.

The above geometric properties of the conjugation action on a compact Lie group (G,Q), asserted in the Maximal Torus Theorem 4.1, generalize to the interesting class of *polar actions*. The study of these actions was pioneered by Szenthe [202, 203] and independently by Palais and Terng [181], see also [182].

Definition 4.8. An isometric action of a compact Lie group G on a complete Riemannian manifold (M, g) is called a *polar action* if there exists a connected, complete, immersed submanifold Σ, called a *section*, that intersects all the G-orbits, and every intersection between Σ and a G-orbit is orthogonal. Moreover, if Σ is flat, the action is called *hyperpolar*.

Exercise 4.9 ($\star$). Let Σ be a section for a polar action on (M, g). Verify that:

(i) The dimension of Σ is equal to the codimension of the principal orbits;
(ii) For any $x \in M$, there exists a section Σ' with $x \in \Sigma'$;
(iii) Sections are totally geodesic submanifolds.

Hint: In order to prove (iii), first verify that a section Σ is totally geodesic on regular points, using properties of integrable Riemannian submersions, see (2.12). Since regular points are dense in Σ by Kleiner's Lemma 3.70, conclude that Σ is totally geodesic.

---

[3]Notice that, under the isomorphism $S^3 \cong \mathrm{SU}(2)$ in Exercise 1.52, the maximal torus (4.1) is precisely the subgroup of (4.2) with $n = 2$ such that $\theta_1 + \theta_2 = 0$.

Remark 4.10. A key property of polar actions is that they can be characterized as the isometric actions whose horizontal distribution (3.20) is integrable on the regular stratum. This is explained in Sect. 5.4, together with other aspects of polar actions that hold in the more general context of *polar foliations*.

The Maximal Torus Theorem 4.1 implies that the conjugation action on a compact Lie group (G,Q) is polar, since maximal tori are sections. Moreover, it is hyperpolar, since tori in (G,Q) are flat, by Proposition 2.26.

A linear polar action on a vector space is called a *polar representation*. It is not difficult to show that the linearization of a polar action, restricted to the normal space to the orbits, is a polar representation [182], cf. Theorem 5.27 (iii).

Proposition 4.11. *If $\mu \colon G \times M \to M$ is a polar action, then the slice representation at each $x \in M$ (see Definition 3.72) is a polar representation of G_x on $v_x G(x)$. Moreover, if Σ is a section through $x \in M$, then $T_x \Sigma$ is a section on $v_x G(x)$.*

In particular, the adjoint representation $\mathrm{Ad} \colon G \times \mathfrak{g} \to \mathfrak{g}$ of (G,Q) is a polar representation. Indeed, it is the slice representation obtained by linearizing at $e \in G$ the (hyperpolar) conjugation action, recall Definition 1.36. Furthermore, a section is given by the Lie algebra $\mathfrak{t}$ of a maximal torus T, since this is the tangent space to a section T for the conjugation action on G. This can be proved directly from the Maximal Torus Theorem 4.1, since the tangent space to the adjoint orbit of $Z \in \mathfrak{t}$ is

$$T_Z\big(\mathrm{Ad}(G)Z\big) = \big\{\tfrac{d}{dt}\mathrm{Ad}(\exp(tX))Z\big|_{t=0} : X \in \mathfrak{g}\big\} = \big\{[X,Z] : X \in \mathfrak{g}\big\}, \tag{4.3}$$

and $0 = Q(X,[Z,Y]) = Q([X,Z],Y)$ for all $Y \in \mathfrak{t}$.

Other examples of polar actions are given by isometric actions whose principal orbits have codimension 1. In fact, by Proposition 3.78, a geodesic that intersects a G-orbit orthogonally intersects all G-orbits orthogonally. Such actions are called *cohomogeneity one actions*[4] and are studied in detail in Chap. 6.

More examples of polar actions can be found among *symmetric spaces*, which are Riemannian manifolds (M,g) such that, for each $p \in M$, there is an isometry I^p that fixes p and reverses geodesics through p. As explained in Exercise 6.32, symmetric spaces are of the form $M = G/H$, where G is the connected component of $\mathrm{Iso}(M,\mathsf{g})$ that contains the identity, and H is the isotropy group of a point $p \in M$; such a pair of groups (G,H) is called a *symmetric pair*. In a symmetric space $M = G/H$, a maximal totally geodesic flat submanifold Σ through $p \in M$ is a section for the subaction of the isotropy group H, which is the left translation $h \cdot gH := hgH$ on G/H, see (6.13). Note that these polar actions are also hyperpolar, as Σ is flat. It can be proved that if (G,H_1) and (G,H_2) are symmetric pairs, then the subaction of H_1 on G/H_2 is also hyperpolar. These actions are called *Hermann actions*, as they were introduced by Hermann [127], see also Thorbergsson [209].

---

[4]The *cohomogeneity* of an action is, by definition, the codimension of its principal orbits.

Remark 4.12. The conjugation action on a Lie group G with bi-invariant metric Q is a particular case of the above hyperpolar actions of isotropy groups of symmetric spaces. In fact, (G, Q) is a symmetric space by Theorem 2.30, which can be written as $G \cong (G \times G)/\Delta G$, see Example 3.39. It is easy to check that conjugation on G corresponds to the subaction of the isotropy group ΔG on $(G \times G)/\Delta G$, and that a maximal torus is a maximal totally geodesic flat in this symmetric space.

It can be shown that the isotropy representation of a symmetric space $M = G/H$ is a polar representation of H. These representations are called *s-representations*. Polar representations were classified by Dadok [72], who proved that any such representation must be *orbit-equivalent*[5] to an *s*-representation.

Podestà and Thorbergsson [186] classified polar actions on compact rank[6] one symmetric spaces (see Example 6.10 for details on these spaces). Besides cohomogeneity one actions, sections for these polar actions are not flat, since they are totally geodesic (see Exercise 4.9) and these spaces have positive sectional curvature. Kollross [146] classified hyperpolar (and cohomogeneity one) actions on compact irreducible symmetric spaces, proving that they are orbit-equivalent to a Hermann action if they have cohomogeneity at least two. It had been conjectured that polar actions on compact irreducible symmetric spaces of rank greater than one are hyperpolar. A partial answer was given by Biliotti [40], who proved it under the extra assumption that the symmetric space is *Hermitian*. The full conjecture was recently settled by Lytchak [155], see Theorem 5.38. Another recent development in the area was the classification of polar manifolds with sec > 0 and cohomogeneity at least two by Fang, Grove and Thorbergsson [84], see Theorem 5.39. For surveys on polar actions, see Thorbergsson [208, 209].

We conclude this section with a discussion of isoparametric submanifolds, which also generalize some of the geometric properties of orbits of the conjugation action.

Definition 4.13. Let L be an immersed submanifold of a Riemannian manifold (M, g). A section ξ of the normal bundle $\nu(L)$ is said to be a *parallel normal field* along L if $\nabla^\nu \xi$ vanishes identically, where ∇^ν is the normal connection.[7] The submanifold L is said to have *flat normal bundle*, if any normal vector can be extended to a locally defined parallel normal field. In addition, L is said to have *globally flat normal bundle*, if the holonomy of the normal bundle $\nu(L)$ is trivial, which means that any normal vector can be extended to a globally defined parallel normal field.

Definition 4.14. A submanifold F of a space form $M(k)$ is called *isoparametric* if its normal bundle is flat and principal curvatures along any parallel normal vector field are constant.

---

[5]Isometric actions of Lie groups G_1 and G_2 on Riemannian manifolds M_1 and M_2 are *orbit-equivalent* if there exists an isometry between M_1 and M_2 that maps G_1-orbits to G_2-orbits.

[6]The *rank* of a symmetric space is the dimension of a maximal totally geodesic flat submanifold.

[7]In other words, $\nabla^\nu \xi$ is the component of $\nabla \xi$ that is normal to L.

It can be shown that principal orbits of a polar action on a Riemannian manifold have globally flat normal bundle, since they are described by an integrable Riemannian submersion (see Remarks 3.79 and 4.10). In particular, equivariant normal vectors turn out to be parallel normal fields. It is not difficult to prove that principal curvatures along equivariant normal vector fields of principal orbits of a polar action are constant (see Proposition 3.78). These two facts together imply the next result.

Proposition 4.15. *Principal orbits of a polar action on Euclidean space are isoparametric submanifolds. In particular, principal orbits of the adjoint action of a compact Lie group G are isoparametric submanifolds of its Lie algebra* $\mathfrak{g}$.

In the particular case of adjoint actions, each regular orbit is a principal orbit (see Theorem 4.36), and its principal curvatures and principal directions can be explicitly computed (see Example 4.29).

It is possible to prove that the normal bundle of an isoparametric submanifold is *globally* flat and hence that each normal vector can be extended to a parallel normal vector field. An important property of isoparametric submanifolds is that sets parallel to an isoparametric submanifold are submanifolds. This means that given an isoparametric submanifold N and a parallel normal vector field ξ, the *endpoint map* $\eta_\xi(x) := \exp_x(\xi)$ is such that $\eta_\xi(N)$ is a submanifold, with dimension possibly lower than $\dim N$. An *isoparametric foliation* $\mathscr{F}$ on $M(k)$ is a partition of $M(k)$ by submanifolds parallel to a given isoparametric submanifold N. It is possible to prove that a leaf L of $\mathscr{F}$ is isoparametric if $\dim L = \dim N$, in which case the partition into parallel submanifolds to L coincides with $\mathscr{F}$.

Remark 4.16. Isoparametric hypersurfaces in space forms have been studied since Cartan [62–65], and remain a very active field of research. In Euclidean and hyperbolic space, they can only be cylinders or umbilic hypersurfaces. In spheres, there are other examples of isoparametric hypersurfaces, many of which are inhomogeneous, see Ferus, Karcher and Münzner [89]. Münzner [170, 171] proved that isoparametric hypersurfaces in spheres separate the ambient into two disk bundles over focal submanifolds and used this to show that the number of principal curvatures can only be 1, 2, 3, 4 or 6, with certain restrictions on the multiplicities. All of these numbers are known to occur. The classification of isoparametric hypersurfaces in spheres is almost complete, except for a last remaining case, see Miyaoka [162] and Siffert [196], and also [69, 210].

A general concept of isoparametric submanifold in higher codimensions was independently introduced by Harle [122], Carter and West [67, 68] and Terng [204], motivated by analogies with orbits of polar actions in $\mathbb{R}^n$. In particular, their study is also connected with Coxeter groups. Regarding the relation of isoparametric submanifolds with group actions, Thorbergsson [206] proved that a compact, irreducible isoparametric submanifold L^n in $\mathbb{R}^{n+k}$ is homogeneous, provided $k \geq 3$ and L^n does not lie in any affine hyperplane of $\mathbb{R}^{n+k}$.

Details on Terng's work about isoparametric submanifolds on space forms and Hilbert spaces can be found in Palais and Terng [182]. Another important reference on isoparametric submanifolds is the book of Berndt, Console and Olmos [32]. More information on the history of isoparametric hypersurfaces and submanifolds can be found in the surveys of Thorbergsson [208, 209], see also Sect. 5.5.

4.2 Normal Slices of Conjugation Actions

In this section, we explain how normal slices for the conjugation action of a compact Lie group (G, Q) can be explicitly described in terms of the collection of maximal tori in G. Notice that the orbits of the conjugation action of G on itself are its conjugacy classes, and the isotropy group of $x \in G$ is its centralizer $G_x = Z_G(\{x\})$.

Lemma 4.17. *Let G_x^0 be the connected component of $G_x = \{g \in G : gxg^{-1} = x\}$ that contains the identity. Then $x \in G_x^0$.*

Proof. Since G is compact, there exists $X \in \mathfrak{g}$ such that $\exp(X) = x$, see Theorem 2.27. Thus, $x = \exp(X) = \exp(tX)\exp(X)\exp(-tX) = \exp(tX)x\exp(-tX)$, so $\exp(tX) \in G_x^0$ for all $t \in \mathbb{R}$ and, in particular, $x = \exp(X) \in G_x^0$. $\qquad\square$

Proposition 4.18. *Let $\mathscr{T}$ be the collection of maximal tori in G, and let $\Lambda(x)$ be the subcollection of maximal tori that contain $x \in G$. Then $\bigcap_{T \in \mathscr{T}} T = Z(G)$ is the center of G, and $\bigcup_{T \in \Lambda(x)} T = G_x^0$ is the identity component of the isotropy group G_x.*

Proof. Item (iii) of the Maximal Torus Theorem 4.1 implies that $G = \bigcup_{T \in \mathscr{T}} T$ and hence $Z(G) = Z(\bigcup_{T \in \mathscr{T}} T) = \bigcap_{T \in \mathscr{T}} Z_G(T) = \bigcap_{T \in \mathscr{T}} T$, by Remark 4.3. Furthermore, it is clear that $\bigcup_{T \in \Lambda(x)} T \subset G_x^0$. For the converse inclusion, let $g \in G_x^0$ and T' be a maximal torus in G_x^0 that contains g. By Lemma 4.17, we have that $x \in G_x^0$, and hence trivially $x \in Z(G_x^0)$. As proved above, $Z(G_x^0)$ is the intersection of all the maximal tori in G_x^0, hence $x \in T'$. Letting T be a maximal torus in G such that $T' \subset T$, we have $g, x \in T$ and hence $g \in \bigcup_{T \in \Lambda(x)} T$. $\qquad\square$

Slice Theorem 4.19. *Let (G, Q) be a compact connected Lie group acting on itself by conjugation, and let S_x be a normal slice at $x \in G$ with radius $\varepsilon > 0$.*

(i) *$S_x = \bigcup_{\sigma \in \Lambda_\varepsilon(x)} \sigma$, where $\Lambda_\varepsilon(x) := \{B_\varepsilon(x) \cap T : T \in \Lambda(x)\}$;*

(ii) *$S_y \subset S_x$ for all $y \in S_x$;*

(iii) *Denoting by $\mathscr{F}$ the homogeneous foliation of G by orbits of the conjugation action, see (3.4), there exists an isoparametric foliation $\widehat{\mathscr{F}}$ of a neighborhood of the origin $0 \in T_x S_x$, such that $\exp_x(\widehat{\mathscr{F}}) = \mathscr{F} \cap S_x$.*

Proof. It follows from Proposition 4.18 that $G_x^0 \cap B_\varepsilon(x) = \bigcup_{\sigma \in \Lambda_\varepsilon(x)} \sigma$. Given $\sigma \in \Lambda_\varepsilon(x)$, the minimal geodesic segment γ joining $y \in \sigma$ to x is contained in σ, since every maximal torus is totally geodesic in G. By the Maximal Torus Theorem 4.1, the conjugation orbit $G(x)$ is orthogonal to the maximal torus σ, hence γ is orthogonal to $G(x)$, so $y \in S_x$. Thus, $\bigcup_{\sigma \in \Lambda_\varepsilon(x)} \sigma \subset S_x$. Moreover, note that $\dim S_x = \dim G - \dim G/G_x = \dim G_x$, from which it follows that

$$G_x^0 \cap B_\varepsilon(x) = \bigcup_{\sigma \in \Lambda_\varepsilon(x)} \sigma = S_x. \tag{4.4}$$

Item (ii) follows from $G_y \subset G_x$ and (4.4). In order to prove (iii), notice that (4.4) implies that there is a neighborhood U of $e \in G_x^0$ such that $L_x(U) = xU = S_x$. Set $\widetilde{\mathscr{F}}_x := \{G_x(y) : y \in S_x\}$ and $\widetilde{\mathscr{F}}_e := \{G_x(y) : y \in U\}$. Since $x \in Z(G_x)$, we have $\widetilde{\mathscr{F}}_x = L_x(\widetilde{\mathscr{F}}_e)$. By the properties of a slice of an isometric action, $\widetilde{\mathscr{F}}_x = \mathscr{F} \cap S_x$. Define $\widehat{\mathscr{F}}_e := \{\mathrm{Ad}(G_x)X : X \in (B_\varepsilon(0) \cap \mathfrak{g}_x)\}$, where $\mathfrak{g}_x$ is the Lie algebra of G_x, and $\widehat{\mathscr{F}} := dL_x(\widehat{\mathscr{F}}_e)$. Clearly, $\widetilde{\mathscr{F}}_e = \exp(\widehat{\mathscr{F}}_e)$. Combining these observations and (2.2),

$$\mathscr{F} \cap S_x = \widetilde{\mathscr{F}}_x = L_x(\widetilde{\mathscr{F}}_e) = L_x\left(\exp\left(\widehat{\mathscr{F}}_e\right)\right) = \exp_x\left(dL_x(\widehat{\mathscr{F}}_e)\right) = \exp_x\left(\widehat{\mathscr{F}}\right).$$

Finally, $\widehat{\mathscr{F}}$ is isoparametric, as $\widehat{\mathscr{F}}_e$ is isoparametric and L_x is an isometry. □

4.3 Roots of a Compact Lie Group

Roots play a fundamental role in the theory of compact Lie groups. They allow, for instance, to classify compact simple Lie groups (see Sect. 4.5) and are also related to principal curvatures of principal orbits of the adjoint action (see Example 4.29). *Roots* are defined in Theorem 4.23.

For the sake of motivation, we first state a result that follows directly from Theorem 4.23 and Remark 4.27, which explains the importance of roots in understanding the adjoint representation, as it splits into rotations on appropriate subspaces.

Proposition 4.20. *Let G be a connected compact nonabelian Lie group with Lie algebra $\mathfrak{g}$ and $T \subset G$ a maximal torus with Lie algebra $\mathfrak{t}$. There exist 2-dimensional subspaces $V_j \subset \mathfrak{g}$ and linear functionals $\alpha_j \colon \mathfrak{t} \to \mathbb{R}$ such that:*

(i) *$\mathfrak{g} = \mathfrak{t} \oplus V_1 \oplus \cdots \oplus V_z$;*
(ii) *If Q is a bi-invariant metric on G, considering a Q-orthonormal basis of V_j,*

$$\mathrm{Ad}\big(\exp(X)\big)\big|_{V_j} = \begin{pmatrix} \cos\alpha_j(X) & -\sin\alpha_j(X) \\ \sin\alpha_j(X) & \cos\alpha_j(X) \end{pmatrix}, \quad \textit{for all } X \in \mathfrak{t};$$

(iii) *The above decomposition is unique, and $\alpha_j \neq \alpha_k$ if $j \neq k$.*

Each linear functional $\alpha \colon \mathfrak{t} \to \mathbb{R}$ above is the root associated to the space V_α, see the definition in Theorem 4.23. Before proving this main theorem, we need to state the following auxiliary result from linear algebra.

Lemma 4.21. *Let $W \subset \mathbb{C}^n$ be a complex subspace. Then W is invariant by complex conjugation (i.e., $W = \overline{W}$) if and only if $W = V \oplus iV$ for a real subspace $V \subset \mathbb{R}^n$.*

Remark 4.22. If $\{v_j\}$ is a real basis for a real subspace $V \subset \mathbb{R}^n$, then $\{v_j\}$ is a complex basis for the complex subspace $W = V \oplus iV$. In particular, $\dim_{\mathbb{C}} W = \dim_{\mathbb{R}} V$.

Notice that if G is a Lie group with Lie algebra $\mathfrak{g}$, the complexification of $\mathfrak{g}$,

$$\mathfrak{g}_{\mathbb{C}} := \mathfrak{g} \otimes_{\mathbb{R}} \mathbb{C}, \tag{4.5}$$

is a (complex) Lie algebra where $[\cdot, \cdot] \colon \mathfrak{g}_{\mathbb{C}} \times \mathfrak{g}_{\mathbb{C}} \to \mathfrak{g}_{\mathbb{C}}$ is the canonical extension of the Lie bracket of $\mathfrak{g}$. As a real vector space, $\mathfrak{g}_{\mathbb{C}} = \mathfrak{g} \oplus i\mathfrak{g}$. We are now ready to prove the main result in this section, based on Duistermaat and Kolk [79, pp. 170–171].

Theorem 4.23. *Let G be a connected compact nonabelian Lie group and $T \subset G$ a maximal torus, with Lie algebras $\mathfrak{g}$ and $\mathfrak{t}$, respectively.*

(i) *For each $X \in \mathfrak{g}$, the linear map $\mathrm{ad}(X) \colon \mathfrak{g}_{\mathbb{C}} \to \mathfrak{g}_{\mathbb{C}}$ is diagonalizable with purely imaginary eigenvalues;*

(ii) *There exists a unique (up to permutations) decomposition of $\mathfrak{g}_{\mathbb{C}}$ in subspaces*

$$\mathfrak{g}_\alpha := \{Y \in \mathfrak{g}_{\mathbb{C}} : [X,Y] = i\alpha(X)Y, \text{ for all } X \in \mathfrak{t}\},$$

where $\alpha \colon \mathfrak{t} \to \mathbb{R}$ is a linear functional called root. *More precisely,*

$$\mathfrak{g}_{\mathbb{C}} = \mathfrak{g}_0 \oplus \bigoplus_{\alpha \in R} \mathfrak{g}_\alpha,$$

where R denotes the set of roots, also called root system;

(iii) *$\mathfrak{g}_0 = \mathfrak{t} \oplus i\mathfrak{t}$ and $\overline{\mathfrak{g}_\alpha} = \mathfrak{g}_{-\alpha}$. In particular, $\alpha \in R$ implies $-\alpha \in R$;*

(iv) *$\dim_{\mathbb{C}} \mathfrak{g}_\alpha = 1$ and $\dim V_\alpha = 2$, where $V_\alpha := (\mathfrak{g}_\alpha \oplus \mathfrak{g}_{-\alpha}) \cap \mathfrak{g}$;*

(v) *$\mathfrak{g}_{\lambda\alpha} = 0$ if $\lambda \in \mathbb{C} \setminus \{-1, 0, 1\}$;*

(vi) *If $e_2 + ie_1$ is the vector that spans $\mathfrak{g}_\alpha$, where $e_1, e_2 \in \mathfrak{g}$, then:*

(vi-a) *$\{e_1, e_2\}$ is a basis of V_α;*

(vi-b) *$[X, e_1] = \alpha(X)e_2$ and $[X, e_2] = -\alpha(X)e_1$, for all $X \in \mathfrak{t}$;*

(vi-c) *$Q(e_1, e_2) = 0$ and $\|e_1\| = \|e_2\|$, with respect to any bi-invariant metric Q;*

(vi-d) *In the Q-orthonormal basis $\left\{ \frac{e_1}{\|e_1\|}, \frac{e_2}{\|e_2\|} \right\}$,*

$$\mathrm{Ad}\big(\exp(X)\big)\big|_{V_\alpha} = \begin{pmatrix} \cos\alpha(X) & -\sin\alpha(X) \\ \sin\alpha(X) & \cos\alpha(X) \end{pmatrix};$$

(vii) *Let $\alpha^\vee \in \mathfrak{t}$ be the coroot of α, defined to be the vector such that $\alpha(\alpha^\vee) = 2$ and $\alpha^\vee$ is Q-orthogonal to $\ker\alpha$.*

(vii-a) *$\alpha^\vee$ is orthogonal to $\ker\alpha$ with respect to any bi-invariant metric;*

(vii-b) $\mathfrak{g}^{(\alpha)} := \operatorname{span}_{\mathbb{R}}\{\alpha^{\vee}\} \oplus V_{\alpha}$ *is a Lie algebra isomorphic to* $\mathfrak{su}(2)$, *and the corresponding Lie group* $G^{(\alpha)} := \exp(\mathfrak{g}^{(\alpha)})$ *is isomorphic to either* SU(2) *or* SO(3);

(vii-c) *For each* $g \in G^{(\alpha)}$, $\operatorname{Ad}(g)\big|_{\ker \alpha} = \operatorname{id}$;

(vii-d) *There exists* $w \in G^{(\alpha)}$, *such that* $\operatorname{Ad}(w)\alpha^{\vee} = -\alpha^{\vee}$.

Proof. Using the Jordan normal form, there exists a decomposition of $\mathfrak{g}_{\mathbb{C}}$ in subspaces $\mathfrak{g}_j$ such that $\operatorname{ad}(X)|_{\mathfrak{g}_j} = c_j I + N_j$ where N_j is a *nilpotent matrix.*[8] Therefore

$$\exp\left(t\operatorname{ad}(X)\right)\big|_{\mathfrak{g}_j} = \exp(tc_j) \sum_{k=0}^{m_j-1} \frac{t^k}{k!} N_j^k. \tag{4.6}$$

On the other hand, using (1.12), compactness of $\operatorname{Ad}(G)$ implies that $\exp(t\operatorname{ad}(X))$ is bounded. Together with (4.6), this implies that c_j is a purely imaginary number and $N_j = 0$, proving (i). Item (ii) follows from the next two claims.

Claim 4.24. There exists a decomposition $\mathfrak{g}_{\mathbb{C}} = \mathfrak{g}_0 \oplus \mathfrak{g}_1 \oplus \cdots \oplus \mathfrak{g}_z$ and linear functionals $\alpha_j \colon \mathfrak{t} \to \mathbb{R}$, such that for each $Y \in \mathfrak{g}_j$, $[X,Y] = i\alpha_j(X)Y$ for all $X \in \mathfrak{t}$. In addition, $\alpha_j \neq \alpha_k$ if $j \neq k$.

In order to verify this claim, consider a basis $\{X_1,\dots,X_r\}$ of $\mathfrak{t}$. On the one hand, from (i), each $\operatorname{ad}(X_l)$ is diagonalizable. On the other hand, $\operatorname{ad}(X_l)$ commutes with $\operatorname{ad}(X_k)$ for all k,l, since $[X_l, X_k] = 0$. Thus, there is a decomposition $\mathfrak{g}_{\mathbb{C}} = \bigoplus_j \mathfrak{g}_j$ in common eigenspaces of the operators $\operatorname{ad}(X_l)$. Define

$$\alpha_j\left(\sum_{l=1}^{r} x_l X_l\right) := \sum_{l=1}^{r} x_l \alpha_{jl},$$

where $[X_l, Y] = i\alpha_{jl}Y$ for all $Y \in \mathfrak{g}_j$. By construction, $\mathfrak{g}_j \oplus \mathfrak{g}_k$ is not an eigenspace of all operators $\operatorname{ad}(X_l)$, hence $\alpha_j \neq \alpha_k$ if $j \neq k$.

Claim 4.25. If $[X,Y] = i\beta(X)Y$ for all $X \in \mathfrak{t}$, then there exists j such that $Y \in \mathfrak{g}_j$ and $\beta = \alpha_j$, where $\mathfrak{g}_j$ and α_j were defined in Claim 4.24.

In order to prove this claim, let $Y = Y_0 + Y_1 + \cdots + Y_z$, with $Y_j \in \mathfrak{g}_j$. Then $\sum_{j=1}^{z} i\beta(X)Y_j = [X,Y] = \sum_{j=1}^{z} i\alpha_j(X)Y_j$. Since $\mathfrak{g}_{\mathbb{C}} = \mathfrak{g}_0 \oplus \mathfrak{g}_1 \oplus \cdots \oplus \mathfrak{g}_z$, we have that $\beta(X) = \alpha_j(X)$. Thus, Claim 4.25 follows from the fact that $\alpha_j \neq \alpha_k$ if $j \neq k$.

Item (iii) follows from Lemma 4.21 and the fact that $\mathfrak{t}$ is a maximal abelian Lie subalgebra of $\mathfrak{g}$.

With regard to (iv) and (v), since the set of roots R is finite, we can choose a vector $X \in \ker \alpha$ such that $\beta(X) \neq 0$, for all roots β that are not multiples of α.

---

[8]This means that $N_j^m = 0$ if $m = m_j$, and $N_j^m \neq 0$ if $m < m_j$

Note that the Lie algebra of the identity connected component of the isotropy group G_X^0 with respect to the adjoint action is

$$\mathfrak{g}_X = \mathfrak{t} \oplus \bigoplus_{\lambda \neq 0} \left(\mathfrak{g}_{\lambda\alpha} \oplus \mathfrak{g}_{-\lambda\alpha} \right) \cap \mathfrak{g}. \tag{4.7}$$

This implies that $Z(\mathfrak{g}_X) = \ker\alpha$, where $Z(\mathfrak{g}_X)$ is the Lie algebra of $Z(G_X^0)$, see Corollary 1.57. Set $G' := G_X^0/Z(G_X^0)$ and note that the Lie algebra of the compact connected Lie group G' is $\mathfrak{g}_X/Z(\mathfrak{g}_X)$. It follows from (4.7) that $\mathfrak{t}/\ker\alpha$ is a maximal abelian Lie subalgebra of $\mathfrak{g}_X/Z(\mathfrak{g}_X)$. Thus, G' has rank one and hence $\mathfrak{g}_X/Z(\mathfrak{g}_X) \cong \mathfrak{su}(2)$ by Theorem 4.6. The fact that $\dim \mathfrak{g}_X/Z(\mathfrak{g}_X) = 3$ implies that

$$\dim \bigoplus_{\lambda \neq 0} \left(\mathfrak{g}_{\lambda\alpha} \oplus \mathfrak{g}_{-\lambda\alpha} \right) \cap \mathfrak{g} = 2. \tag{4.8}$$

Items (iv) and (v) follow from (4.8) and Remark 4.22.

As for (vi), since $\mathfrak{g}_\alpha \cap \mathfrak{g}_{-\alpha} = \{0\}$, $e_1 \neq \lambda e_2$. This, and the fact that $\dim V_\alpha = 2$, implies (vi-a). Item (vi-b) follows directly from $[X, e_2 + i e_1] = i\alpha(X)e_2 - \alpha(X)e_1$. Item (vi-c) follows from the fact that ad is skew-symmetric with respect to every bi-invariant metric (see Proposition 2.26). Finally, (vi-d) follows from (vi-b) and from

$$\tfrac{d}{dt}\mathrm{Ad}\left(\exp(tX) \right)e_j\big|_{t=0} = [X, e_j].$$

In order to prove (vii), we first prove that $\mathfrak{g}^{(\alpha)}$ is a Lie algebra and that the definition of $\alpha^\vee$ does not depend on the bi-invariant metric. Let Y be an infinitesimal generator of $\mathfrak{t}$. The Jacobi identity and (vi-b) imply that $[[e_1, e_2], Y] = 0$. Thus, from Claim 4.2, $[e_1, e_2] \in \mathfrak{t}$. Note that $[e_1, e_2] \neq 0$, because otherwise $\ker\alpha \oplus V_\alpha$ would be an abelian subalgebra with dimension greater than $\dim \mathfrak{t}$. Item (vi-b) and the fact that ad is skew-symmetric with respect to every bi-invariant metric imply that $[e_1, e_2]$ is orthogonal to $\ker\alpha$. In particular, the definition of $\alpha^\vee$ does not depend on the bi-invariant metric. To verify that $\mathfrak{g}^{(\alpha)}$ is a Lie algebra, note that

$$[[e_1, e_2], e_1] = \alpha([e_1, e_2])e_2, \quad \text{and} \quad [[e_1, e_2], e_2] = -\alpha([e_1, e_2])e_1.$$

Note also that $\mathfrak{g}^{(\alpha)} \subset \mathfrak{g}_X$, $\mathfrak{g}^{(\alpha)} \cap Z(\mathfrak{g}_X) = 0$ and $\dim \mathfrak{g}^{(\alpha)} = 3 = \dim \mathfrak{g}_X/Z(\mathfrak{g}_X)$, where X was defined in the proof of (iv). Therefore, $\mathfrak{g}^{(\alpha)} \cong \mathfrak{g}_X/Z(\mathfrak{g}_X) \cong \mathfrak{su}(2)$ are isomorphic Lie algebras. In particular, $G^{(\alpha)}$ is a compact Lie group isomorphic to SU(2) or SO(3), see Exercise 1.55, which implies that $\mathrm{Ad}(G^{(\alpha)})\big|_{\mathfrak{g}^{(\alpha)}} \cong$ SO(3). Since spheres are homogeneous in $\mathfrak{g}^{(\alpha)}$, there exists $w \in G^{(\alpha)}$ such that $\mathrm{Ad}(w)\alpha^\vee = -\alpha^\vee$. Finally, as $\mathrm{Ad}(\exp(Y))Z = \exp(\mathrm{ad}(Y))Z = Z$ for $Y \in \mathfrak{g}^{(\alpha)}$ and $Z \in \ker\alpha$, we conclude that $\mathrm{Ad}(g)\big|_{\ker\alpha} = \mathrm{id}$ for all $g \in G^{(\alpha)}$. $\square$

In what follows, we use the same notation from Theorem 4.23.

Definition 4.26. A connected component of $t \setminus \bigcup_{\alpha \in R} \ker \alpha$ is called a *Weyl chamber*, and each hyperplane $\ker \alpha$ is called a *wall*. For a fixed Weyl chamber C, we define the set of *positive roots* as

$$P := \big\{ \alpha \in R : \alpha(X) > 0, \text{ for all } X \in C \big\}.$$

Note that $R = P \cup -P$ and $P \cap -P = \emptyset$, where $-P := \{-\alpha : \alpha \in P\}$. In particular, $|P| = \frac{1}{2}|R|$, that is, P has *exactly half* of the elements of R, cf. Theorem 4.23 (iii).

Instead of defining the set of positive roots P for a given choice of Weyl chamber C, one can instead define a Weyl chamber for a given choice of positive roots. More precisely, a set of roots P satisfying:

(i) $P \cup -P = R$ and $P \cap -P = \emptyset$;
(ii) If $\sum_{\alpha \in P} c_\alpha \alpha = 0$, with $c_\alpha \geq 0$ for all $\alpha \in P$, then $c_\alpha = 0$ for all $\alpha \in P$;

is called a *choice of positive roots*. The associated Weyl chamber is then defined as

$$C := \big\{ X \in t : \alpha(X) > 0, \text{ for all } \alpha \in P \big\}.$$

Thus, there is a correspondence between sets of positive roots and Weyl chambers.

Remark 4.27. It follows from Theorem 4.23 that if G is a connected compact nonabelian Lie group, then $\mathfrak{g} = t \oplus \bigoplus_{\alpha \in P} V_\alpha$, where P is a choice of positive roots, cf. Proposition 4.20. Thus, the number of roots in G is $|R| = \dim G - \operatorname{rank} G$. In particular, the codimension of any maximal torus in G is even.

Remark 4.28. If $X \in C$ belongs to a Weyl chamber, then $\operatorname{Ad}(G)X$ is a regular orbit and its codimension is equal to the dimension of T. Indeed, since $X \notin \ker \alpha$ for all $\alpha \in R$, we conclude that $\mathfrak{g}_X = t$. In Theorem 4.36, we prove that regular orbits of the adjoint action are principal orbits.

It was proved in Proposition 4.15 that principal orbits of the adjoint action of a compact Lie group are isoparametric (recall Definition 4.14). We now compute the corresponding principal curvatures and principal directions, in terms of roots.

Example 4.29. Let (G, Q) be a connected compact Lie group with bi-invariant metric and $T \subset G$ a maximal torus, with Lie algebras $\mathfrak{g}$ and t respectively. Let $Z \in t$ be such that $\operatorname{Ad}(G)Z$ is a principal orbit, and recall that $T_Z(\operatorname{Ad}(G)Z) = \{[X, Z] : X \in \mathfrak{g}\}$, see (4.3). From the Maximal Torus Theorem 4.1, if $N \in t$, then

$$\widehat{N}(\operatorname{Ad}(g)Z) := \operatorname{Ad}(g)N$$

is a well-defined normal field along $\operatorname{Ad}(G)Z$. Differentiating at $g = e$, we have

$$d\widehat{N}_Z([X, Z]) = \big(\overline{\nabla}_{[X,Z]} \widehat{N}\big)_Z = [X, N], \tag{4.9}$$

which proves that $\widehat{N}$ is a parallel normal field along $\mathrm{Ad}(G)Z$, since $Q([X,N],Y) = Q(X,[N,Y]) = 0$ for all $Y \in \mathfrak{t}$. Thus, from (4.9),

$$\mathscr{S}_{\widehat{N}}([X,Z]) = -[X,N], \tag{4.10}$$

where $\mathscr{S}_{\widehat{N}}$ is the shape operator in the direction $\widehat{N}$, see (2.8). If $\{e_1, e_2\}$ is a basis of $V_\alpha \subset \mathfrak{g}$ as in Theorem 4.23, replacing X by e_j in (4.10), it follows that

$$\mathscr{S}_{\widehat{N}}(e_1) = -\frac{\alpha(N)}{\alpha(Z)}e_1, \quad \mathscr{S}_{\widehat{N}}(e_2) = -\frac{\alpha(N)}{\alpha(Z)}e_2.$$

Therefore, V_α are the curvature distributions of the orbit $\mathrm{Ad}(G)Z$, with principal curvature $-\frac{\alpha(N)}{\alpha(Z)}$. In particular, the spaces V_α are pairwise Q-orthogonal.

We conclude this section by computing the roots of $\mathrm{U}(n)$ and $\mathrm{SU}(n)$. Analogous computations for the other classical Lie groups can be found in Fegan [85].

Example 4.30. Consider the maximal torus (4.2) of $\mathrm{U}(n)$, recall Example 4.7. Its Lie algebra $\mathfrak{t} \subset \mathfrak{u}(n)$ consists of the matrices

$$X = \begin{pmatrix} i\theta_1 & & \\ & \ddots & \\ & & i\theta_n \end{pmatrix}, \tag{4.11}$$

where $\theta_k \in \mathbb{R}$. For each $k \neq l$, denote by E_{kl} the matrix with 1 in the (k,l)th entry and -1 in the (l,k)th entry, and by F_{kl} the matrix with i in both (k,l)th and (l,k)th entries, and zeros elsewhere. It is easy to see that E_{kl} and F_{kl} span the complement V of $\mathfrak{t} \subset \mathfrak{u}(n)$, i.e., $\mathfrak{u}(n) = \mathfrak{t} \oplus V$. A direct computation yields:

$$\mathrm{ad}(X)E_{kl} = XE_{kl} - E_{kl}X = (\theta_k - \theta_l)F_{kl},$$
$$\mathrm{ad}(X)F_{kl} = XF_{kl} - F_{kl}X = -(\theta_k - \theta_l)E_{kl}.$$

Thus, $\mathrm{ad}(X)\colon V \to V$ leaves each 2-dimensional subspace $V_{kl} = \mathrm{span}_{\mathbb{R}}\{E_{kl}, F_{kl}\}$, invariant, where it has eigenvalues $\pm i(\theta_k - \theta_l)$. By uniqueness of the decomposition $\mathfrak{u}(n) = \mathfrak{t} \oplus \bigoplus_{\alpha \in P} V_\alpha$, see Theorem 4.23 and Remark 4.27, the roots of $\mathrm{U}(n)$ are:

$$\alpha_{kl}\colon \mathfrak{t} \to \mathbb{R}, \quad \alpha_{kl} := \theta_k^* - \theta_l^*, \quad k \neq l, \tag{4.12}$$

where $\theta_k^*(X) := \theta_k$ for an X as in (4.11). Notice that, since $k \neq l$, there are precisely $n(n-1)$ roots α_{kl}, and $\dim \mathrm{U}(n) = n^2$ and $\mathrm{rank}\,\mathrm{U}(n) = n$, cf. Remark 4.27. It is easy to see that $P = \{\alpha_{kl} : k < l\}$ is a choice of positive roots.

The computation above also yields the roots of $\mathrm{SU}(n)$, since the Lie algebra of its maximal torus is $\mathfrak{t}/Z(\mathfrak{u}(n))$, which can be identified with the subspace $\mathfrak{t}'$ of $\mathfrak{t}$

spanned by elements X as in (4.11) that satisfy $\theta_1 + \cdots + \theta_n = 0$, see Example 4.7. Thus, the restriction of (4.12) to $\mathfrak{t}'$ are the roots of $\mathrm{SU}(n)$. Again, notice that there are $n(n-1)$ roots, and $\dim \mathrm{SU}(n) = n^2 - 1$ and $\mathrm{rank}\,\mathrm{SU}(n) = n-1$, cf. Remark 4.27.

4.4 Weyl Group

Let (G,Q) be a connected compact Lie group with bi-invariant metric and T a maximal torus. In this section, we define the *Weyl group* and use it to study intersections of orbits of the adjoint action with the Lie algebra $\mathfrak{t}$ of the maximal torus T.

Definition 4.31. The *Weyl group* is defined by

$$W := N(T)/T,$$

where $N(T) = \{g \in G : gTg^{-1} = T\}$ is the normalizer of the maximal torus T in G.

The Weyl group W is a finite group, since the connected component of $N(T)$ that contains the identity is T. In fact, recall that $N(T) = \{g \in G : \mathrm{Ad}(g)\mathfrak{t} \subset \mathfrak{t}\}$ and its Lie algebra is $\mathfrak{n} = \{X \in \mathfrak{g} : \mathrm{ad}(X)\mathfrak{t} \subset \mathfrak{t}\}$, see Exercise 1.61. Let $X \in \mathfrak{t}$ be an infinitesimal generator of T and $Y \in \mathfrak{n}$. Then $Q([X,Y],Z) = -Q(Y,[X,Z]) = 0$ for all $Z \in \mathfrak{t}$, hence $[X,Y] = 0$, and so $Y \in \mathfrak{t}$ by Claim 4.2. This proves that $\mathfrak{n} = \mathfrak{t}$ and, in particular, the above claim.

Remark 4.32. Since $T = Z_G(T)$ is its own centralizer (see Remark 4.3), it follows that $W = N_G(T)/Z_G(T)$ is the quotient of the normalizer by the centralizer of T, which is the group of automorphisms of T induced by *inner automorphisms* in G.

Remark 4.33. By the Maximal Torus Theorem 4.1, if T' is another maximal torus in G, then $T' = gTg^{-1}$ for some $g \in G$. Since $N(gTg^{-1}) = gN(T)g^{-1}$, it follows that the corresponding Weyl groups $N(T')/T'$ and $N(T)/T$ are isomorphic.

Exercise 4.34. Consider the maximal torus T of $S^3 \cong \mathrm{SU}(2)$ given by (4.1). Compute $N(T)$ and verify that the Weyl group of $\mathrm{SU}(2)$ is $W \cong \mathbb{Z}_2$.

From the above discussion, we also have that

$$\mu : W \times \mathfrak{t} \to \mathfrak{t}, \quad \mu(wT,X) := \mathrm{Ad}(w)X, \tag{4.13}$$

is a well-defined effective isometric W-action on $\mathfrak{t}$, whose orbits coincide with the intersections of $\mathfrak{t}$ with the orbits of the adjoint action of G on $\mathfrak{g}$. In order to study in more details how these actions are related, it is convenient to simultaneously use roots $\alpha : \mathfrak{t} \to \mathbb{R}$ and their duals $\alpha^* \in \mathfrak{t}$ determined by the bi-invariant metric Q, i.e.,

$\alpha(\cdot) = Q(\alpha^*, \cdot)$. Notice that $\alpha^* = 2\alpha^\vee / \|\alpha^\vee\|^2$, where $\alpha^\vee$ is the coroot associated to α, see Theorem 4.23 (vii). We denote by φ_α the reflection across the wall $\ker\alpha \subset \mathfrak{t}$,

$$\varphi_\alpha : \mathfrak{t} \to \mathfrak{t}, \quad \varphi_\alpha(Z) := Z - 2Q\left(Z, \frac{\alpha^*}{\|\alpha^*\|}\right) \frac{\alpha^*}{\|\alpha^*\|}.$$

In particular, notice that $\varphi_\alpha(Z) = Z - \alpha(Z)\alpha^\vee$ for all $Z \in \mathfrak{t}$.

Item (vii) of Theorem 4.23 implies that the reflections φ_α are elements of the Weyl group W. We prove in Theorem 4.37 that W is generated by these reflections. With this aim, we now establish a preliminary result that has the interesting consequence that the adjoint action does not have exceptional orbits.

Lemma 4.35. *The isotropy group $G_X = \{g \in G : \mathrm{Ad}(g)X = X\}$ of $X \in \mathfrak{t}$ with respect to the adjoint action is a connected Lie group.*

Proof. Given $x \in G_X$, we want to prove that $x \in G_X^0$, i.e., x is in the identity connected component of G_X. Let G_x be its isotropy group with respect to the conjugation action. Then X is in the Lie algebra of G_x, since $t\mathrm{Ad}(x)X = tX$ and hence $x\exp(tX)x^{-1} = \exp(tX)$, which is equivalent to $\exp(-tX)x\exp(tX) = x$.

Let T' be a maximal torus of G_x^0, tangent to X. Since $[Z, X] = 0$ for every Z in the Lie algebra of T', Z is in the Lie algebra of G_X, hence $T' \subset G_X^0$. Since $x \in G_x^0$ (see Lemma 4.17), by the Maximal Torus Theorem 4.1, there exists $g \in G_x^0$ such that $gxg^{-1} \in T'$. Finally, since $g \in G_x^0$, we have that $x = gxg^{-1}$, and hence $x \in G_X^0$. $\qquad\square$

Theorem 4.36. *Every regular orbit of the adjoint action is a principal orbit.*

Proof. Let $X_0 \in \mathfrak{g}$ be a regular point. This implies that $\dim G_{X_0} \leq \dim G_X$ for all $X \in \mathfrak{g}$. If S_{X_0} is a slice at X_0, then $G_Y \subset G_{X_0}$ for each $Y \in S_{X_0}$. Since $\dim G_{X_0} \leq \dim G_Y$ and both are connected, it follows that $G_Y = G_{X_0}$, i.e., $G(X_0)$ is a principal orbit. $\qquad\square$

Theorem 4.37. *Let W be the Weyl group. Then the following hold:*

(i) *If C_1 and C_2 are two Weyl chambers such that $X \in \partial C_1 \cap \partial C_2$, then there exist reflections $\varphi_{\alpha_1}, \ldots, \varphi_{\alpha_n}$ such that $\varphi_{\alpha_i}(X) = X$ and $(\varphi_{\alpha_n} \circ \cdots \circ \varphi_{\alpha_1})(C_2) = C_1$;*

(ii) *The closure $\overline{C}$ of each Weyl chamber C is a fundamental domain for the action of W on $\mathfrak{t}$, i.e., each orbit of the adjoint action intersects $\overline{C}$ exactly once;*

(iii) *W is generated by $\{\varphi_\alpha\}_{\alpha \in R}$, where R is the set of roots.*

Proof. Each isometry $w \in W$ maps regular points to regular points, and hence $w(C_1)$ is a Weyl chamber. Let K_X be the subgroup of W generated by the reflections φ_α such that $\varphi_\alpha(X) = X$. Let $B_{2\varepsilon}(X)$ be an open ball in the intersection of a slice at X with $\mathfrak{t}$, and consider two regular points p_1 and p_2 such that $p_i \in C_i$ and $p_i \in B_\varepsilon(X)$. Define $f : K_X \to \mathbb{R}$ as $f(\varphi_\alpha) := \|\varphi_\alpha(p_2) - p_1\|$, and let $w_0 \in K_X$ be a minimum of f. If $w_0(p_2) \notin C_1$, then the straight line segment joining $w_0(p_2)$ to p_1 intersects a wall $\ker\beta$. Since this intersection is contained in $B_\varepsilon(X)$, we have $X \in \ker\beta$ and hence $\varphi_\beta \in K_X$. Therefore $f(\varphi_\beta w_0) < f(w_0)$, contradicting the fact that w_0 is a minimum of f. Thus $w_0(p_2) \in C_1$, concluding the proof of (i).

As for (ii), the Maximal Torus Theorem 4.1 implies that each orbit of the adjoint action intersects t. Composing with reflections φ_α, we conclude that each regular orbit intersects C at least once, and every singular orbit intersects ∂C at least once. We claim that each principal orbit intersects C *exactly* once. Set

$$H := \{h \in G : \mathrm{Ad}(h)X \in C, \text{ for all } X \in C\},$$

and assume that there exists $X \in C$ such that the orbit $\mathrm{Ad}(G)X$ meets C more than once, i.e., there is $g \in G$ such that $X \neq \mathrm{Ad}(g)X \in C$. Clearly $g \notin T$, and it is not difficult to prove that $g \in H$. Since $\mathrm{Ad}(G)$ is compact, g has finite order[9] n. Since $g \in H$, we have that $Y := X + \mathrm{Ad}(g)X + \cdots + \mathrm{Ad}(g^{n-1})X \in C$, and clearly $\mathrm{Ad}(g)Y = Y$. Thus, $g \in G_Y$. On the other hand, since $Y \in C$, it follows from Remark 4.28 and Lemma 4.35 that $G_Y = G_Y^0 = T$, contradicting the fact that $g \notin T$.

Similarly, we claim that each singular orbit of the adjoint action intersects ∂C exactly once. Suppose there are $0 \neq X \in \partial C$ and $g \in G$ such that $X \neq \mathrm{Ad}(g)X \in \partial C$. Let $\gamma \colon [0,1] \to C$ be a curve with $\gamma(0) = X$ and $\gamma(t) \in C$ for $0 < t \leq 1$. Composing with reflections φ_α if necessary, observe that $\mathrm{Ad}(g)\gamma|_{(0,1]} \subset C$. Therefore, if t is sufficiently small, $\gamma(t) \neq \mathrm{Ad}(g)\gamma(t) \in C$, contradicting the fact proved above that each principal orbit of the adjoint action intersects C exactly once.

It only remains to prove (iii). For any $g \in W$, there exist reflections $\varphi_{\alpha_1}, \ldots, \varphi_{\alpha_n}$ such that $(\varphi_{\alpha_n} \circ \cdots \circ \varphi_{\alpha_1} \circ g)(C) \subset C$. From (ii), we have that, for all $X \in C$,

$$(\varphi_{\alpha_n} \circ \cdots \circ \varphi_{\alpha_1} \circ g)(X) \in \{\mathrm{Ad}(G)X \cap C\} = \{X\}$$

Therefore, $(\varphi_{\alpha_n} \circ \cdots \circ \varphi_{\alpha_1} \circ g)|_t$ is the identity, concluding the proof. □

Remark 4.38. Let $\mathscr{F} = \{\mathrm{Ad}(G)X\}_{X \in \mathfrak{g}}$ be the (homogeneous) singular foliation determined by the orbits of the adjoint action. The above result implies that $\mathscr{F} \cap t$ is invariant under reflections across the walls of Weyl chambers in t.

Remark 4.39. Notice that the orbit space $\mathfrak{g}/G$ of the adjoint action can be identified with the closure of a Weyl chamber, which can in turn be identified with the orbit space t/W of the Weyl group action (4.13) on t.

Remark 4.40. The above theorem implies that the Weyl group permutes the Weyl chambers and that the cardinality $|W|$ is equal to the number of Weyl chambers.

Exercise 4.41. From Exercises 4.5 and 4.34, the great circle T in (4.1) is a maximal torus in $\mathrm{SU}(2)$, and the Weyl group is $W \cong \mathbb{Z}_2$. Prove that its Lie algebra $t = \mathbb{R}$ has two Weyl chambers $C_\pm = \mathbb{R}_\pm$, which are half lines, and W acts by reflection across the origin. Conclude that the orbit space of the adjoint action of $\mathrm{SU}(2)$ on $\mathfrak{su}(2) \cong \mathbb{R}^3$ is a half line, and relate this with Exercises 1.9 and 1.55.

---

[9]This means that there exists n such that $g^n = e$ and $g^{n-1} \neq e$.

Example 4.42. The Weyl group of SU(n) is the group $\mathfrak{S}_n$ of permutations in n letters. This can be proved by computing the normalizer of a maximal torus, see Example 4.7. This group acts on an element X as in (4.11) by permuting its coordinates $\theta_1, \ldots, \theta_n$. By Remark 4.40, it follows that there are $n!$ Weyl chambers.

Let us consider SU(3), which has rank 2 and Weyl group $\mathfrak{S}_3$. The Lie algebra of a maximal torus is identified with $\mathfrak{t}' = \{(\theta_1, \theta_2, \theta_3) \in \mathbb{R}^3 : \theta_1 + \theta_2 + \theta_3 = 0\}$ and $P = \{\alpha_{12}, \alpha_{13}, \alpha_{23}\}$ is a choice of positive roots, where $\alpha_{kl} = \theta_k^* - \theta_l^*$, see Example 4.30. The restriction of α_{kl} to $\mathfrak{t}'$ is clearly $(\alpha_{kl})|_{\mathfrak{t}'} = \theta_k^* - \theta_l^* = 2\theta_k^* + \theta_m^*$, where $\{k, l, m\} = \{1, 2, 3\}$. Introducing orthonormal coordinates (x, y) on the plane $\mathfrak{t}' \subset \mathbb{R}^3$, elementary computations show that the walls $\ker(\alpha_{kl})|_{\mathfrak{t}'}$ in $\mathfrak{t}' \cong \mathbb{R}^2$ are the lines spanned by the vectors $(1, 0)$, $\left(-\frac{1}{2}, \frac{\sqrt{3}}{2}\right)$ and $\left(-\frac{1}{2}, -\frac{\sqrt{3}}{2}\right)$, and hence each of the 6 Weyl chambers is a wedge of angle $\pi/3$. In these coordinates, the duals $\left((\alpha_{kl})|_{\mathfrak{t}'}\right)^*$ are the vectors $\left(\frac{3}{2}, -\frac{\sqrt{3}}{2}\right)$, $\left(\frac{3}{2}, \frac{\sqrt{3}}{2}\right)$ and $(0, \sqrt{3})$. The Weyl group $\mathfrak{S}_3$ is clearly the symmetry group generated by reflections across these walls, cf. Theorem 4.37. By the above discussion, the orbit space of the adjoint action of SU(3) on its Lie algebra $\mathfrak{su}(3)$ is a 2-dimensional (closed) wedge of angle $\pi/3$.

Remark 4.43. The definition of Weyl group can be extended to the general context of polar actions $\mu : G \times M \to M$, where it is also referred to as *polar group*. If Σ is a section for such an action, then the Weyl group is defined as $W := N/Z$, where

$$N = \{g \in G : \mu(g, x) \in \Sigma \text{ for all } x \in \Sigma\} \text{ and } Z = \{g \in G : \mu(g, x) = x \text{ for all } x \in \Sigma\},$$

cf. Remark 4.32. It can be proved that W is a discrete subgroup of $N(H)/H$, where H is the isotropy group of a principal G-orbit. Similarly to the case (4.13) of the conjugation action of a compact Lie group on itself, the Weyl group has an effective isometric action on Σ, whose orbits coincide with the intersections of Σ with the G-orbits on M. In particular, the orbit spaces M/G and Σ/W are isometric.

The Weyl group of a polar action is also finite, provided that the orbits and sections are compact. In order to see this, let $x \in M$ be in a principal G-orbit, and note that a slice S_x at x locally coincides with the section Σ_x through x. The isotropy G_x coincides with the isotropy group W_x of the Weyl group action. By Exercise 3.77, the slice representation is trivial, and hence $W_x = \{e\}$. By the compactness assumptions, it follows that $W(x) \cong W/W_x$ is finite and hence W is finite.

4.5 Dynkin Diagrams

In this section, we briefly discuss Dynkin diagrams and the classification of compact simple Lie groups, without giving proofs, which can be found in standard textbooks such as [58, 90, 121, 126, 191, 195].

We henceforth assume that (G, Q) is a connected compact *semisimple* Lie group with bi-invariant metric and, as before, we denote by T a maximal torus in G and by R the set of roots. Moreover, if $\alpha, \beta \in R$ are roots, we set $Q(\alpha, \beta) := Q(\alpha^*, \beta^*)$, where $\alpha^*, \beta^* \in \mathfrak{t}$ are their duals, i.e., $\alpha(\cdot) = Q(\alpha^*, \cdot)$ and $\beta(\cdot) = Q(\beta^*, \cdot)$.

Proposition 4.44. *Let* $\alpha_1, \alpha_2 \in R$ *be roots with* $\alpha_1 \neq \pm\alpha_2$, *and denote by* $\sigma = \mathrm{span}\{\alpha_1^*, \alpha_2^*\}$ *the plane in* $\mathfrak{t}$ *spanned by their duals. There are exactly* $2m$ *duals of roots* $\alpha_j^* \in \sigma$, *where* $m \in \{2, 3, 4, 6\}$, *and the normalized vectors* $\alpha_j^*/\|\alpha_j^*\|$ *are uniformly distributed along the unit circle of* σ. *Moreover, if* $\alpha_j^*, \alpha_k^* \in \sigma$ *are such that* $Q(\alpha_j, \alpha_k) \leq 0$, *then* $Q(\alpha_j, \alpha_k) = \|\alpha_j\|\|\alpha_k\| \cos(\pi - \phi_{jk})$, *where:*

   (i) $\phi_{jk} = \pi/2$, *and* $\|\alpha_j\| = \|\alpha_k\|$;
  (ii) $\phi_{jk} = \pi/3$, *and* $\|\alpha_j\| = \|\alpha_k\|$;
 (iii) $\phi_{jk} = \pi/4$, *and* $\|\alpha_j\| = \sqrt{2}\|\alpha_k\|$ *or* $\|\alpha_k\| = \sqrt{2}\|\alpha_j\|$;
 (iv) $\phi_{jk} = \pi/6$, *and* $\|\alpha_j\| = \sqrt{3}\|\alpha_k\|$ *or* $\|\alpha_k\| = \sqrt{3}\|\alpha_j\|$.

Remark 4.45. Part of the above rigidity in the geometric arrangement of $\alpha_j^* \in \sigma$ can be explained as follows. The reflections φ_α on σ generate a finite group $F \subset O(2)$ of linear isometries, whose intersection with $SO(2) \cong S^1$ is an index 2 subgroup, which must be isomorphic to $\mathbb{Z}_m$ for some $m \in \mathbb{N}$. This subgroup is hence generated by a rotation of angle $2\pi/m$ on the plane σ. The trace of this rotation matrix is $2\cos(2\pi/m)$, and (from representation theory) this number must be an integer, so the only possible values for m are 2, 3, 4, and 6.

Definition 4.46. A positive root α is called *simple* if it cannot be written as the sum of two other positive roots.

Theorem 4.47. *Let* P *be a choice of positive roots in* R.

  (i) *The subset* $S \subset P$ *of simple roots is a basis of* $\mathfrak{t}^*$;
 (ii) *Each* $\alpha \in R$ *is a linear combination of elements of* S *with integer coefficients, either all nonnegative or all nonpositive;*
(iii) *If* $\alpha, \beta \in S$, *then* $Q(\alpha, \beta) \leq 0$.

The set S *is called a* base *of the root system* R.

Definition 4.48. The *Dynkin diagram* of R with respect to a base S is a directed graph with one vertex for each simple root $\alpha_j \in S$. If $j \neq k$, there are either 0, 1, 2, or 3 edges between α_j and α_k, which are oriented pointing from the larger root, say α_j, to the smaller root, say α_k, according to the cases in Proposition 4.44. Namely, if $Q(\alpha_j, \alpha_k) = \|\alpha_j\|\|\alpha_k\| \cos(\pi - \phi_{jk})$, we proceed as follows:

  (i) If $\phi_{jk} = \pi/2$, there are no edges;

$$\overset{\textstyle\bullet}{\alpha_j} \qquad\qquad \overset{\textstyle\bullet}{\alpha_k}$$

(ii) If $\phi_{jk} = \pi/3$, there is 1 edge;

$$\underset{\alpha_j}{\bullet} \underline{\qquad} \underset{\alpha_k}{\bullet}$$

(iii) If $\phi_{jk} = \pi/4$, and $\|\alpha_j\| = \sqrt{2}\|\alpha_k\|$, there are 2 edges;

$$\underset{\alpha_j}{\bullet} \Longrightarrow \underset{\alpha_k}{\bullet}$$

(iv) If $\phi_{jk} = \pi/6$, and $\|\alpha_j\| = \sqrt{3}\|\alpha_k\|$, there are 3 edges.

$$\underset{\alpha_j}{\bullet} \Longrightarrow \underset{\alpha_k}{\bullet}$$

It is possible to prove that the Dynkin diagram of R does not depend on the choice of bi-invariant metric Q, nor on the choice of positive roots P.

Example 4.49. Let us determine the Dynkin diagram of $SO(7)$. From Example 4.7, the Lie algebra t of a maximal torus in $SO(7)$ is formed by matrices of the form

$$X = \begin{pmatrix} 0 & \theta_1 & & & & & \\ -\theta_1 & 0 & & & & & \\ & & 0 & \theta_2 & & & \\ & & -\theta_2 & 0 & & & \\ & & & & 0 & \theta_3 & \\ & & & & -\theta_3 & 0 & \\ & & & & & & 0 \end{pmatrix}.$$

The same type of argument used in Example 4.30 implies that the roots of $SO(7)$ are $\pm\theta_k^*$ and $\theta_k^* \pm \theta_l^*$, $k \neq l$, where $\theta_k^*(X) = \theta_k$ for X as above. A possible choice of positive roots is θ_k^* and $\theta_k^* \pm \theta_l^*$, $k < l$. For this choice, the basis of simple roots is $S = \{\alpha_1, \alpha_2, \alpha_3\}$, where

$$\alpha_1 := \theta_1^* - \theta_2^*, \qquad \alpha_2 := \theta_2^* - \theta_3^*, \qquad \alpha_3 := \theta_3^*.$$

Consider the bi-invariant metric $Q(X,Y) := \operatorname{tr} XY^t$ of $SO(7)$. It is easy to check that $Q(\theta_k^*, \theta_l^*) = \delta_{kl}$, hence $\|\alpha_1\| = \|\alpha_2\| = \sqrt{2}$, $\|\alpha_3\| = 1$ and moreover, $\phi_{12} = \pi/3$, $\phi_{13} = \pi/2$ and $\phi_{23} = \pi/4$. Therefore, the Dynkin diagram of $SO(7)$ is:

$$\underset{\alpha_1}{\bullet} \underline{\qquad} \underset{\alpha_2}{\bullet} \Longrightarrow \underset{\alpha_3}{\bullet}.$$

Exercise 4.50 (⋆). Use the calculus of roots of $SU(n)$ in Example 4.30 to prove that the Dynkin diagram of $SU(3)$ is

$$\bullet \!\!-\!\!\!-\!\!\!-\!\! \bullet$$

Hint: Use Examples 4.42 and 4.49.

The classification of (connected) compact simple Lie groups, can be reduced to the classification of Dynkin diagrams, due to the next result. This classification of diagrams is purely combinatorial, and we simply state its outcome, relating the diagrams with the corresponding groups. For the proofs, we refer the reader to Fulton and Harris [90].

Theorem 4.51. *If G_1 and G_2 are (connected) simply-connected compact semisimple Lie groups with the same Dynkin diagram, then G_1 and G_2 are isomorphic.*

Theorem 4.52. *The possible Dynkin diagrams of a connected, compact simple Lie group are:*

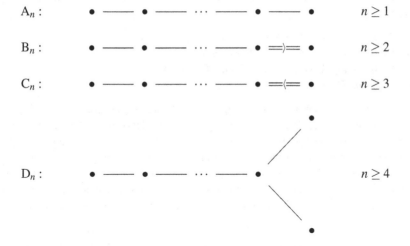

where n denotes the number of vertices, and also:

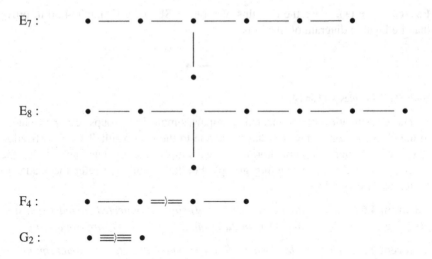

Each of the above is the Dynkin diagram of a connected compact simple Lie group, whose rank is equal to the subindex used. Regarding classical Lie groups of rank n,

  (i) A_n is the Dynkin diagram of $SU(n+1)$, for $n \geq 1$;
 (ii) B_n is the Dynkin diagram of $SO(2n+1)$, for $n \geq 2$;
(iii) C_n is the Dynkin diagram of $Sp(n)$, for $n \geq 3$;
 (iv) D_n is the Dynkin diagram of $SO(2n)$, for $n \geq 4$.

The remaining Dynkin diagrams E_6, E_7, E_8, F_4 and G_2 are the Dynkin diagrams of the so-called *exceptional* Lie groups, which are denoted by the same symbol. Despite the terminology, these groups make several natural appearances in geometric problems (see Example 6.10 (v) and Table 6.1), especially in connection with *Cayley numbers*.[10] Their exceptionality is mostly due to not being part of an infinite family of groups with increasing dimension, as the classical Lie groups. For details on the construction of such exceptional Lie groups, we refer the reader to [26, 123].

Remark 4.53. According to the above result, the Lie groups $SO(7)$ and $Sp(3)$ are nonisomorphic, but each has 3 simple roots, $\{\alpha_1, \alpha_2, \alpha_3\}$ and $\{\beta_1, \beta_2, \beta_3\}$, with

$$\frac{\langle \alpha_j, \alpha_k \rangle}{\|\alpha_j\| \|\alpha_k\|} = \frac{\langle \beta_j, \beta_k \rangle}{\|\beta_j\| \|\beta_k\|} = \cos(\pi - \phi_{jk})$$

where ϕ_{jk} are the angles $\pi/2$, $\pi/3$ and $\pi/4$. Indeed, their Dynkin diagrams are the same up to the orientation of the edges. In particular, the orbit spaces of the

---

[10]The *Cayley numbers* $\mathbb{C}a$, or *octonions*, form one of the 4 real normed division algebras, together with real numbers $\mathbb{R}$, complex numbers $\mathbb{C}$ and quaternions $\mathbb{H}$.

adjoint actions of SO(7) and Sp(3) are *isometric*, despite these representations being nonequivalent. This interesting phenomenon alludes to a recent research trend that studies when a representation (or, more generally, an isometric action) can be reconstructed from its orbit space together with some additional data, see [18, 102, 118].

Chapter 5
Polar Foliations

In this chapter, we give a survey on some results from the theory of polar foliations, also called singular Riemannian foliations with sections. This theory is a generalization of the classical theory of adjoint actions (presented in Chap. 4), and several results in this chapter are extensions of its results. Since our main goal is to provide a flavor of this new field, to not lose sight of the big picture, we only give some sketches of proofs without going into technical details.

5.1 Definitions and First Examples

We start by stating the definition of singular Riemannian foliation, see Molino [165].

Definition 5.1. Let $\mathscr{F}$ be a partition of a complete Riemannian manifold M into connected immersed submanifolds, called *leaves*. We say that:

(i) $\mathscr{F}$ is *singular* if the module $\mathscr{X}_\mathscr{F}$ of smooth vector fields on M that are tangent at each point to the corresponding leaf acts transitively on each leaf, in the sense that, for each leaf L and each $v \in T_pL$, there exists $X \in \mathscr{X}_\mathscr{F}$ with $X(p) = v$;
(ii) $\mathscr{F}$ is *transnormal* if a geodesic that is perpendicular to a leaf at one point is *horizontal*, i.e., remains perpendicular to all leaves it intersects.

If $\mathscr{F}$ satisfies (i), then it is called a *singular foliation* of M. If $\mathscr{F}$ satisfies (i) and (ii), then it is called a *singular Riemannian foliation* of M.

Let $\mathscr{F}$ be a singular Riemannian foliation of a complete Riemannian manifold M. A leaf L of $\mathscr{F}$ (and each point in L) is called *regular* if the dimension of L is maximal; otherwise, it is called *singular*. If all the leaves of $\mathscr{F}$ are regular, then $\mathscr{F}$ is called a *Riemannian foliation*. The *codimension* of a singular foliation is defined as the codimension of its regular leaves.

© Springer International Publishing Switzerland 2015   109
M.M. Alexandrino, R.G. Bettiol, *Lie Groups and Geometric Aspects
of Isometric Actions*, DOI 10.1007/978-3-319-16613-1_5

Definition 5.2. Let $\mathscr{F}$ be a singular Riemannian foliation on a complete Riemannian manifold M. Then $\mathscr{F}$ is a *polar foliation*, also called a *singular Riemannian foliation with sections*, if for each regular point p, the set $\Sigma = \exp_p(\nu_p L_p)$ is a complete immersed submanifold that intersects all leaves orthogonally. In this case, Σ is called a *section*.

Example 5.3. Using Killing vector fields, one can prove that the partition (3.4) of a Riemannian manifold M into orbits of an isometric action is a singular Riemannian foliation (recall Proposition 3.78). Moreover, the partition of M into orbits of a polar action (see Definition 4.8) is a polar foliation. A foliation of a space form by isoparametric submanifolds (see Definition 4.14) is also a polar foliation, and is called an *isoparametric foliation*.

Terng and Thorbergsson [205] introduced the concept of equifocal submanifolds with flat sections in symmetric spaces, generalizing the definition of isoparametric submanifolds in Euclidean space. We now give a slightly more general definition of equifocal submanifolds in Riemannian manifolds.

Definition 5.4. A connected immersed submanifold L of a complete Riemannian manifold M is *equifocal* if it satisfies:

  (i) The normal bundle $\nu(L)$ is flat;
 (ii) L admits sections, i.e., for each $p \in L$, the set $\Sigma = \exp_p(\nu_p L_p)$ is a complete immersed totally geodesic submanifold;
(iii) For each $p \in L$, there exists a neighborhood $U \subset L$ such that, for each parallel normal field ξ along U, the derivative of the *endpoint map* $\eta_\xi : U \to M$, $\eta_\xi(x) := \exp_x(\xi)$, has constant rank.

Terng and Thorbergsson [205] proved that any partition of a simply-connected compact symmetric space M into submanifolds parallel to an equifocal submanifold L with flat sections is a polar foliation.

As we prove in Theorem 5.26, regular leaves of polar foliations are equifocal submanifolds. Conversely, a closed embedded equifocal submanifold is a leaf of some polar foliation, see Remark 5.66.

We conclude this section with the following well-known fact about equifocal submanifolds of the round sphere.

Proposition 5.5. *Any equifocal submanifold of the round sphere S^n is an isoparametric submanifold of S^n.*

Proof (Sketch). Let ξ be a unit normal parallel field along an equifocal submanifold L of S^n. The proof is divided into two steps. We first establish a relation between principal curvatures and the kernel of endpoint maps. Next, we use this relation and equifocality of L to conclude that the set of principal curvatures with respect to ξ at the point x is actually independent of $x \in L$, i.e., L is isoparametric, see Definition 4.14.

Claim 5.6. For any fixed $x \in L$, each principal curvature λ_i corresponds bijectively to a radius t_i (up to integer multiples of π) by the equation $\lambda_i = \cot t_i$, so that $\ker d\eta_{t_i \xi} \subset T_x L$ is the eigenspace of $\mathscr{S}_\xi$ corresponding to the eigenvalue λ_i.

To prove the above, consider the geodesic $\gamma_x(t) := \exp_x(t\xi)$ and a variation $\alpha(s,t) = \exp_{\beta(s)}(t\xi)$, where β is a curve in L with $\beta(0) = x$. We are interested in the zeros of L-Jacobi fields $J(t) = \frac{\partial}{\partial s}\alpha(s,t)|_{s=0}$ along γ_x. Since the normal bundle of L is flat, it is possible to prove that the L-Jacobi fields satisfy

$$\begin{cases} J''(t) + J(t) = 0 \\ J'(0) + \mathscr{S}_\xi J(0) = 0. \end{cases} \tag{5.1}$$

Let $e_i(t)$ be parallel normal fields along the geodesic $\alpha(0,t)$, such that $\mathscr{S}_\xi(e_i(0)) = \lambda_i e_i(0)$. Writing $J(t) = \sum_i a_i(t)e_i(t)$, we conclude that if J is a solution of (5.1), then $a_i(t) = a_i(0)(\cos t - \lambda_i \sin t)$. In particular, $a_i(t_i) = 0$ *if and only if* $a_i(0) = 0$ or if $\lambda_i = \cot t_i$.

Given one such principal curvature λ_i corresponding to a principal direction $e_i \in T_x L$, it follows from Claim 5.6 that $e_i \in \ker d\eta_{t_i \xi}(x)$. Since L is equifocal, $\dim \ker d\eta_{t_i \xi}$ is constant. Thus, by Claim 5.6, λ_i is also a principal curvature (with the same multiplicity) at any other point of L, concluding the proof. $\qquad \square$

5.2 Holonomy and Orbifolds

In this section, we recall some preliminary facts about Riemannian foliations, most of which can also be found in Molino [165]. In the particular case of partition of the space into orbits of an proper isometric actions, some of the results and remarks of this section were presented in Chap. 3; e.g., recall Proposition 3.78, Remark 3.79 and Exercise 3.81. We start with two equivalent characterizations of Riemannian foliations.

Proposition 5.7. *Let $\mathscr{F}$ be a foliation on a complete Riemannian manifold (M, g). The following statements are equivalent:*

(i) *$\mathscr{F}$ is a Riemannian foliation;*
(ii) *For each $q \in M$, there exists a neighborhood U of q in M (called* simple open subset*), a Riemannian manifold (σ, b) and a Riemannian submersion $f : (U, g) \to (\sigma, b)$ such that the connected components of $\mathscr{F} \cap U$ (called* plaques*) are preimages of f;*
(iii) *Let g_T be the transverse metric, i.e., $g_T = N^*g$ where $N_p : T_p M \to v_p L$ is the orthogonal projection onto the normal space. Then the Lie derivative $\mathscr{L}_X g_T$ vanishes for each $X \in \mathscr{X}_{\mathscr{F}}$, see Definition 5.1.*

With the above notion of plaques, we have the following:

Proposition 5.8. *Let $\mathscr{F}$ be a (singular) foliation on a complete Riemannian manifold (M,g). Then $\mathscr{F}$ is a (singular) Riemannian foliation if and only if its leaves are locally equidistant, i.e., for each $q \in M$ there exists a tubular neighborhood*

$$\mathrm{Tub}_\varepsilon(P_q) = \{x \in M : \mathrm{dist}(x,P_q) < \varepsilon\}$$

of a plaque P_q of L_q, such that for $x \in \mathrm{Tub}(P_q)$, each plaque P_x contained in $\mathrm{Tub}(P_q)$ is contained in the cylinder of radius $\mathrm{dist}(x,P_q)$ and axis P_q.

An important property of Riemannian foliations is *(Bott) equifocality*. In order to understand this concept, we need some preliminary definitions. A *Bott* or *basic connection* ∇ of a foliation $\mathscr{F}$ is a connection on the normal bundle of the leaves, with $\nabla_X Y = [X,Y]^{\nu\mathscr{F}}$ whenever $X \in \mathscr{X}_\mathscr{F}$ and Y is a section of the normal bundle $\nu\mathscr{F}$ of the foliation. Here, $(\cdot)^{\nu\mathscr{F}}$ denotes projection onto $\nu\mathscr{F}$. A *normal foliated vector field* is a normal field parallel with respect to the Bott connection. If we consider a local submersion f that describes the plaques of $\mathscr{F}$ in a neighborhood of a point of L, then a normal foliated vector field is a normal projectable/basic vector field with respect to f. In this context, *Bott equifocality* means that if ξ is a normal parallel vector field (with respect to the Bott connection) along a curve $\beta: [0,1] \to L$, then the curve $t \mapsto \exp_{\beta(t)}(\xi)$ is contained in the leaf $L_{\exp_{\beta(0)}(\xi)}$.

Remark 5.9. Bott equifocality also holds for singular Riemannian foliations, and implies that one can reconstruct the (singular) foliation by taking all parallel submanifolds of a (regular) leaf with trivial holonomy, see Alexandrino and Töben [20].

The above allows us to introduce the concept of parallel translation (with respect to the Bott connection) of horizontal segments of geodesic.

Definition 5.10. Let $\beta: [a,b] \to L$ be a piecewise smooth curve and $\gamma: [0,1] \to M$ be a segment of a horizontal geodesic such that $\gamma(0) = \beta(a)$. Let ξ_0 be a vector of the normal space $\nu_{\beta(a)}L$, such that $\exp_{\gamma(0)}(\xi_0) = \gamma(1)$, and let $\xi: [a,b] \to \nu L$ be the parallel translation of ξ_0 with respect to the Bott connection along β. Define $\|_\beta(\gamma) := \widehat{\gamma}$, where $\widehat{\gamma}: [0,1] \to M$ is the geodesic segment $s \mapsto \widehat{\gamma}(s) = \exp_{\beta(b)}(s\,\xi(b))$.

Due to equifocality of $\mathscr{F}$, there is an alternative definition of holonomy map of a Riemannian foliation.

Definition 5.11. Let $\beta: [0,1] \to L$ be a piecewise smooth curve and let

$$S_{\beta(i)} := \{\exp_{\beta(i)}(\xi) : \xi \in \nu_{\beta(i)}L, \|\xi\| < \varepsilon\} \tag{5.2}$$

be a *slice* at $\beta(i)$, for $i = 0, 1$, and $\varepsilon > 0$ small. The *holonomy map* is defined by

$$\varphi_{[\beta]}: S_{\beta(0)} \to S_{\beta(1)}, \quad \varphi_{[\beta]}(x) := \|_\beta \gamma(r), \tag{5.3}$$

where $\gamma\colon [0,r] \to S_{\beta(0)}$ is the minimal geodesic segment joining $\beta(0)$ to x. Since the Bott connection is locally flat, the parallel translation depends only on the homotopy class fixing endpoints $[\beta]$.

Remark 5.12. The holonomy map of a foliation $\mathscr{F}$ is usually defined as follows. Consider a partition $0 = t_0 < t_1 < \ldots < t_k = 1$ such that $\beta\big((t_{i-1}, t_i)\big)$ is contained in a simple open set U_i defined by a trivialization of $\mathscr{F}$. Then $\{U_1, \ldots U_k\}$ is called a chain of simple open sets. Sliding along the plaques in U_k, one can define a diffeomorphism φ_k from an open neighborhood of the transverse manifold $S_{\beta(t_{k-1})}$ onto an open set of $S_{\beta(t_k)}$. In the same way, sliding along the plaques U_{k-1}, one defines a diffeomorphism φ_{k-1} from an open neighborhood of $S_{\beta(t_{k-2})}$ onto an open neighborhood of $S_{\beta(t_{k-1})}$. After some steps, we obtain a diffeomorphism φ_1 from an open neighborhood $S_{\beta(t_0)}$ onto a neighborhood of $S_{\beta(t_1)}$. Finally, the germ of the diffeomorphism $\varphi_k \circ \varphi_{k-1} \circ \cdots \circ \varphi_1$ is the holonomy map $\varphi_{[\beta]}$. It is possible to prove that the latter does not depend on the chain of simple open sets used in its construction, but only on the homotopy class fixing endpoints of β.

Remark 5.13. Note that, since the holonomy map depends only on the homotopy class, there is a group homomorphism $\varphi\colon \pi_1(L_0, x_0) \to \mathrm{Diff}_{x_0}(S_{x_0})$, from the fundamental group of the leaf L_0 to the group of germs at x_0 of local diffeomorphisms of S_{x_0}. The image of this homomorphism is the *holonomy group* of L_0 at x_0.

Remark 5.14. Consider a proper G-action μ on a manifold M, and the homogeneous foliation (3.4). Assume that G is connected and that all G-orbits are regular. Using Remarks 3.60 and 5.12, it is possible to check that for each holonomy map $\varphi_{[\beta]}$, there exists g such that $\mu(g, \cdot)|_{S_{x_0}} = \varphi_{[\beta]}$. In particular, the holonomy group coincides with the image of the slice representation of G_{x_0} in S_{x_0} (see Remark 3.61).

We now mention some facts about *pseudogroups* and *orbifolds*, related to Riemannian foliations. More details can be found in Salem [165, Appendix D] or Moerdijk and Mrčun [163].

Definition 5.15. Let Σ be a Riemannian manifold, not necessarily connected. A *pseudogroup* W of isometries of Σ is a collection of (local) isometries $w\colon U \to W$, where U and V are open subsets of Σ, such that:

(i)   If $w \in W$, then $w^{-1} \in W$;
(ii) If $w\colon U \to V$ and $\widetilde{w}\colon \widetilde{U} \to \widetilde{V}$ are in W, then $(\widetilde{w} \circ w)\colon w^{-1}(\widetilde{U} \cap V) \to \widetilde{V}$ is in W;
(iii) If $w\colon U \to V$ is in W, then its restriction to each open subset $\widetilde{U} \subset U$ is in W;
(iv) If $w\colon U \to V$ is an isometry between open subsets of Σ that coincides in a neighborhood of each point of U with an element of W, then $w \in W$.

Definition 5.16. Let A be a family of local isometries of Σ containing the identity. The pseudogroup obtained by taking inverses of the elements of A, restrictions to open subsets, compositions, and unions is called the *pseudogroup generated* by A.

Definition 5.17. Let W and Φ be pseudogroups of isometries of Riemannian manifolds Σ and σ, respectively. An *equivalence* between W and Φ is a maximal collection Ψ of isometries between open subsets of Σ and open subsets of σ such that:

(i) The domains of the elements of Ψ form an open cover of Σ and the ranges of the elements of Ψ form an open cover of σ;

(ii) If $\psi \in \Psi$, $w \in W$, and $\varphi \in \Phi$, then $(\varphi \circ \psi \circ w) \in \Psi$.

(iii) If $\psi_\alpha, \psi_\beta \in \Psi$, $w \in W$, and $\varphi \in \Phi$, then

$$\left(\psi_\beta \circ w \circ \psi_\alpha^{-1}\right) \in \Phi \quad \text{and} \quad \left(\psi_\alpha^{-1} \circ \varphi \circ \psi_\beta\right) \in W.$$

Example 5.18. An important example of a Riemannian pseudogroup is the holonomy pseudogroup of a Riemannian foliation. Let $\mathscr{F}$ be a Riemannian foliation of codimension k of a Riemannian manifold (M, g). Then $\mathscr{F}$ can be described by an open cover $\{U_i\}$ of M with Riemannian submersions $f_i \colon (U_i, g) \to (\sigma_i, b)$, where σ_i is a submanifold of dimension k, such that there are isometries $w_{ij} \colon f_i(U_i \cap U_j) \to f_j(U_j \cap U_i)$ with $f_j = w_{ij} \circ f_i$. The elements w_{ij} acting on $\Sigma = \bigsqcup \sigma_i$ generate a pseudogroup of isometries of Σ, called the *holonomy pseudogroup of $\mathscr{F}$*.

Definition 5.19. A k-dimensional *Riemannian orbifold* is an equivalence class of pseudogroups W of isometries on a k-dimensional Riemannian manifold Σ, such that:

(i) The orbit space Σ/W is Hausdorff;

(ii) For each $x \in \Sigma$, there exists an open neighborhood U of x in Σ such that the restriction of W to U is generated by a finite group of diffeomorphisms of U.

The orbit space Σ/W is the underlying topological space of the orbifold.

If, in addition, W is a group of isometries and Σ is connected, then Σ/W is called a *good* Riemannian orbifold.

Remark 5.20. Alternatively, a *good Riemannian orbifold* Σ/W can be defined as the orbit space of a discrete group of isometries W acting properly on a connected Riemannian manifold Σ. It is possible to prove that if two good Riemannian orbifolds are isometric, then they are equivalent. More generally, as suggested by Lytchak, one can define a *Riemannian orbifold* as a metric space X where each point $x \in X$ has a neighborhood U isometric to a quotient M/Γ, where M is a Riemannian manifold and Γ is a finite group of isometries, see Gorodski [101].

An important example of Riemannian orbifold is the space of leaves $M/\mathscr{F}$, where M is a Riemannian manifold and $\mathscr{F}$ is a Riemannian foliation of M with closed embedded leaves. Here, $M/\mathscr{F}$ is isomorphic (in fact, isometric) to Σ/W, where Σ and W were defined in Example 5.18. The following is a converse result.

Proposition 5.21. *Every Riemannian orbifold Σ/W is the space of leaves of a Riemannian foliation with compacts leaves.*

Proof (Sketch). Given a good Riemannian orbifold Σ/W, we describe a right $U(n)$-action on a manifold M such that $\Sigma/W = M/U(n)$. The remaining details in the proof can be found in Moerdijk and Mrčun [163, Prop. 2.23].

The metric $g(\cdot,\cdot)$ of $T\Sigma$ induces a Hermitian metric on the complex vector bundle $E := T\Sigma \otimes \mathbb{C}$, given by the complexification of the tangent bundle $T\Sigma$. Let $U(E_p)$ denote the set of unitary frames of E_p. Then $U(E) := \bigcup_{p \in \Sigma} U(E_p)$ is called the *unitary frame bundle* of E, and is a principal bundle with fiber $U(n)$ and base Σ. The underlying free right action $\widehat{\mu}_1 : U(E) \times U(n) \to U(E)$, see Proposition 3.33, is defined as $\widehat{\mu}_1(\xi_p, g) := \xi_p \circ g$, where the unitary frame ξ_p is identified with a unitary linear map between $\mathbb{C}^n$ and E_p.

Consider the action $\widehat{\mu}_2 : W \times U(E) \to U(E)$ given by $\widehat{\mu}_2(w, \xi) := dw \cdot \xi$ for each unitary frame ξ and $w \in W$. This is a free proper left action, hence $M := U(E)/W$ is a manifold. Since the actions $\widehat{\mu}_1$ and $\widehat{\mu}_2$ commute, the action $\widehat{\mu}_1$ descends to a proper right action $\mu : M \times U(n) \to M$. One can then prove that Σ/W is equivalent to the orbit space of this action, i.e.,

$$\Sigma/W = M/U(n). \tag{5.4}$$

With the appropriate metric on $M = U(E)/W$, the equality in (5.4) can be interpreted not only as an equivalence but also as an isometry between Riemannian orbifolds.

Let us briefly explain the construction of this metric. Note that if ξ_p is an orthogonal frame of $T_p\Sigma$, then ξ_p is a unitary frame of E_p. This fact can be used to prove that the orthogonal frame bundle $O(T\Sigma)$ is contained in the unitary frame bundle $U(E)$, and, moreover, it is a subbundle of $U(E)$. The Riemannian connection on $T\Sigma$ induces a linear connection $\widehat{\mathscr{H}}$ on $O(T\Sigma)$.[1] Using $\widehat{\mu}_1$, we can extend the linear connection $\widehat{\mathscr{H}}$ on $O(T\Sigma)$ to a linear connection on $U(E)$. Note that there is an induced metric $\widehat{g}$ on $U(E)$, for which the bundle projection $\pi : U(E) \to \Sigma$ is a Riemannian submersion and the linear connection $\widehat{\mathscr{H}}$ is orthogonal to the orbits of $\widehat{\mu}_1$. Since isometries preserve parallel transport, we conclude that $d(\widehat{\mu}_2)^w(\widehat{\mathscr{H}}_{\widehat{x}}) = \widehat{\mathscr{H}}_{(\widehat{\mu}_2)^w(\widehat{x})}$ and hence the distribution $\widehat{\mathscr{H}}$ descends to a distribution $\mathscr{H}$ on $M = U(E)/W$, transverse to the orbits of μ. We can define a metric on $\mathscr{H}$, i.e., a *transverse metric*, by setting $g_{[p]}([X],[Y]) := \widehat{g}_p(X,Y)$, where X, Y are vectors tangent to $\widehat{\mathscr{H}}_p$ and $[\cdot] : U(E) \to U(E)/W$ is the quotient map. Finally, one can extend this transverse metric to a Riemannian metric on M, by requiring that $\mathscr{H}$ be orthogonal to the orbits and averaging an arbitrary metric along the orbits. $\square$

---

[1] The *linear connection* on $O(T\Sigma)$ is defined as follows. For a fixed frame ξ_p, set $\widehat{\mathscr{H}}_{\xi_p}$ as the space of all vectors $\widehat{\alpha}'(0)$, where $\widehat{\alpha}$ is the parallel transport of ξ_p along a curve $\alpha : [0,1] \to \Sigma$ with $\alpha(0) = p$.

5.3   Surgery and Suspension of Homomorphisms

The examples of polar foliations presented in the first section have a symmetric space as ambient. Simple ways to construct new polar foliations on nonsymmetric spaces are to consider a polar foliation $\mathscr{F}$ with compact leaves on a manifold M and change either the leaf metric locally, or the transverse metric on a tubular neighborhood of a regular leaf L with trivial holonomy.[2] After a generic perturbation of its metric, M becomes nonsymmetric. In this section, we explain some techniques that allow to construct more general examples of polar foliations, namely via *surgery* and *suspension of homomorphisms*.

Let us first discuss how *surgery* can be used to construct polar foliations, see Alexandrino and Töben [19]. Let $\mathscr{F}_i$ be a polar foliation with codimension k and compact leaves on a complete Riemannian manifold M_i, for $i = 1, 2$. Suppose that there exist regular leaves with trivial holonomy $L_1 \in \mathscr{F}_1$ and $L_2 \in \mathscr{F}_2$ that are diffeomorphic. Since L_i has trivial holonomy, there exist trivializations $\psi_i \colon \mathrm{Tub}_{3\varepsilon}(L_i) \to L_1 \times B_{3\varepsilon}$, where $\mathrm{Tub}_{3\varepsilon}(L_i)$ is the tubular neighborhood of L_i with radius 3ε, and $B_{3\varepsilon}$ is a ball in the Euclidean k-dimensional space.

Define $\tau \colon (L_1 \times B_\varepsilon) \setminus \overline{(L_1 \times B_{\varepsilon/2})} \to (L_1 \times B_{2\varepsilon}) \setminus \overline{(L_1 \times B_\varepsilon)}$ as the inversion in the cylinder of radius ε and axis L_1. Define also $\phi \colon \mathrm{Tub}_\varepsilon(L_1) \setminus \overline{\mathrm{Tub}_{\varepsilon/2}(L_1)} \to \mathrm{Tub}_{2\varepsilon}(L_2) \setminus \overline{\mathrm{Tub}_\varepsilon(L_2)}$ by $\phi := \psi_2^{-1} \circ \tau \circ \psi_1$. We now change the transverse metric of M_i in the tubular neighborhoods of L_i, such that the restriction of ϕ to each section in $\mathrm{Tub}_\varepsilon(L_1) \setminus \overline{\mathrm{Tub}_{\varepsilon/2}(L_1)}$ is an isometry. This can be done, for instance, by taking $\frac{1}{\|x\|^2} g_0$ as the transverse metric on $\mathrm{Tub}_{2\varepsilon}(L_i)$ and keeping the original transverse metric g outside $\mathrm{Tub}_{3\varepsilon}(L_i)$, using partitions of unity with two appropriate functions. Finally, define $\tilde{M}$ as:

$$\tilde{M} := \left(M \setminus \overline{\mathrm{Tub}_{\varepsilon/2}(L_1)} \right) \sqcup_\phi \left(M \setminus \overline{\mathrm{Tub}_\varepsilon(L_2)} \right).$$

This new manifold has a singular foliation $\widetilde{\mathscr{F}}$, and leaves are locally equidistant with respect to the transverse metric that already exists. To conclude the construction, we only have to define a tangential metric to the leaves with partitions of unity. Note that if Σ_i is a section of $\mathscr{F}_i$, then the connected sum $\Sigma_1 \# \Sigma_2$ is a section of $\widetilde{\mathscr{F}}$.

Let us now explore an example of polar foliation constructed with *suspension of homomorphisms*, extracted from Alexandrino [6]. Other examples can be found in Alexandrino [4]. We start by recalling the method of suspensions, see Molino [165, p. 28–29, 96–97] for details. Let Q and V be Riemannian manifolds of dimension p and n respectively, and consider a homomorphism $\rho \colon \pi_1(Q, q_0) \to \mathrm{Iso}(V)$. Let $\widehat{P} \colon \widehat{Q} \to Q$ be the projection of the universal covering of Q. There is a natural action of $\pi_1(Q, q_0)$ on $\tilde{M} = \widehat{Q} \times V$, given by

$$(\widehat{q}, v) \cdot [\alpha] := \left(\widehat{q} \cdot [\alpha], \rho(\alpha^{-1}) \cdot v \right),$$

---

[2]This is possible since the set of regular leaves is open and dense in M.

where $\widehat{q} \cdot [\alpha]$ denotes the deck transformation associated to $[\alpha]$ applied to a point $\widehat{q} \in \widehat{Q}$. We denote by M the set of orbits of this action, and by $\Pi : \widetilde{M} \to M$ the canonical projection. We claim that M is a smooth manifold. Indeed, given a simple open neighborhood $U_j \subset Q$, consider the bijection

$$\Psi_j : \Pi\big(\widehat{P}^{-1}(U_j) \times V\big) \to U_j \times V, \quad \Psi_j(\Pi(\widehat{q}, v)) := \big(\widehat{P}(\widehat{q}) \times \{v\}\big).$$

If $U_i \cap U_j \neq \emptyset$ is connected, we can see that $\big(\Psi_i \cap \Psi_j^{-1}\big)(q, v) = \big(q, \rho([\alpha]^{-1})v\big)$ for a fixed $[\alpha]$. Hence, there exists a unique manifold structure on M for which Ψ_j are local diffeomorphisms. Let

$$P : M \to Q, \quad P(\Pi(\widehat{q}, v)) := \widehat{P}(\widehat{q}).$$

It follows that M is the total space of a fiber bundle, and P is its projection over the base Q. In addition, the fiber is V and the structural group is the image of ρ.

Finally, define $\mathscr{F} := \big\{\Pi(\widehat{Q}, v)\big\}$, i.e., the projection of the trivial foliation defined as the product of $\widehat{Q}$ with each v. It is possible to prove that this is a foliation transverse to the fibers. Furthermore, this foliation is a Riemannian foliation whose transverse metric coincides with the metric on V.

Example 5.22. In what follows, we construct a polar foliation such that the intersection of a local section with the closure of a regular leaf is an orbit of an action of a subgroup of isometries of the local section. This isometric action is not polar. Therefore, *there exists a polar foliation $\mathscr{F}$ such that the partition formed by the closures of the leaves is a nonpolar singular Riemannian foliation.*

Set $V := \mathbb{R}^2 \times \mathbb{C} \times \mathbb{C}$ and $\widehat{\mathscr{F}_0}$ the singular foliation of codimension 5 on V, whose leaves are products of points in $\mathbb{C} \times \mathbb{C}$ with circles in $\mathbb{R}^2$ centered at the origin. It is easy to see that the foliation $\widehat{\mathscr{F}_0}$ is polar. Let Q be the circle S^1 and $\alpha \notin \mathbb{Q}$ be an irrational number. Define the homomorphism

$$\rho : \pi_1(Q, q_0) \to \mathrm{Iso}(V), \quad \rho(n) := \big((x, z_1, z_2) \mapsto (x, e^{in\alpha} \cdot z_1, e^{in\alpha} \cdot z_2)\big).$$

Finally, set $\mathscr{F} := \Pi\big(\widehat{Q} \times \widehat{\mathscr{F}_0}\big)$. It turns out that $\mathscr{F}$ is a polar foliation and the intersection of the section $\Pi(0 \times \mathbb{R} \times \mathbb{C} \times \mathbb{C})$ with the closure of a regular leaf is an orbit of an isometric action on the section. This isometric action is not polar, since the Hopf action is not polar (see Remark 3.90).

5.4 Differential and Geometric Aspects of Polar Foliations

In this section, we describe a few results from the theory of polar foliations. The following was proposed by Palais and Terng [182, Rem 5.6.8]:

Palais-Terng Conjecture 5.23. *Let G be an isometric action of a compact Lie group on M. If the distribution of normal spaces to regular orbits is integrable, then the G-action is polar.*

This conjecture was proved by Heintze, Liu and Olmos [124]. In particular, they proved that the set of regular points is dense in each section. A result in Alexandrino [6] gives a positive answer to this conjecture in the case of singular Riemannian foliations.

Theorem 5.24. *Let $\mathscr{F}$ be a singular Riemannian foliation on a complete Riemannian manifold M. If the distribution of normal spaces to the regular leaves is integrable, then $\mathscr{F}$ is a polar foliation. In addition, the set of regular points is open and dense in each section.*

Proof (Sketch). Let p be a regular point. The first step is to verify that if γ is a geodesic orthogonal to the regular leaf L_p, then singular points are isolated along γ. In the case where $\mathscr{F}$ is the partition of M into orbits of an isometric action, the reader can check this claim using the slice representation, or Kleiner's Lemma 3.70.

The second step in the proof is to ensure that, for each regular point x, there exists a totally geodesic disk $D_\delta(x)$ orthogonal to the leaves. Its radius δ depends only on the convex neighborhood centered at x, and not on the tubular neighborhood of a plaque at x.

For the third step, note that we can extend $D_\delta(p)$ along a horizontal geodesic γ by gluing totally geodesic disks along γ. Let Σ_{r_1} denote the extensions constructed along every geodesic γ with length r_1 such that $\gamma'(0) \in T_p D_\delta(p)$.

In the fourth step, one checks that Σ_{r_1} is covered by a finite number of disks D_i. This implies that Σ_{r_1} is an immersed submanifold. Indeed, let D_i^* be the lift of D_i to the Grassmannian bundle $\mathrm{Gr}_k TM$. Since D_i are totally geodesic, if D_i^* and D_j^* have a common point, then they coincide in a neighborhood of this point. Thus, the union $\Sigma_{r_1}^*$ of D_i^*'s is an embedded submanifold in $\mathrm{Gr}_k TM$ that projects to the immersed submanifold Σ_{r_1}.

Finally, considering $r_n \to \infty$ and applying the Hopf-Rinow Theorem 2.9, one can check that the projection Σ of $\Sigma^* := \bigcup_{r_i} \Sigma_{r_i}^*$ is a complete immersed totally geodesic submanifold. $\square$

Remark 5.25. Particular cases of the above result where proved by Molino and Pierrot [166], Boualem [49] and Lytchak and Thorbergsson [156]. Szenthe [203] also worked on this conjecture for isometric actions.

Henceforth, $\mathscr{F}$ denotes a polar foliation on a complete Riemannian manifold M. The following result of Alexandrino [4] relates polar foliations to equifocal submanifolds, see also Alexandrino and Töben [20].

Theorem 5.26. *Let L be a regular leaf of a polar foliation $\mathscr{F}$ of a complete Riemannian manifold M.*

(i) *L is equifocal. In particular, the union of regular leaves that have trivial normal holonomy is an open and dense set in M, provided that all leaves are compact;*

(ii) *Let β be a smooth curve on L and ξ be a parallel normal field to L along β. Then the curve $\eta_\xi \circ \beta$ is in a leaf of $\mathscr{F}$;*

(iii) *Suppose that L has trivial holonomy and let Ξ denote the set of all parallel normal fields on L. Then $\mathscr{F} = \{\eta_\xi(L)\}_{\xi \in \Xi}$.*

Proof (Sketch). We first explain why the image $\eta_\xi(U)$ of the endpoint map is contained in a leaf, for the appropriate basic vector ξ. Let x be a singular point and x_0 be a regular point in a small tubular neighborhood $\mathrm{Tub}(P_x)$. Let ξ be the parallel normal vector field of the plaque P_{x_0} with $\exp(\xi_{x_0}) = x$. Assume that the geodesic $\gamma(t) := \exp(t\,\xi_{x_0})$ lies in $\mathrm{Tub}(P_x)$ for all $t \in [0,2]$, and $x = \gamma(1)$ is the only singular point along γ. Define $\alpha(s,t) := \exp(t\xi_{c(s)})$ and $J(t) = \frac{\partial}{\partial s}\alpha(s,t)|_{s=0}$, where $c(s)$ is a curve in P_{x_0} with $c(0) = x_0$. We want to check that the Jacobi field J is always tangential to the leaves, and hence that the curve $\alpha(\cdot,t)$ is contained in $L_{\gamma(t)}$.

Since Σ is totally geodesic, we have that $T_{\gamma(t)}\Sigma$, as well as $(T_{\gamma(t)}\Sigma)^\perp$, are families of parallel subspaces along γ which are invariant under $R(\gamma,\cdot)\dot\gamma$. Note that $J(0), J'(0) \in (T_{\gamma(0)}\Sigma)^\perp$. These facts and the Jabobi equation (2.5) imply that J is orthogonal to Σ for $0 \le t \le 2$.

Since the singular points are isolated in γ, we have that $J(t)$ is tangent to the leaves $L_{\gamma(t)}$ for $t \ne 1$ and hence the curve $\alpha(\cdot,t)$ is contained in $L_{\gamma(t)}$ for $t \ne 1$. It remains to prove that $J(1)$ is tangent to $L_{\gamma(1)}$. Set

$$h(s,t) := \mathrm{dist}\big(\alpha(s,t), P_x\big) - \mathrm{dist}\big(\alpha(0,t), P_x\big).$$

The above discussion implies that $h(\cdot,t) = 0$ for $t \ne 1$ and hence, since h is continuous, $h(\cdot,1) = 0$. This implies that $J(1)$ is tangent to $L_{\gamma(1)}$, and $\alpha(\cdot,1)$ is contained in $L_{\gamma(1)}$. As we know that $\alpha(\cdot,t) \subset L_{\gamma(t)}$, it follows that $\eta_{t\xi}(U) \subset L_{\gamma(t)}$.

Let us now explain why $\mathrm{d}\eta_{t\xi}$ has constant rank for t close to 1. The classical theory implies that, for $t < 1$, $\eta_{t\xi}(U)$ is an open subset of $L_{\gamma(t)}$ and $\eta_{t\xi} : U \to \eta_{t\xi}(U) \subset L_{\gamma(t)}$ is a diffeomorphism. If $t = 1$, we have that the rank of $\mathrm{d}\eta_\xi$ is constant on U, because the closest-point projection $\pi : \mathrm{Tub}(P_x) \to P_x$ is a submersion and $\pi|_U = \eta_\xi|_U$. Finally, consider $t > 1$. The fact that π is a submersion and $\pi|_U = \eta_\xi|_U$ imply that $\ker \mathrm{d}\eta_{t\xi}$ is tangent to the slice. On the other hand, $\eta_{t\xi}$ is a diffeomorphism when restricted to the slice. In fact, one can check that the foliation restricted to a small slice is invariant under geodesic involutions (i.e., invariant under the maps $\exp_x \circ(-\,\mathrm{id}) \circ \exp_x^{-1}$). Thus, $\eta_{t\xi}|_U$ is a local diffeomorphism. It is not difficult to see that it is bijective, and hence a diffeomorphism.[3]

---

[3] Alternatively, one can prove that $\mathrm{d}\eta_{t\xi}$ has constant rank for t close to 1 using ideas presented in the proof of Theorem 5.63.

We have sketched the proof of the result for *small* ξ's. The general case can be proved by composing endpoint maps, i.e., $\eta_\xi := \eta_{\xi_n} \circ \ldots \circ \eta_{\xi_0}$, where ξ_i are the appropriate small vectors.   □

Note that this result guarantees that, given a regular leaf L with trivial holonomy, we can reconstruct $\mathscr{F}$ by taking submanifolds parallel to L.

In order to state the next result, we recall the concepts of slice and local section. Given $x \in M$, let $\mathrm{Tub}(P_x)$ be a tubular neighborhood of a plaque P_x that contains x. The connected component of $\exp_x(\nu P_x) \cap \mathrm{Tub}(P_x)$ that contains x is called a *slice* at x, and is usually denoted S_x. A *local section* σ (centered at x) of a section Σ is a connected component $\mathrm{Tub}(P_x) \cap \Sigma$ that contains x.

Let us recall some results about the local structure of $\mathscr{F}$, in particular about the structure of the set of singular points in a local section. The next result of Alexandrino [4] is a generalization of Theorem 4.19.

Slice Theorem 5.27. *Let $\mathscr{F}$ be a polar foliation of a complete Riemannian manifold M, and let S_x be a slice at a singular point x.*

(i) *$S_x = \bigcup_{\sigma \in \Lambda(x)} \sigma$, where $\Lambda(x)$ is the set of local sections σ centered at x;*
(ii) *$S_y \subset S_x$ for all $y \in S_x$;*
(iii) *$\mathscr{F}|_{S_x}$ is a polar foliation of S_x with the metric induced from M;*
(iv) *$\mathscr{F}|_{S_x}$ is diffeomorphic to an isoparametric foliation of an open subset of $\mathbb{R}^n$, where $n = \dim S_x$.*

Proof (Sketch). In order to prove (i), first note that $S_x \supset \bigcup_{\sigma \in \Lambda(x)} \sigma$. In fact, let $\sigma \in \Lambda(x)$. Since σ is totally geodesic and orthogonal to L_x, we have that $\sigma \subset S_x$.

Let σ be a local section. In what follows, we show that:

$$\text{If } \sigma \cap S_x \neq \emptyset, \text{ then } x \in \sigma \text{ and } \sigma \subset S_x. \tag{5.5}$$

Since $S_x \supset \bigcup_{\sigma \in \Lambda(x)} \sigma$, it suffices to check that $x \in \sigma$. We claim that if the intersection of a local section σ with a slice S_x contains a regular point, then $x \in \sigma$. Indeed, let $p \in S_x \cap \sigma$ be a regular point and γ be a geodesic segment orthogonal to L_x, that joins x to p. Since $\mathscr{F}$ is a polar foliation, γ is orthogonal to L_p and belongs to the local section σ, hence $x \in \sigma$. Let us now consider the remaining case, in which $\sigma \cap S_x$ may not contain regular points. The previous claim and the fact that the regular points are dense in σ imply that there exists a slice $S_{\tilde{x}}$ with $\tilde{x} \in L_x$, such that $\tilde{x} \in \sigma$ and $\sigma \subset S_{\tilde{x}}$. Since $S_x \cap S_{\tilde{x}} \neq \emptyset$, we conclude that $x = \tilde{x}$.

In order to verify that $S_x \subset \bigcup_{\sigma \in \Lambda(x)} \sigma$, let $y \in S_x$ and σ be a local section that contains y. It follows from (5.5) that $x \in \sigma$, and hence $y \in \bigcup_{\sigma \in \Lambda(x)} \sigma$.

As for (ii), let $y \in S_x$ and σ be a local section that contains y. It follows from (5.5) that $\sigma \subset S_x$. Since, by (i), S_y is a union of local sections that contain y, we conclude that $S_y \subset S_x$. Item (iii) follows easily from (ii).

Finally let us sketch the proof of item (iv). Set $\mathscr{F}_x := \mathscr{F}|_{S_x}$. From the previous items, $\mathscr{F}_x$ is a polar foliation on S_x. Consider a *geodesic homothetic transformation* h_λ with respect to x, i.e., $h_\lambda(\gamma(\varepsilon)) = \gamma(\lambda \varepsilon)$ for all horizontal unit speed geodesics γ starting at x. Due to equifocality, $\mathscr{F}_x$ is invariant under h_λ. We also have that the sections of $\mathscr{F}_x$ are invariant under h_λ. Therefore, $\mathscr{F}_x$ is a polar foliation with respect to the metric $g_\lambda := \frac{1}{\lambda^2} h_\lambda^* g$. On the other hand, as $\lambda \searrow 0$, the metric g_λ converges smoothly to the metric g_0 of the tangent space[4] at x to the slice S_x. Thus, $\mathscr{F}_x$ is a polar foliation with respect to the flat metric g_0, and hence $\mathscr{F}_x$ is a polar foliation on $\mathbb{R}^n$, after the appropriate identification.[5]

Recall that the regular leaves of $\mathscr{F}_x$ are equifocal. A standard computation shows that equifocal submanifolds in Euclidean spaces are isoparametric, cf. Proposition 5.5. Therefore, each regular leaf L is isoparametric. Note that, by equifocality, the isoparametric foliation of L coincides with $\mathscr{F}_x$, concluding the proof. $\qquad\square$

From Theorem 5.27 (iv), it is not difficult to derive the following consequence, see Alexandrino [4].

Corollary 5.28. *Let σ be a local section. The set of singular points of $\mathscr{F}$ contained in σ is a finite union of totally geodesic hypersurfaces. These hypersurfaces are diffeomorphic to focal hyperplanes contained in a section of an isoparametric foliation of an open set of Euclidean space.*

We call the set of singular points of $\mathscr{F}$ contained in σ the *singular stratification of the local section σ*. Let M_{reg} denote the set of regular points in M. A *Weyl chamber* of a local section σ is the closure in σ of a connected component of $M_{\text{reg}} \cap \sigma$, cf. Definition 4.26. It is possible to prove that a Weyl chamber of a local section is a convex set.

Theorem 5.26 allows us to define the *singular holonomy map*, which is very useful to study $\mathscr{F}$. The following is also in Alexandrino [4].

Proposition 5.29. *Let $\mathscr{F}$ be a polar foliation on a complete Riemannian manifold M, and $x_0, x_1 \in L_x$. Let $\beta : [0,1] \to L_p$ be a smooth curve contained in a regular leaf L_p, such that $\beta(i) \in S_{x_i}$, where S_{x_i} is the slice at x_i for $i = 0, 1$. Let σ_i be a local section contained in S_{x_i} that contains $\beta(i)$ and x_i, $i = 0, 1$. Finally, let $[\beta]$ denote the homotopy class of β. Then there exists an isometry $\varphi_{[\beta]} : U_0 \to U_1$, where the domain U_0 and range U_1 are contained in σ_0 and σ_1 respectively, satisfying:*

(i) $x_0 \in U_0$;
(ii) $\varphi_{[\beta]}(y) \in L_y$ *for each $y \in U_0$;*
(iii) $d\varphi_{[\beta]} \xi(0) = \xi(1)$, *where $\xi(s)$ is a parallel normal field along $\beta(s)$.*

Such an isometry is called the *singular holonomy map along β*. We remark that, in the definition of the singular holonomy map, singular points can be contained in the domain U_0. If the domain U_0 and the range U_1 are sufficiently small, and L_x is regular, then the singular holonomy map coincides with the usual holonomy map along β.

---

[4]The metric g_0 is such that $\exp_x$ becomes an isometry.

[5]I.e., by composing the inverse of the normal exponential map $\exp_x : B_\varepsilon(0) \cap T_x S_x \to S_x$ with an isometry between $T_x S_x$ and $\mathbb{R}^n$.

Theorem 5.27 establishes a relation between polar foliations and isoparametric foliations. Similarly, as in the theory of isoparametric submanifolds, it is natural to ask whether a (generalized) Weyl group action can be defined on σ. The following definitions and results deal with this question.

Definition 5.30. The pseudosubgroup $W(\sigma)$ generated by all singular holonomy maps $\varphi_{[\beta]}$ such that $\beta(0)$ and $\beta(1)$ belong to the same local section σ is called the *generalized Weyl pseudogroup* of σ. Define $W(\Sigma)$ for a section Σ analogously.

Proposition 5.31. *Let σ be a local section. The reflections across the hypersurfaces of the singular stratification of the local section σ leave $\mathscr{F}|_\sigma$ invariant. Moreover, these reflections are elements of $W(\sigma)$.*

Remark 5.32. Compare the above result, in Alexandrino [4], with Theorem 4.37 (i) and Remark 4.38.

Using suspensions, one can construct an example of polar foliation such that $W(\sigma)$ is larger than the pseudogroup generated by the reflections across the hypersurfaces of the singular stratification of σ. On the other hand, a sufficient condition to ensure that both pseudogroups coincide is that the leaves of $\mathscr{F}$ have trivial normal holonomy and are compact. Thus, it is natural to ask under which conditions the normal holonomy of regular leaves is trivial. The next result, in Alexandrino and Töben [19], Alexandrino [7, 9], and Lytchak [154], deals with this question.

Theorem 5.33. *Let $\mathscr{F}$ be a polar foliation on a simply-connected manifold M.*

  (i) *Each regular leaf has trivial holonomy;*
 (ii) *The leaves of $\mathscr{F}$ are closed and embedded;*
(iii) *$M/\mathscr{F}$ is a simply-connected good Coxeter orbifold;[6]*
 (iv) *Let Σ be a section of $\mathscr{F}$ and $\Pi\colon M \to M/\mathscr{F}$ the canonical projection. Denote by Ω a connected component of the set of regular points in Σ. Then $\Pi\colon \Omega \to M_{\mathrm{reg}}/\mathscr{F}$ and $\Pi\colon \overline{\Omega} \to M/\mathscr{F}$ are homeomorphisms, where M_{reg} denotes the set of regular points in M, and Ω is convex.[7]*

Remark 5.34. Compare the above result with Theorems 4.36 and 4.37.

Proof (Sketch). In what follows we sketch the proof following Alexandrino [9].

Let $\pi_\Sigma\colon \widetilde{\Sigma} \to \Sigma$ be the Riemannian universal covering, $p \in \Sigma$ a regular point, and $\widetilde{p} \in \widetilde{\Sigma}$ be such that $\pi_\Sigma(\widetilde{p}) = p$. Let $w \in W(\Sigma)$ and consider a curve $c\colon [0,1] \to \Sigma$ with $c(0) = p$ and $c(1) = w(p)$. There exists an isometry $\widetilde{w}_{[c]}$ on $\widetilde{\Sigma}$ such that $\pi_\Sigma \circ \widetilde{w}_{[c]} = w \circ \pi_\Sigma$ and $\widetilde{w}_{[c]}(\widetilde{p}) = \widetilde{c}(1)$, where $\widetilde{c}$ is the horizontal lift of c starting at $\widetilde{p}$. The isometry $\widetilde{w}_{[c]}$ defined above is called *a lift of w along c*, and clearly depends on

---

[6]I.e., a quotient of a simply-connected Riemannian manifold by a *reflection group in the classical sense*, i.e., a discrete group of isometries generated by reflections acting properly and effectively.

[7]I.e., for any two points p and q in Ω, every minimal geodesic segment between p and q lies entirely in Ω.

the homotopy class $[c]$. Let $\widetilde{W}$ be the group generated by all $\widetilde{w}$, called the *lifted Weyl group*. It is proved in [9] that there exists a surjective homomorphism

$$\widetilde{\rho} \colon \pi_1(M) \to \widetilde{W}/\widetilde{\Gamma}, \tag{5.6}$$

where $\widetilde{\Gamma}$ is a normal subgroup, called *lifted reflection group*, that is constructed as follows. Let H be a wall of the Weyl chamber and $\widetilde{H} \subset \widetilde{\Sigma}$ be such that $\pi_\Sigma(\widetilde{H}) = H$. Each reflection in $\widetilde{H}$ is an element of $\widetilde{W}$, and $\widetilde{\Gamma}$ is the group generated by all these reflections. It follows from Davis [76, Lemma 1.1] that $\widetilde{\Gamma}$ is a *reflection group in the classical sense*. In particular, a Weyl chamber C of $\widetilde{\Gamma}$ is a fundamental domain of the action of $\widetilde{\Gamma}$, see [76, Thm 4.1]. Along this proof, M is simply-connected, and hence (5.6) implies that $\widetilde{W} = \widetilde{\Gamma}$.

We first prove (i), i.e., show that each regular leaf has trivial holonomy. If this was not true, there would be a regular point $p \in \Sigma$ and $w \in W(\Sigma)$ such that $w(p) = p$ and $w \neq \mathrm{id}$. Let $\delta \colon [0,1] \to \Sigma$ be an arbitrary curve starting at $\delta(0) = p$. Define a lift of w by $\widetilde{w}(\delta(1)) := \widetilde{w \circ \delta}(1)$, i.e., the lift of w along the constant curve $c = p$. Note that $\widetilde{w}(\widetilde{p}) = \widetilde{p}$ and $\widetilde{w} \neq \mathrm{id}$. Since $\widetilde{\Gamma} = \widetilde{W}$ is a reflection group, the only points that can be fixed by elements of $\widetilde{\Gamma}$ are singular, i.e., points that are mapped by π_Σ to singular points of Σ. Therefore, $\widetilde{w}(\widetilde{p}) \neq \widetilde{p}$, a contradiction.

Now we prove (ii), i.e., show that the leaves of $\mathscr{F}$ are closed and embedded. Due to equifocality, it suffices to prove that the leaves are embedded. If L_p is not embedded, then there would be a sequence $\{w_n\} \subset W(\Sigma)$ such that $w_n(p) \to p$. Define $\widetilde{w}_n(\delta(1)) := (c_n * \widetilde{w_n \circ \delta})(1)$, where c_n are short curves contained in a simply-connected neighborhood of p and $*$ denotes concatenation of curves. Then one can check that $\widetilde{w}_n(\widetilde{p}) \to \widetilde{p}$. Since $\widetilde{\Gamma} = \widetilde{W}$ is a reflection group in the classical sense, $\{\widetilde{w}_n(\widetilde{p})\}$ cannot converge to $\widetilde{p}$, giving the desired contradiction.

We continue by proving (iii), i.e, that $M/\mathscr{F}$ is a simply-connected good Coxeter orbifold. For each loop c in Σ, consider the isometry $\widetilde{\mathrm{id}}_{[c]}$, i.e., the lift of the identity along c. In this way, we identify $\pi_1(\Sigma)$ with a subgroup of $\widetilde{W}$. It is not difficult to check that this is a normal subgroup and $W(\Sigma) = \widetilde{W}/\pi_1(\Sigma)$. Since $\Sigma = \widetilde{\Sigma}/\pi_1(\Sigma)$, we have that $\Sigma/W(\Sigma) = \widetilde{\Sigma}/\widetilde{W}$. As $\widetilde{W} = \widetilde{\Gamma}$ is a reflection group, we conclude that $M/\mathscr{F} = \Sigma/W(\Sigma)$ is the good Coxeter orbifold $\widetilde{\Sigma}/\widetilde{\Gamma}$. In order to check that $\widetilde{\Sigma}/\widetilde{\Gamma}$ is simply-connected, consider a homotopy $\widetilde{c}_s$ in $\widetilde{\Sigma}$ between a loop $\widetilde{c}$ contained in $C = \widetilde{\Sigma}/\widetilde{\Gamma} \subset \widetilde{\Sigma}$ and the constant curve $\widetilde{c}(0)$. Then $\rho(\widetilde{c}_s)$ is an homotopy in $C = \widetilde{\Sigma}/\widetilde{\Gamma}$ between $\widetilde{c}$ and $\widetilde{c}(0)$, where $\rho \colon \widetilde{\Sigma} \to \widetilde{\Sigma}/\widetilde{\Gamma}$ is the quotient map, so the conclusion follows from the fact that $\widetilde{\Sigma}$ is simply-connected.

Finally, we prove item (iv), i.e., that $\overline{\Omega}$ is a fundamental domain of the action of $W(\Sigma)$ on Σ. Up to composing with reflections across the walls (that are elements of $W(\Sigma)$), we may assume that each orbit intersects $\overline{\Omega}$ at least once. Suppose, by contradiction, that there exist two points $x, y \in \overline{\Omega}$ that belong to the same leaf. Then there is an isometry $w \in W(\Sigma)$ such that $w(x) = y$. Joining x and y with a curve c contained in $\overline{\Omega}$ and constructing an isometry $\widetilde{w}_{[c]} \in \widetilde{W}$ that maps a point $\widetilde{x}$ to a point

$\widetilde{y}$, both inside the fundamental domain C of the action of $\widetilde{\Gamma} = \widetilde{W}$ on $\widetilde{\Sigma}$, we obtain a contradiction. As we proved that $\overline{\Omega} = \Sigma/W(\Sigma) = C = \widetilde{\Sigma}/\widetilde{\Gamma}$, we conclude that $\overline{\Omega}$ is convex.   □

The following is an important consequence of Theorem 5.33.

Corollary 5.35. *Orbits of a polar action on a simply-connected complete Riemannian manifold are closed and embedded. Moreover, there are no exceptional orbits.*

Remark 5.36. As remarked in [9], Theorem 5.33 also holds for a more general class of foliations, namely infinitesimally polar foliations. In particular, a proper isometric action on a simply-connected manifold whose orbit space is a good orbifold has no exceptional orbits. Apart from polar actions, other examples of isometrics actions whose orbit spaces are good orbifolds are given by variationally complete actions, see Lytchack [154].

Remark 5.37. As remarked in [14], the results in [9] and Theorem 5.33 can be used to provide elementary proofs of the existence of closed geodesics on leaf spaces of polar foliations on compact simply-connected spaces. More generally, we have the following result if $\mathscr{F}$ is a singular Riemannian foliation with closed embedded leaves on a simply-connected Riemannian manifold M and $M/\mathscr{F}$ is a compact orbifold, then $M/\mathscr{F}$ admits a nontrivial closed geodesic.

We now discuss very recent results in the theory of polar foliations and actions. The following result of Lytchak [155] gives a positive answer to an important conjecture in polar actions on symmetric spaces:

Theorem 5.38. *Let M be a simply-connected, irreducible symmetric space with sec ≥ 0, and let $\mathscr{F}$ be a polar foliation on M of codimension at least 3. Then either all the leaves of $\mathscr{F}$ are points, or $\mathscr{F}$ is hyperpolar, or the symmetric space has rank one.*

Similar techniques to the ones used by Lytchak [155] were independently employed by Fang, Grove and Thorbergsson [84], who recently proved the following classification result.

Theorem 5.39. *A polar action on a compact simply-connected manifold with sec > 0 and of cohomogeneity ≥ 2 is equivariantly diffeomorphic to a polar action on a compact rank one symmetric space (CROSS).*[8]

Remark 5.40. Theorem 6.30, due to Verdiani [218] and Grove, Wilking and Ziller [114], complements the above result with regard to manifolds with sec > 0 and cohomogeneity 1 (which are automatically polar). In this case, there are several manifolds different from a CROSS that support an invariant metric with sec > 0.

---

[8]For details on compact rank one symmetric spaces, see Example 6.10.

Remark 5.41. As proved in Mucha [169], there exist examples of hyperpolar actions with cohomogeneity ≥ 2 on locally irreducible compact homogeneous spaces with sec ≥ 0 which are not homeomorphic to symmetric spaces.

Grove and Ziller [118] associate to a polar foliation without exceptional orbits (that they call *Coxeter polar action*), a graph of groups $G(C)$, where C is a chamber of a fixed section. The following fundamental reconstruction result is proved using this notion.

Theorem 5.42. *A Coxeter polar action (M, G) is determined by its polar data $(C, G(C))$. More precisely, there is a canonical construction of a Coxeter polar G-manifold $M(D)$ from the polar data $D = (C, G(C))$, and if D is the polar data for a Coxeter polar G-manifold M, then $M(D)$ is equivariantly diffeomorphic to M.*

This result was recently used by Gozzi [103] to classify polar actions in manifolds of dimension ≤ 5. In Chap. 6, we discuss the above result in more details in the particular case of cohomogeneity one actions, see Remark 6.38.

Since we have seen several examples of theorems about polar foliations and actions in spaces with sec ≥ 0, is natural to wonder what happens in the dual case of sec ≤ 0. The following was obtained by Töben [214].

Theorem 5.43. *Let $\mathscr{F}$ be a polar foliation on a Hadamard manifold. Then, up to diffeomorphisms, $\mathscr{F}$ is the product foliation of a compact isoparametric foliation and the trivial foliation of $\mathbb{R}^n$ with only one leaf.*

Remark 5.44. It is a classical fact that there are no nontrivial Killing vector fields in compact negatively curved spaces, see Petersen [183, p. 191]. Töben [214] proved that there are no polar foliations with closed leaves on such spaces, and Lytchak [153] extended this result proving that there are not even singular Riemannian foliations.

To conclude this section, we discuss a result that describes the behavior of polar foliations whose leaves are not embedded. Molino [165] proved that if M is compact, then the closure of the leaves of a (regular) Riemannian foliation forms a partition of M that is a singular Riemannian foliation. He also proved that leaf closures are orbits of a locally constant sheaf of germs of (transversal) Killing fields. Furthermore, he proposed the following conjecture in the same book:

Molino Conjecture 5.45. *Let $\mathscr{F}$ be a singular Riemannian foliation of a compact Riemannian manifold M. Then the closures of the leaves of $\mathscr{F}$ forms a partition of M that is also a singular Riemannian foliation.*

Molino [165, Thm 6.2] proved that the closure of the leaves is a *transnormal system*, however, it still remains an open problem to prove that $\mathscr{F}$ is a singular foliation. The next result, in Alexandrino [6], gives a positive answer to the Molino Conjecture 5.45 if $\mathscr{F}$ is polar.

Theorem 5.46. *Let $\mathscr{F}$ be a polar foliation on a complete Riemannian manifold M.*

(i) *The closures $\{\overline{L}\}_{L \in \mathscr{F}}$ of the leaves of $\mathscr{F}$ form a partition of M which is a singular Riemannian foliation;*

(ii) *Each $x \in M$ is contained in a homogeneous submanifold $\mathscr{O}_x$ (possibly 0-dimensional). If we fix a local section σ that contains x, then $\mathscr{O}_x$ is a connected component of an orbit of the closure of the Weyl pseudogroup of σ;*

(iii) *If $x \in \overline{L}$, then a neighborhood of x in $\overline{L}$ is the product of the homogeneous submanifold $\mathscr{O}_x$ with plaques with the same dimension of the plaque P_x;*

(iv) *If x is a singular point and T is the intersection of the slice S_x with the singular stratum that contains x, then the normal connection of T in S_x is flat;*

(v) *Let x be a singular point and T as in (iv). Consider v a parallel normal vector field along T, $y \in T$ and $z = \exp_y(v)$. Then $\mathscr{O}_z = \eta_v(\mathscr{O}_y)$.*

Proof (Sketch). Let x and σ be as in the statement. It follows from the definition of Weyl pseudogroups that the intersection of the leaf L_x with the local section σ coincides with the orbit of x under the Weyl pseudogroup. This observation and results of Salem [165, Appendix D] then imply that $\mathscr{O}_x$ is a connected component of an orbit of the closure of the Weyl pseudogroup of σ. The rest of (ii), i.e., that $\mathscr{O}_x$ does not depend on σ, can be proved using that the Weyl pseudogroups of different local sections are conjugate to each other.

Note that the partition of σ into orbits of the closure of the Weyl pseudogroup is a singular Riemannian foliation. This fact and (ii) imply that $\{\overline{L}\}_{L \in \mathscr{F}}$ is a transnormal system. Item (iv) can be proved using that walls of the singular stratification of the local section σ are totally geodesic in σ. In addition, the singular stratification is diffeomorphic to a product of simplicial cones with a subspace, which can be identified with the intersection of σ with the stratum that contains x, called the *local minimal stratum*. The above fact is also the key to prove (v). Roughly speaking, the orbit $\mathscr{O}_z$ must be equidistant to the walls of the singular stratification of the local section and therefore parallel to the local minimal stratum. Finally, combining (iv) and (v), one can prove that the partition $\{\overline{L}\}_{L \in \mathscr{F}}$ is singular.   □

The above theorem was illustrated in Example 5.22.

Remark 5.47. Lytchak [154] recently proved that polar foliations on simply-connected complete Riemannian manifolds have closed and embedded leaves, see also Theorem 5.33. Therefore, the above result is of interest only in the case where the ambient space is *not* simply-connected.

5.5   Transnormal and Isoparametric Maps

In the previous section, polar foliations were presented as a natural generalization of isoparametric foliations on Euclidean spaces, in particular of orbits of adjoint actions. In this section, we consider a different generalization, related to partitions by preimages of certain special maps, called *transnormal* and *isoparametric* maps. This approach was already considered by Cartan [62–65], Harle [122], Terng [204] and Wang [224].

We start by discussing the definition of transnormal functions and isoparametric functions; for historical context, see Thorbergsson [208, 209].

Definition 5.48. A smooth function $f: M \to \mathbb{R}$ on a complete Riemannian manifold M is called *transnormal* if

$$\|\operatorname{grad} f\|^2 = b \circ f \qquad (5.7)$$

for some C^2 function $b: f(M) \to \mathbb{R}$. Moreover, if

$$\Delta f = a \circ f \qquad (5.8)$$

for some continuous function $a: f(M) \to \mathbb{R}$, then f is called *isoparametric*.[9]

Geometrically, (5.7) implies that the level sets of f are equidistant, i.e., the corresponding partition of M is trasnormal (recall Definition 5.1), while (5.8) implies that the regular level sets have constant mean curvature:

Exercise 5.49. Let $f: M \to \mathbb{R}$ be a transnormal function satisfying (5.7). Let $I = (c_0, c_1)$ be an interval without singular values of f, and set $N := f^{-1}(I)$.

(i) Show that $f: N \to \left(I, \frac{1}{b}\mathrm{d}x^2\right)$ is a Riemannian submersion;
(ii) Prove that integral curves of $\operatorname{grad} f$, parametrized to have unit speed, are geodesics;
(iii) Let $\gamma(t)$ be one such geodesic and set $c(t) := f(\gamma(t))$. Denote by H_t the mean curvature of the hypersurface $f^{-1}(c(t)) \subset N$ at $\gamma(t)$. Prove that:

$$\Delta f(\gamma(t)) = (f \circ \gamma)''(t) - (f \circ \gamma)'(t) H_t, \qquad (5.9)$$

with the conventions that $\Delta f = \operatorname{tr}(\operatorname{Hess} f)$ is the (nonpositive) Laplacian and the mean curvature of a hypersurface is given by (2.9). Conclude that the hypersurfaces $f^{-1}(c(t))$ have constant mean curvature if f is isoparametric.

It is natural to wonder which of the above geometric properties of *regular* level sets remain true for *critical* level sets. This is answered by the following result of Wang [224].

---

[9] According to Thorbergsson [209], the term *isoparametric hypersurface* was introduced by Levi-Civita and $\|\operatorname{grad} f\|^2$ and Δf were called the *differential parameters* of f. We stress that, in modern references, the term *isoparametric hypersurface* is used for level sets of isoparametric functions, see Remark 5.58.

Theorem 5.50. *Let M be a connected complete Riemannian manifold and $f : M \to \mathbb{R}$ be a transnormal function.*

 (i) *The critical level sets of f are submanifolds of M;*
 (ii) *The focal manifolds of regular level sets are components of critical level sets of f;*
(iii) *There are at most two critical level sets, and each regular level set is a geometric tube over the latter.*

Remark 5.51. It also follows from the proof of the above result that the regular level sets are equifocal. Miyaoka [161] has recently improved this by showing that if M admits a transnormal function, then either M is diffeomorphic to a vector bundle over a submanifold, or M is diffeomorphic to a union of disk bundles over two submanifolds. This is analogous to a property of orbits of a cohomogeneity one action, see Proposition 6.33.

Remark 5.52. Interesting aspects of submanifold geometry of the level sets of isoparametric functions on compact manifolds were studied by Ni [174], Wang [224], and Ge and Tang [95, 96]. For example, it is proved that focal submanifolds of an isoparametric function are minimal, and that there exists at least one minimal hypersurface level set. These results are also analogous to properties of orbits of a cohomogeneity one action, see Remark 6.43, and can be alternatively deduced as an application of the mean curvature flow on polar foliations, see [17].

Let us now illustrate the above discussion of isoparametric functions with some standard examples.

Exercise 5.53. Consider the Euclidean space $\mathbb{R}^{n+1} = \{(x_1, \ldots, x_{n+1})\}$. Let

$$f(x) = x_1^2 + \cdots + x_{k+1}^2,$$

with $0 \le k \le n$. Check that $f : \mathbb{R}^{n+1} \to \mathbb{R}$ is an isoparametric function, such that:

 (i) $\|\operatorname{grad} f\|^2 = 4f^2$;
 (ii) $\Delta f = 2(k+1)$;
(iii) The level set $f^{-1}(c) = S^k(\sqrt{c}) \times \mathbb{R}^{n-k}$, $c > 0$, has principal curvatures $\lambda_1 = \frac{1}{c}$ and $\lambda_2 = 0$, with multiplicities k and $n - k$ respectively.

Exercise 5.54 ($\star$). Consider the unit round sphere $S^{n+1} = \{x_1^2 + \cdots + x_{n+2}^2 = 1\}$. Let

$$\widetilde{f}(x) := x_1^2 + \cdots + x_{k+1}^2,$$

with $0 \le k \le n$, and set $f := \widetilde{f}|_{S^{n+1}}$. Check that $f : S^{n+1} \to \mathbb{R}$ is an isoparametric function, such that:

 (i) $\|\operatorname{grad} f\|^2 = 4f - 4f^2$;
 (ii) $\Delta f = 2(k+1) - 2(n+2)f$;

(iii) The level set $f^{-1}(c) = S^k(\sqrt{c}) \times S^{n-k}(\sqrt{1-c})$, $0 < c < 1$, has principal curvatures $\lambda_1 = \frac{1-c}{\sqrt{c-c^2}}$ and $\lambda_2 = \frac{-c}{\sqrt{c-c^2}}$, with multiplicities k and $n-k$ respectively.

Remark 5.55. Let us make some geometric observations, in the particular case $k = 1$ and $n = 2$, about the foliation described in Exercise 5.54. This foliation of $S^3 \subset \mathbb{R}^4$ by level sets of $f(x) = x_1^2 + x_2^2$ is studied in detail in Example 6.48, from the cohomogeneity one viewpoint. Here, we note that the critical level sets $f^{-1}(1)$ and $f^{-1}(0)$ are respectively the great circles in S^3 obtained by intersecting S^3 with the linear subspaces $\mathbb{R}^2 = \{(x_1,x_2,0,0)\}$ and $\mathbb{R}^2 = \{(0,0,x_3,x_4)\}$. The points $(1,0,0,0) \in f^{-1}(1)$ and $(0,0,1,0) \in f^{-1}(0)$ are joined by the unit speed geodesic segment $\gamma(t) = (\cos t, 0, \sin t, 0)$, $0 \leq t \leq \frac{\pi}{2}$, in $S^3 \subset \mathbb{R}^4$, which is perpendicular to $f^{-1}(c)$ for each $0 \leq c \leq 1$ and is hence a section for this polar foliation. The value of f along $\gamma(t)$ is $c(t) := f(\gamma(t)) = \cos^2 t$, whose preimage is the torus $f^{-1}(c(t)) = S^1(\cos t) \times S^1(\sin t)$, with principal curvatures $\lambda_1 = \tan t$ and $\lambda_2 = -\cot t$. Observe that this also allows to compute its mean curvature $H_t = \lambda_1 + \lambda_2 = \tan t - \cot t$ from (5.9), using Exercise 5.54 (ii), and H_t vanishes only when $t = \frac{\pi}{4}$, see Remark 6.49.

All of the above observations have direct analogues for general k and n, regarding the foliation of S^{n+1} given by $f^{-1}(c(t)) = S^k(\cos t) \times S^{n-k}(\sin t)$, $0 \leq t \leq \frac{\pi}{2}$. The singular leaves S^k and S^{n-k} correspond respectively to $t = 0$ and $t = \frac{\pi}{2}$, and are the intersection of $S^{n+1} \subset \mathbb{R}^{n+2}$ with each of the summands in the orthogonal direct sum decomposition $\mathbb{R}^{n+2} = \mathbb{R}^{k+1} \oplus \mathbb{R}^{n-k+1}$, see Exercise 6.50.

Exercise 5.56 ($\star$). Consider the hyperbolic space $H^{n+1} = \{-x_1^2 + \cdots + x_{n+2}^2 = -1\}$ given by the hyperboloid model where the Riemannian metric is induced by the Lorentz metric $g_L = -dx_1^2 + \cdots + dx_{n+2}^2$ on $\mathbb{R}^{n+2}$. Let

$$\widetilde{f}(x) := -x_1^2 + \cdots + x_{k+1}^2,$$

with $2 \leq k \leq n$, and set $f := \widetilde{f}|_{H^{n+1}}$. Check that $f \colon H^{n+1} \to \mathbb{R}$ is an isoparametric function, such that:

(i) $\|\text{grad } f\|^2 = 4f + 4f^2$;
(ii) $\Delta f = 2(k+1) + 2(n+2)f$;
(iii) The level set $f^{-1}(-c) = H^k(\sqrt{c}) \times S^{n-k}(\sqrt{c-1})$, $c > 1$, has principal curvatures $\lambda_1 = \frac{c-1}{\sqrt{c^2-c}}$ and $\lambda_2 = \frac{c}{\sqrt{c^2-c}}$, with multiplicities k and $n-k$ respectively.

Remark 5.57. Similarly to Remark 5.55, consider the particular case $k = 2$ and $n = 3$ in Exercise 5.56, which gives a foliation of $H^4 \subset (\mathbb{R}^5, g_L)$ by level sets of $f(x) = -x_1^2 + x_2^2 + x_3^2$. Note that the critical level set $f^{-1}(-1)$ is the totally geodesic H^2 in H^3 obtained by intersecting H^3 with the linear subspace $\mathbb{R}^3 = \{(x_1,x_2,x_3,0,0)\}$. The unit speed geodesic $\gamma(t) = (\cosh t, 0, 0, \sinh t, 0)$, $t \geq 0$, in $H^4 \subset \mathbb{R}^5$ is perpendicular to $f^{-1}(-c)$ for each $c \geq 1$ and is hence a section for this polar foliation. The value of f along $\gamma(t)$ is $f(\gamma(t)) = -\cosh^2 t$, so that setting $c(t) := \cosh^2 t$, $t \geq 0$, the preimage $f^{-1}(-c(t)) = H^2(\cosh t) \times S^1(\sinh t)$ has principal curvatures $\lambda_1 = \tanh t$ and $\lambda_2 = \coth t$. Observe that this also allows to compute its mean curvature $H_t = \lambda_1 + \lambda_2 = \tanh t + \coth t$ from (5.9), using Exercise 5.56 (ii).

All of the above observations have direct analogues for general k and n, regarding the foliation of H^{n+1} given by $f^{-1}(-c(t)) = H^k(\cosh t) \times S^{n-k}(\sinh t)$, $t \geq 0$. The singular leaf H^k corresponds to $t = 0$, and is the intersection of $H^{n+1} \subset (\mathbb{R}^{n+2}, g_L)$ with the first summand in the direct sum decomposition $\mathbb{R}^{n+2} = \mathbb{R}^{k+1} \oplus \mathbb{R}^{n-k+1}$. These are also orbits of a cohomogeneity one action on H^{n+1}, see Remark 6.51.

Remark 5.58. Note that regular level sets of the isoparametric functions on space forms considered in Exercises 5.53, 5.54 and 5.56 are *isoparametric hypersurfaces* in the classical sense, i.e., hypersurfaces with constant principal curvatures. Cartan [62] proved that hypersurfaces in space forms are isoparametric if and only if they are regular level sets of isoparametric functions, see also Remark 4.16. On the other hand, Wang [223] found an example of isoparametric function f on a complex projective space, whose regular level sets do not have constant principal curvatures. As pointed out above, in modern references, the term *isoparametric submanifold* is used for this more general concept of regular level sets of an isoparametric function.

Wang [224] also remarked that level sets of transnormal functions on $\mathbb{R}^{n+1}$ and S^{n+1} are isoparametric hypersurfaces. This follows from equifocality of the regular level sets and Proposition 5.5. In general, however, transnormal functions are not necessarily isoparametric. Consider, for instance, $f: \mathbb{R}^{n+1} \to \mathbb{R}$ given by $f(x) = \cos(\|x\|)$. A direct computation gives $\|\text{grad } f\|^2 = 1 - f^2$, so f is transnormal; and $\Delta f = -f + (1-n)\sin(\|x\|)/\|x\|$, so f is not isoparametric. Moreover, note that the level sets of f are the same as those of $g(x) = \|x\|^2$, which is isoparametric (see Example 5.53).

Remark 5.59. The above is not true for hyperbolic spaces, i.e., there exist transnormal functions on H^{n+1} whose level sets are not isoparametric hypersurfaces. To see this, let N be a nonisoparametric hypersurface in H^{n+1}, with principal curvatures between -1 and 1. Let N_t be the parallel hypersurface to N at distance t, which is well-defined for all $t > 0$, since N does not have focal points. The function $f: H^{n+1} \to \mathbb{R}$ defined by $f(N_t) = t$ is transnormal, giving the desired example.

We now go in the direction of generalizing the above discussion to higher codimension, i.e., looking for a notion of map $H: M \to \mathbb{R}^q$ whose regular level sets are leaves of a polar foliation of codimension q. Inspired by Definition 5.48 and Exercise 5.49, we can infer some properties that this map should fulfill. If M_0 is the principal stratum (i.e., the set of leaves with trivial holonomy), then $H|_{M_0}: M_0 \to (\mathbb{R}^q, \tilde{g})$ should be a Riemannian submersion, where $\tilde{g}$ is an appropriate metric on $\mathbb{R}^q$. Like in Exercise 5.49, it should be possible to write $\tilde{g}$ in terms of the gradients of the components h_i of the map H. Since the regular level sets should be leaves of a polar foliations, the distribution given by their normal spaces ought to be integrable; in particular, near the regular level sets, the set spanned by grad h_i should be integrable. Therefore, we should expect some condition on the brackets [grad h_i, grad h_j], by the Frobenius Theorem A.19. Altogether, it is natural to expect that our candidate $H: M \to \mathbb{R}^q$ to a *transnormal map* is *an integrable Riemannian submersion with singularities*. In addition, if we want H to be *isoparametric*, i.e., the mean curvature vector of regular level sets to be basic and have constant length, then some condition similar to (5.8) should hold. The above discussion leads to the following definition of transnormal and isoparametric maps, introduced by Harle [122] and Terng [204]:

Definition 5.60. Let M^{n+q} be a complete Riemannian manifold. A smooth map $H = (h_1, \ldots, h_q): M^{n+q} \to \mathbb{R}^q$ is called *transnormal* if

(i) H has a regular value;
(ii) For each regular value c, there exist a neighborhood V of $H^{-1}(c)$ in M and smooth functions b_{ij} on $H(V)$ such that, for every $x \in V$,

$$\langle \operatorname{grad} h_i(x), \operatorname{grad} h_j(x) \rangle = (b_{ij} \circ H)(x); \qquad (5.10)$$

(iii) There exists a sufficiently small neighborhood of each regular level set such that [grad h_i, grad h_j] is a linear combination of grad $h_1, \ldots$, grad h_q, with coefficients that are functions of H, for all i and j.

Moreover, a transnormal map H is said to be *isoparametric* if V can be chosen to be the whole manifold M and $\Delta h_i = a_i \circ H$, where a_i are smooth functions.

Exercise 5.61. Show that a map $H = (h_1, \ldots, h_q): M^{n+q} \to \mathbb{R}^q$ is *transnormal* if and only if it has a regular value, and for each regular value c, there exists a neighborhood V of $H^{-1}(c)$ in M such that $H|_V: V \to H(V)$ is an integrable Riemannian submersion, where the metric of $H(V)$ is given by the inverse matrix of $B = (b_{ij})$.

Exercise 5.62. Let $H: \mathrm{SU}(3) \to \mathbb{C} \cong \mathbb{R}^2$ be the trace map $H(A) := \operatorname{tr}(A)$. Consider the foliation $\mathscr{F}$ given by the partition of $\mathrm{SU}(3)$ into the orbits of the conjugation action. Check that leaves of $\mathscr{F}$ are preimages of H and that H is a transnormal map.

A submanifold in a space form is isoparametric if and only if it is the preimage of some isoparametric map, see [182]. Therefore, it is natural to ask under which conditions this equivalence holds in the general case of Riemannian manifolds. In the rest of this section we deal with this question.

First, we present a generalization of Theorem 5.50, due to Alexandrino [3], for analytic transnormal maps on analytic Riemannian manifolds (e.g., symmetric spaces).

Theorem 5.63. *Let $H: M \to \mathbb{R}^q$ be an analytic transnormal map on a real analytic complete Riemannian manifold M. Let c be a regular value and $L \subset H^{-1}(c)$ be a connected component of $H^{-1}(c)$. Denote by Ξ the set of all parallel normal fields along L.*

(i) $\mathcal{F}_{c,L} = \{\eta_\xi(L)\}_{\xi \in \Xi}$ *is a polar foliation with embedded leaves;*

(ii) *For each regular value $\widehat{c}$, the connected components of $H^{-1}(\widehat{c})$ are equifocal submanifolds and leaves of $\mathcal{F}_{c,L}$;*

(iii) *$\mathcal{F}_{c,L}$ is independent of the choice of c and L, i.e., for another regular value $\widetilde{c}$ and connected component $\widetilde{L} \subset H^{-1}(\widetilde{c})$, $\mathcal{F}_{c,L} = \mathcal{F}_{\widetilde{c},\widetilde{L}}$.*

Proof (Sketch). Let us briefly sketch the proofs of (i) and (ii), from [3]. Consider the submanifold L defined in the theorem. The first step is to construct a candidate section Σ, which is a totally geodesic submanifold orthogonal to L and to the regular level sets of H. This can be done using the analyticity of the map H and arguments presented in the proof of Theorem 5.24.

The second step is to check that L is an equifocal submanifold. In other words, for a parallel normal field ξ along L and $p \in L$, we have to show that the rank of $d\eta_\xi$ is constant on a neighborhood U of p in L. The main idea is to show that the focal distances and the multiplicity of focal points are constant, using a Morse index argument.

More precisely, set $g(x,t) := h_i \circ \eta_{t\xi}(p) - h_i \circ \eta_{t\xi}(x)$. The map H is an integrable Riemannian submersion on a neighborhood of the regular level set, and ξ is a projectable field of this submersion. Therefore $g(x,t) = 0$ for every small t and for all $x \in L$. Since $g(x,t)$ is analytic, $g(x,t) = 0$ for all t. Thus, we proved that $\eta_{t\xi}(L) \subset H^{-1}(d)$. Using the fact that H is a Riemannian submersion with singularities, one can also prove the following:

Claim 5.64. *Let $L_t := \eta_{t\xi}(L)$ and d be the value such that $L_t \subset H^{-1}(d)$. Then $\eta_{t\xi} : L \to L_t$ is a diffeomorphism when d is a regular value.*

Due to Claim 5.64, we can suppose without loss of generality that p belongs to a convex normal neighborhood V of $\eta_\xi(p)$. For a small neighborhood $U \subset L_p$ of p, we can define, for each $x \in U$, the geodesic segment $\gamma_x(t) := \exp_x(t\xi), t \in [-\varepsilon, 1 + \varepsilon]$. Note that $\gamma_x(t) \in V$ if U and $\varepsilon > 0$ are sufficiently small.

Using the Jacobi equation and the fact that Σ is totally geodesic, it is possible to prove that the focal points of U along $\gamma_x(t)$ are of tangential type, i.e., come from variations of basic vector fields. We then have

$$m(\gamma_x) = \sum_i \dim \ker d\eta_{t_i(x)\xi}(x), \tag{5.11}$$

where $t_i(x)$ denotes the distance between the focal points on γ_x and x, and $m(\gamma_x)$ denotes the number of focal points on γ_x counted with multiplicities. Since H is analytic, the critical points of H on γ_x are isolated and, without loss of generality, we may assume that $\gamma_p(1)$ is the only critical point of H on γ_p. It follows from the above claims that the map $\eta_{t\xi}$ may only fail to be a diffeomorphism if $t = 1$. Therefore, (5.11) simplifies to $m(\gamma_x) = \dim \ker d\eta_\xi(x)$. On the other hand, from the Morse Index Theorem, $m(\gamma_x) \geq m(\gamma_p)$, for all x in some neighborhood of p in L. Thus, since $\dim \ker d\eta_\xi(x) \leq \dim \ker d\eta_\xi(p)$, we conclude that $\dim \ker d\eta_\xi$ is constant on U, completing the second step of the proof.

The third and last step is to prove that the partition $\mathscr{F} := \{\eta_\xi(L)\}_{\xi \in \Xi}$ is a polar foliation. It is interesting to note that in this part we do not need analyticity of M or the existence of H, but only that L is equifocal and Σ has some special properties, see Remark 5.66 below.

Since the derivative of the endpoint map has constant rank, $\eta_\xi(U)$ is an embedded manifold (for small U). These small submanifolds are the plaques of the foliation. In what follows, we claim that part of the Slice Theorem holds for these plaques:

Claim 5.65. Let S_x be the slice at $x \in \eta_\xi(U)$. Then

$$S_x = \bigcup_{\sigma \in \Lambda(x)} \sigma,$$

where $\Lambda(x)$ is the set of local sections that contain x.

Claim 5.65 almost completes the proof that $\mathscr{F}$ is a polar foliation. In fact, for another $\eta_{\tilde{\xi}}(\widetilde{U})$ and for $\tilde{x} \in \eta_{\tilde{\xi}}(\widetilde{U}) \cap S_x$, it can be used to show that $S_{\tilde{x}} \subset S_x$. From this, one can concludes that $\eta_{\tilde{\xi}}(\widetilde{U})$ is contained in the boundary of a tube over $\eta_\xi(U)$. In other words, we conclude that $\eta_{\tilde{\xi}}(\widetilde{U})$ and $\eta_\xi(U)$ are equidistant, which implies that $\mathscr{F}$ is a transnormal system. Finally, $S_{\tilde{x}} \subset S_x$ implies that $v(S_x) \subset v(S_{\tilde{x}})$, hence $\mathscr{F}$ is a singular foliation. $\qquad \square$

Remark 5.66. As remarked in [5], the above proof can be slightly adapted to give an alternative proof of a result of Töben [213], that can be reformulated as follows. Let L be a closed embedded equifocal submanifold, and suppose that the subset of regular points[10] of each section Σ is open and dense. Define Ξ as the set of all parallel normal fields along L. Then $\mathscr{F} := \{\eta_\xi(L)\}_{\xi \in \Xi}$ is a polar foliation.

It is also known (see Carter and West [68] and Terng [204]) that given an isoparametric submanifold L of $\mathbb{R}^{n+k}$, one can construct a polynomial isoparametric map from $\mathbb{R}^{n+k}$ into $\mathbb{R}^k$ that has L as a level set. Therefore, it is natural to look for conditions under which a polar foliation can be described with preimages of transnormal maps. It follows from Theorem 5.27 and from the classical theory of isoparametric foliations that this is always locally true, as stated in the following observation of Alexandrino [4].

Proposition 5.67. *Let $\mathscr{F}$ be a polar foliation on a complete Riemannian manifold M. Each $p \in M$ has a neighborhood U such that the plaques of $\mathscr{F} \cap U$ are preimages of a transnormal map.*

Since there are examples of polar foliations with nontrivial holonomy (see Alexandrino [4]), there are examples of polar foliations that are not given as

---

[10]Here, a point of a section is called regular if there exists only one local section that contains p, i.e., if two local sections σ and $\tilde{\sigma}$ contain p, then they have the same germ at p.

preimages of transnormal maps. The next result, by Alexandrino and Gorodski [13] and Alexandrino [7], gives sufficient conditions under which a polar foliation can be described by a transnormal map. Thus, it can be considered as a converse to Theorem 5.63.

Theorem 5.68. *Let $\mathscr{F}$ be a polar foliation on a complete simply-connected Riemannian manifold M. Assume that leaves of $\mathscr{F}$ are closed and embedded, and $\mathscr{F}$ admits a flat section of dimension n. Then the leaves of $\mathscr{F}$ are given by the level sets of a transnormal map $H: M \to \mathbb{R}^n$.*

Remark 5.69. Heintze, Liu and Olmos [124] proved the above result with the additional assumption that M is a simply-connected symmetric space of compact type.

If we allow the image of H to be contained in a general manifold instead of $\mathbb{R}^n$, we have the following more general result, that can be found in [10, 12].

Theorem 5.70. *Let $\mathscr{F}$ be a singular Riemannian foliation on a simply-connected complete Riemannian manifold M. Assume that $M/\mathscr{F}$ is a good orbifold Σ/W, e.g., if $\mathscr{F}$ is polar. Then there exists a smooth map $H: M \to \Sigma$ such that the leaves of $\mathscr{F}$ coincide with the level sets of H, and such that $H(M)$ is isometric to $M/\mathscr{F}$.*

Remark 5.71. The above result was first proved for the case of polar foliations in Alexandrino [10], and as remarked in Alexandrino, Briquet and Töben [12], one can easily use pseudogroups to obtain the above formulation.

Proof (Sketch). For the sake of brevity, we sketch the proof when $\mathscr{F}$ is a polar foliation. In what follows, we use the notation introduced in the proof of Theorem 5.33. The first step is to construct a $\widetilde{\Gamma}$-equivariant map $G: \widetilde{\Sigma} \to \widetilde{\Sigma}$ with special properties. More precisely, let $\widetilde{\Sigma}$ be the universal Riemannian covering space of a section Σ and $\widetilde{\Gamma}$ be the lifted reflection group. Fix a chamber $C \subset \widetilde{\Sigma}$ and denote by $\pi: \widetilde{\Sigma} \to \widetilde{\Sigma}/\widetilde{\Gamma} = C$ the canonical projection. Then there exists a smooth $\widetilde{\Gamma}$-equivariant map $G: \widetilde{\Sigma} \to \widetilde{\Sigma}$ such that

(i) $G|_C: C \to C$ is a homeomorphism;
(ii) If K is a connected component of a set of points with the same orbit types for the $\widetilde{\Gamma}$-action on $\widetilde{\Sigma}$, then the restriction $G|_K$ is a diffeomorphism;
(iii) The composition $G \circ \pi: \widetilde{\Sigma} \to \widetilde{\Sigma}$ is smooth.

An easy example that illustrates the above claim is given by $\widetilde{\Sigma} = \mathbb{R}$, $\widetilde{\Gamma} = \{\pm \mathrm{id}\}$, $C = [0, \infty)$, $G(x) := h(|x|)\frac{x}{|x|}$ $(x \neq 0)$ and $G(0) = 0$, for an appropriate choice of function h. Let us outline the proof of this claim, without going into details. First, one defines a new metric $\widetilde{g}$ such that normal slices to the strata are flat and totally geodesic. Then the desired map G can be defined as the composition of $\widetilde{\Gamma}$-equivariant maps F_i that are constructed using the exponential map with respect to $\widetilde{g}$. Due to the properties of $\widetilde{g}$, it is possible to verify that G is smooth.

We now return to the proof of the main result. If follows from (5.6) that, if $\pi_1(M)$ is trivial, then $\widetilde{W} = \widetilde{\Gamma}$ and $W(\Sigma) = \widetilde{\Gamma}/\pi_1(\Sigma)$. This fact and the existence of the map

G imply that there exists a smooth $W(\Sigma)$-invariant map $H\colon \Sigma \to \Sigma$ such that the following diagram commutes:

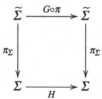

where $\pi_\Sigma\colon \widetilde{\Sigma} \to \Sigma$ is the Riemannian covering map. Due to a result of Alexandrino and Gorodski [13] concerning extensions of basic functions and forms, H admits a smooth extension $\widetilde{H}\colon M \to \Sigma$ that turns out to be the desired smooth $\mathscr{F}$-invariant map. $\qquad\Box$

For the particular case of polar actions, the above theorem can be reformulated as follows.

Corollary 5.72. *Consider a polar action on a simply-connected complete Riemannian manifold M. Then there exists an isoparametric map $H\colon M \to \Sigma$, where Σ is a section, such that the orbits of the action coincide with the level sets of H. In addition, $H(M)$ is a fundamental domain for the action of the Weyl group on Σ.*

Remark 5.73. Recently Radeschi [189] used Clifford systems to construct infinitely many new examples of nonhomogenous (and nonpolar) singular Riemannian foliations on spheres S^{2l-1}, that are preimages of maps $H\colon S^{2l-1} \to \mathbb{R}^{m+1}$.

5.6 Perspectives

In this last section, we survey on some other results regarding singular Riemannian foliations, and future research perspectives in the area.

We start by briefly discussing a few others classes of singular Riemannian foliations. Lytchak and Thorbergsson [156] introduced the class of singular Riemannian foliation without horizontal conjugate points, in order to generalize variationally complete isometric actions. A singular Riemannian foliation $\mathscr{F}$ is said to be *without horizontally conjugate points*, or simply *variationally complete*, if for every leaf L, every L-Jacobi field tangential to a leaf other than L is vertical. They proved that a singular Riemannian foliation without horizontal conjugate points in a complete Riemannian manifold with $\sec \geq 0$ admits flat sections, i.e., are hyperpolar foliations. Subsequently, the same authors proved in [157] that the quotient space of a variationally complete group action is a good Riemannian orbifold. This result was also proved to hold for singular Riemannian foliations without horizontal conjugate points.

In the same paper, the concept of infinitesimally polar foliation is introduced. A singular Riemannian foliation $\mathscr{F}$ is called *infinitesimally polar* if the intersection of $\mathscr{F}$ with each slice is diffeomorphic to a polar foliation on an Euclidean space, i.e., an isoparametric foliation. In fact, this diffeomorphism is given as composition of the exponential map with a linear map. It is also proved in [157] that $\mathscr{F}$ is an infinitesimally polar foliation if and only if, for each point $x \in M$, there exists a neighborhood U of x such that the leaf space of the restricted foliation $\mathscr{F}|_U$ is an orbifold. Since the leaf space of an isoparametric foliation is a Coxeter orbifold, they concluded that the leaf space $U/(\mathscr{F}|_U)$ is an orbifold if and only if it is a Coxeter orbifold. They also proved that being infinitesimally polar is equivalent to having locally bounded transverse curvature.

Typical examples of infinitesimally polar foliations are polar foliations (see [4]), singular Riemannian foliations without horizontal conjugate points (see [156, 157]) and singular Riemannian foliations of codimension 1 or 2 (see [157]). In particular, a partition into orbits of an isometric action is infinitesimally polar if the action is polar, or variationally complete, or has cohomogeneity ≤ 2.

Lytchak [154] continued to study infinitesimally foliations and generalized the blow up introduced by Töben [213], proving that a singular Riemannian foliation admits a resolution preserving the transverse geometry if and only if it is infinitesimally polar. He also deduced that singular Riemannian foliations on simply-connected manifolds that either have sections or no horizontal conjugate points have closed leaves.

We now discuss a few results about general singular Riemannian foliations. An important conjecture regarding open nonnegatively curved spaces has been recently proved by Wilking [230], using Riemannian foliations. More precisely, the *dual foliation* of a singular Riemannian foliation is used to show that the metric projection onto the soul, also called Sharafutdinov projection, is smooth. In the same paper, Wilking proved that the dual foliation to a singular Riemannian foliation $\mathscr{F}$ of a manifold with sec > 0 has only one leaf. In other words, any two points can be joined by a piecewise smooth horizontal curve. Wilking's dualization was also used in the above mentioned work of Lytchak [155]; recall Theorem 5.38.

An important question in the theory of singular Riemannian foliations is the classification of all such objects in the Euclidean space or the sphere. In spheres, Riemannian foliations were classified by Grove and Gromoll [108] when leaves have codimension ≤ 3, and singular Riemannian foliations with leaves of dimension ≤ 3 were classified by Radeschi [188].

As mentioned above, the concept of equifocality is also valid for singular Riemannian foliations. In fact, Alexandrino and Töben [20] proved that the regular leaves of a singular Riemannian foliation are Bott equifocal, i.e., the endpoint map of a normal foliated vector field has constant rank. This implies that it is possible to reconstruct the singular foliation by taking all parallel submanifolds to a regular leaf with trivial holonomy. In addition, the endpoint map of a normal foliated vector field on a leaf with trivial holonomy is a covering map.

Equifocality is a very important concept. It was used in Alexandrino [8] to prove that each singular Riemannian foliation $\mathscr{F}$ on a compact manifold M can be desingularized to a Riemannian foliation $\widehat{\mathscr{F}}$ of $\widehat{M}$ after successive blow ups, generalizing a result of Molino [164]. It was remarked by Lytchak that if the desingularization preserves transverse geometry, then the foliation should be an infinitesimal foliation. One sees in [8] that *part* of the transverse geometry is preserved. Using this fact, it was proved in [8] that if the leaves of $\mathscr{F}$ are compact, then for each small $\varepsilon > 0$, the regular foliation $\widehat{\mathscr{F}}$ can be chosen so that the desingularization map $\widehat{\rho} : \widehat{M} \to M$ that projects leaves of $\widehat{\mathscr{F}}$ to leaves of $\mathscr{F}$ induces an ε-isometry between $\widehat{M}/\widehat{\mathscr{F}}$ and $M/\mathscr{F}$. This implies that $\widehat{M}/\widehat{\mathscr{F}}$ is a Gromov-Hausdorff limit of a sequence $\widehat{M}_\varepsilon/\widehat{\mathscr{F}}_\varepsilon$ of Riemannian orbifolds, that is, $\text{dist}_{GH}\left(\widehat{M}_\varepsilon/\widehat{\mathscr{F}}_\varepsilon, M/\mathscr{F}\right) < \varepsilon$. It actually gives more information, since the desingularization induces a map $\rho_\varepsilon : \widehat{M}_\varepsilon/\widehat{\mathscr{F}}_\varepsilon \to M/\mathscr{F}$ that is smooth, i.e., pull-back of basic smooth functions[11] are basic smooth functions. We recall that, in general, the ε-isometries used in Gromov-Haudsdorff convergence are not necessarily continuous.

As proved by Alexandrino and Radeschi [16], the desingularization turns out to be an important tool to study the smooth structure of leaf spaces. In fact, desingularization is used in [16] to prove smoothness of isometric flows in orbit spaces. This result implies that isometric actions on compact orbit spaces are smooth, indicating the possibility of developing a theory of actions on quotient spaces, in the spirit of Chap. 3. In the same paper, smoothness of isometric flows in the quotient was used to give a positive answer to the Molino Conjecture 5.45 in the case of orbit-like foliations, which is the class of singular Riemannian foliations originally studied by Molino [165]. For more information on the smooth structure of leaf spaces see Alexandrino and Lytchak [15] and Alexandrino and Radeschi [18], where an analogue of the Myers-Steenrod Theorem 2.12 for the leaf spaces is proved.

Finally, we mention that recent progress regarding *tautness* of singular foliations was made by Wiesendorf [228]. Among other results, it is shown that tautness of a singular Riemannian foliation is a property of the leaf space.

Further details on singular Riemannian foliations, especially regarding nonpolar foliations, can be found in Alexandrino, Briquet and Töben [12], while more on (regular) Riemannian foliations and Riemannian submersions can be found in the books of Gromoll and Walschap [105] and Molino [165].

---

[11] Here, a basic function is a function constant along the leaves.

Chapter 6
Low Cohomogeneity Actions and Positive Curvature

In this chapter, we study special types of isometric group actions on compact Riemannian manifolds, namely those with *low cohomogeneity*. The *cohomogeneity* of an action is the codimension of its principal orbits, and can also be regarded as the dimension of its orbit space. Low cohomogeneity is an indication that there are few orbit types, and that the original space has many symmetries. In this situation, it is possible to study many geometric features that are not at reach in the general case. Throughout this chapter, we emphasize connections between the geometry and topology of manifolds with low cohomogeneity stemming from curvature positivity conditions, such as positive (sec > 0) and nonnegative (sec ≥ 0) sectional curvature.

After briefly introducing a deformation technique that is used throughout this chapter, we start with the case of *cohomogeneity zero*, that corresponds to homogeneous spaces. We give a quick introduction with examples, stating the classification of positively curved closed simply-connected homogeneous spaces. We then proceed to *cohomogeneity one*, studying the basic structure and exploring it via many examples (ranging from compact rank one symmetric spaces to exotic Kervaire spheres). Finally, in the last section, we give a survey of how manifolds with low cohomogeneity can be used to better understand the notions of positive and nonnegative sectional curvature, in the spirit of the so-called Grove program.

6.1 Cheeger Deformation

This section contains some preliminary material on a metric deformation technique that is a very important tool in the remainder of this chapter. This deformation consists of rescaling a Riemannian metric in the directions tangent to the orbits of an isometric action, generalizing the construction of *Berger spheres* (see Example 6.16). This method was developed by Cheeger [70] to construct metrics of nonnegative sectional curvature on the connected sum of two compact rank one

© Springer International Publishing Switzerland 2015
M.M. Alexandrino, R.G. Bettiol, *Lie Groups and Geometric Aspects of Isometric Actions*, DOI 10.1007/978-3-319-16613-1_6

symmetric spaces, and was later systematically studied by Müter in his PhD thesis [172]. A summary of the results in [172] is provided by Ziller [234, 236]. In what follows, we discuss some of these key results. Since its inception, this technique has proved of foundational importance and continues to influence research in the area, see [34–36, 115, 184, 192, 194].

Let (M, g) be a Riemannian manifold and G a compact Lie group acting isometrically on M. Fix a bi-invariant metric Q on G, and endow $M \times G$ with the product metric $\left(g \oplus \frac{1}{t} Q\right)$. Note that there is a free isometric G-action on $M \times G$ given by:

$$h \cdot (p, g) := (hp, hg), \quad p \in M, \ g, h \in G. \tag{6.1}$$

The orbit space of this action is easily seen to be diffeomorphic to M. Namely, consider the principal G-bundle $G \to G \to \{e\}$ whose base is a point. Using the G-action on M, we may construct the associated bundle $M \times_G G$ to this principal G-bundle (see Theorem 3.51), which is precisely the orbit space of (6.1). Since $M \times_G G$ is a fiber bundle whose fiber is M and base is a point, we have the desired diffeomorphism, cf. Exercise 3.56. The quotient map can be explicitly computed to be the submersion:

$$\rho : M \times G \to M, \quad \rho(p, g) = g^{-1} p. \tag{6.2}$$

Since the G-action (6.1) is free and isometric, by Exercise 3.81, there is a unique Riemannian metric g_t on M such that $\rho \colon \left(M \times G, \left(g \oplus \frac{1}{t} Q\right)\right) \to (M, g_t)$ is a Riemannian submersion. Note that the original G-action on (M, g) is also isometric on (M, g_t) for $t > 0$. Indeed, the isometric G-action on $\left(M \times G, \left(g \oplus \frac{1}{t} Q\right)\right)$ given by

$$k * (p, g) := \left(p, g k^{-1}\right), \quad p \in M, \ g, k \in G \tag{6.3}$$

commutes with (6.1), and hence descends to an isometric G-action on the corresponding orbit space (M, g_t), see Exercise 3.40. Observe that $k\rho(p, g) = kg^{-1} p = \rho(k * (p, g))$, hence the G-action induced by (6.3) is the original G-action on M.

The 1-parameter family of metrics g_t on M varies smoothly with t, and extends smoothly to $t = 0$, with $g_0 = g$, see Proposition 6.3. Thus, $g_t, t \geq 0$, is a deformation of g by other G-invariant metrics on M, called the *Cheeger deformation* of g. One of the main reasons why Cheeger deformations are useful is that, in addition to preserving the isometric G-action, they preserve the curvature condition $\sec \geq 0$.

Proposition 6.1. *If* $\sec_{g_0} \geq 0$, *then* $\sec_{g_t} \geq 0$, *for all* $t \geq 0$.

Proof. From Proposition 2.26, we have that (G, Q) has $\sec \geq 0$. Thus, if $\sec_{g_0} \geq 0$, then $M \times G$ with the product metric $\left(g \oplus \frac{1}{t} Q\right)$ also has $\sec \geq 0$. From the Gray-O'Neill formula (2.14), it follows that also $\sec_{g_t} \geq 0$, for all $t \geq 0$.   □

Remark 6.2. Although Cheeger deformations preserve $\sec \geq 0$, they might not preserve other lower curvature bounds such as $\sec \geq k > 0$, see Example 6.5.

In order to have more geometric intuition on Cheeger deformations, we represent g_t in terms of g, to analyze how the metrics evolve, following the formalism of Müter [172] and Ziller [234]. For each $p \in M$, denote by $\mathfrak{g}_p$ the Lie algebra of the isotropy G_p. Fix the Q-orthogonal splitting $\mathfrak{g} = \mathfrak{g}_p \oplus \mathfrak{m}_p$, and identify $\mathfrak{m}_p$ with the tangent space $T_p G(p)$ to the G-orbit through p via action fields, i.e., we identify each $X \in \mathfrak{m}_p$ with

$$X_p^* = \tfrac{\mathrm{d}}{\mathrm{d}s} \exp(sX)p\big|_{s=0} \in T_p G(p),$$

recall (3.11). This determines a g-orthogonal splitting $T_p M = \mathscr{V}_p \oplus \mathscr{H}_p$, as described in (3.19) and (3.20), in *vertical* and *horizontal* spaces, respectively

$$\mathscr{V}_p := T_p G(p) = \{X_p^* : X \in \mathfrak{m}_p\} \quad \text{and} \quad \mathscr{H}_p := \{v \in T_p M : g(v, \mathscr{V}_p) = 0\}. \quad (6.4)$$

It is not difficult to check that this g-orthogonal splitting is also g_t-orthogonal for all t, that is, vertical (respectively horizontal) directions of g are also vertical (respectively horizontal) for g_t, for all $t > 0$. For each $t \geq 0$, there is a unique Q-symmetric automorphism $P_t \colon \mathfrak{m}_p \to \mathfrak{m}_p$ such that

$$Q(P_t(X), Y) = g_t(X_p^*, Y_p^*), \quad X, Y \in \mathfrak{m}_p. \quad (6.5)$$

In addition, let $C_t \colon T_p M \to T_p M$ be the g-symmetric automorphism that relates the original metric g and its Cheeger deformation g_t, i.e., such that

$$g(C_t(X), Y) = g_t(X, Y), \quad X, Y \in T_p M. \quad (6.6)$$

We now give an explicit formula for these automorphisms in terms of P_0.

Proposition 6.3. *The symmetric automorphisms $P_t \colon \mathfrak{m}_p \to \mathfrak{m}_p$ and $C_t \colon T_p M \to T_p M$ for $t \geq 0$ are determined by P_0 in the following way:*

$$\begin{aligned}
P_t(X) &= (P_0^{-1} + t\,\mathrm{id})^{-1}(X) = P_0\,(\mathrm{id} + tP_0)^{-1}(X), & X &\in \mathfrak{m}_p, \\
C_t(X) &= \left((\mathrm{id} + tP_0)^{-1}(X_\mathfrak{m})\right)_p^* + X^{\mathscr{H}}, & X &\in T_p M,
\end{aligned} \quad (6.7)$$

where $X^{\mathscr{V}}$ and $X^{\mathscr{H}}$ are the vertical and horizontal parts of $X \in T_p M$ respectively, and $X_\mathfrak{m}$ is the unique vector in $\mathfrak{m}_p$ such that $(X_\mathfrak{m})_p^* = X^{\mathscr{V}}$.

Proof. It is easy to see that the derivative $\mathrm{d}\rho_{(p,e)} \colon T_p M \times \mathfrak{g} \to T_p M$ of (6.2) satisfies

$$\mathrm{d}\rho_{(p,e)}(X^*, Y) = X_p^* - Y_p^*, \quad X, Y \in \mathfrak{g}. \quad (6.8)$$

The vertical space[1] at (p,e) for ρ is formed by the pairs $(Z^*,Z) \in T_pM \times \mathfrak{g}$. Thus, the vector $(X^*,Y) \in T_pM \times \mathfrak{g}$ is $\left(\mathfrak{g} \oplus \frac{1}{t}Q\right)$-orthogonal to all pairs (Z^*,Z), and hence horizontal for ρ, if and only if for all $Z \in \mathfrak{g}$,

$$0 = g(X^*,Z^*) + \tfrac{1}{t}Q(Y,Z) = Q(P_0X_\mathrm{m},Z) + Q(\tfrac{1}{t}Y,Z) = Q\left(P_0X_\mathrm{m} + \tfrac{1}{t}Y,Z\right),$$

that is, $Y = -tP_0X_\mathrm{m}$. Here, for $X \in \mathfrak{g}$, we denote by X_m the component of X in $\mathfrak{m}_p$. In particular, the horizontal lift with respect to the Riemannian submersion ρ of $X^* \in T_pM$ is given by $(W^*, -tP_0W_\mathrm{m}) \in T_pM \times \mathfrak{g}$ where $W \in \mathfrak{g}$ satisfies the equation $X^* = d\rho_{(p,e)}(W^*, -tP_0W_\mathrm{m})$. Using (6.8), this becomes

$$X_p^* = W_p^* + (tP_0W_\mathrm{m})_p^* = \left((\mathrm{id} + tP_0)W_\mathrm{m}\right)_p^*, \tag{6.9}$$

to which $W = (\mathrm{id} + tP_0)^{-1}X \in \mathfrak{m}$ provides a solution. Therefore, the horizontal lift of $X^* \in T_pM$ is $\left(((\mathrm{id} + tP_0)^{-1}X)^*, -tP_0(\mathrm{id} + tP_0)^{-1}X\right)$. Notice that, since

$$P_0(\mathrm{id} + tP_0)^{-1} = P_0\left(P_0(P_0^{-1} + t\,\mathrm{id})\right)^{-1} = P_0(P_0^{-1} + t\,\mathrm{id})^{-1}P_0^{-1} = (P_0^{-1} + t\,\mathrm{id})^{-1},$$

the horizontal lift of $X^* \in T_pM$ can be rewritten as:

$$\left((P_0^{-1}(P_0^{-1} + t\,\mathrm{id})^{-1}X)_p^*, -t(P_0^{-1} + t\,\mathrm{id})^{-1}X\right) \in T_pM \times \mathfrak{g}. \tag{6.10}$$

Rewriting it in this way is convenient for the computations in the remainder of this proof, see (6.12). Analogously, the horizontal lift of a general vector $V \in T_pM$ that has vertical and horizontal parts $V = V^{\mathcal{V}} + V^{\mathcal{H}} = (V_\mathrm{m})_p^* + V_p^{\mathcal{H}} \in T_pM$, is given by

$$\left((P_0^{-1}(P_0^{-1} + t\,\mathrm{id})^{-1}V_\mathrm{m})_p^* + V_p^{\mathcal{H}}, -t(P_0^{-1} + t\,\mathrm{id})^{-1}V_\mathrm{m}\right) \in T_pM \times \mathfrak{g}, \tag{6.11}$$

since it is horizontal for ρ and mapped to V by $d\rho_{(p,e)}$.

Since $\rho \colon \left(M \times G, \left(\mathfrak{g} \oplus \frac{1}{t}Q\right)\right) \to (M, g_t)$ is a Riemannian submersion, the square length of $X^* \in T_pM$ with respect to g_t equals the square length of its horizontal lift (6.10) with respect to $\left(\mathfrak{g} \oplus \frac{1}{t}Q\right)$. Thus, we have that for all $X \in \mathfrak{m}$,

$$\begin{aligned}
g_t(X_p^*, X_p^*) &= g\left((P_0^{-1}(P_0^{-1} + t\,\mathrm{id})^{-1}X)_p^*, (P_0^{-1}(P_0^{-1} + t\,\mathrm{id})^{-1}X)_p^*\right) \\
&\quad + \tfrac{1}{t}Q\left(t(P_0^{-1} + t\,\mathrm{id})^{-1}X, t(P_0^{-1} + t\,\mathrm{id})^{-1}X\right) \\
&= Q\left((P_0^{-1} + t\,\mathrm{id})^{-1}X, P_0^{-1}(P_0^{-1} + t\,\mathrm{id})^{-1}X\right) \\
&\quad + tQ\left((P_0^{-1} + t\,\mathrm{id})^{-1}X, (P_0^{-1} + t\,\mathrm{id})^{-1}X\right)
\end{aligned}$$

---

[1] Note that vertical/horizontal spaces for the submersion $\rho \colon M \times G \to M$ are subspaces of $T_pM \times \mathfrak{g}$, while vertical/horizontal spaces for the G-action on M are subspaces of T_pM.

$$= Q\big((P_0^{-1}+t\,\mathrm{id})^{-1}X, P_0^{-1}(P_0^{-1}+t\,\mathrm{id})^{-1}X + t(P_0^{-1}+t\,\mathrm{id})^{-1}X\big)$$
$$= Q\big((P_0^{-1}+t\,\mathrm{id})^{-1}X, (P_0^{-1}+t\,\mathrm{id})(P_0^{-1}+t\,\mathrm{id})^{-1}X\big)$$
$$= Q\big((P_0^{-1}+t\,\mathrm{id})^{-1}X, X\big), \tag{6.12}$$

which, according to the definition (6.5), proves that $P_t = (P_0^{-1}+t\,\mathrm{id})^{-1}$. The formula for C_t follows immediately from the above and (6.6). □

Remark 6.4. Notice that the horizontal lift (6.11) of $V \in T_pM$ is quite complicated, while the horizontal lift of $C_t^{-1}V = \big((\mathrm{id}+tP_0)V_{\mathrm{m}}\big)_p^* + V^{\mathscr{H}}$ is equal to $(V, -tP_0V_{\mathrm{m}})$, which is much simpler to use for computations.

The above gives insight on how the geometry of g_t changes with t. If P_0 has eigenvalues λ_i, then C_t has eigenvalues $\frac{1}{1+t\lambda_i}$ corresponding to the vertical directions and eigenvalues 1 in the horizontal directions. In other words, *as t grows, the metric g_t shrinks in the direction of the orbits and remains unchanged in the orthogonal directions*. Note that the speed in which the orbits shrink may vary with the orbit.

An easy consequence of the above observations is that (M, g_t) converges in the Gromov-Hausdorff sense to the orbit space M/G as $t \nearrow +\infty$. In fact, horizontal directions remain unchanged, while vertical directions collapse. This observation is useful to study the geometry of orbit spaces of isometric actions, since it provides a natural approximation of such objects by smooth Riemannian manifolds, cf. Remark 3.80.

Let us informally describe two concrete examples of Cheeger deformations before proceeding with the theory. All of the assertions below can be made precise using the formulas in Exercise 6.62.

Example 6.5. Consider the round sphere (S^2, g) with the rotation action by S^1. The orbit space is the closed interval $[0, \pi]$, and (S^2, g_t) can be visualized as spheres that get "skinnier" as t gets large, since the parallels (orbits of the rotation action) are being shrunk. For $t \gg 1$, (S^2, g_t) looks like a cylinder of height $[0, \pi]$ and very small radius, with spherical caps at the ends, i.e., near the poles. All such manifolds have positive curvature, but the curvature becomes arbitrarily large near the poles and arbitrarily small everywhere else.

As a second example, consider the flat plane $\mathbb{R}^2$ with the rotation action by S^1. The orbit space is a half line $[0, +\infty)$, and $(\mathbb{R}^2, g_t)$, $t > 0$, are positively curved surfaces that, as $t \nearrow +\infty$, get asymptotically close to a cylinder with shrinking radius. Intuitively, one can picture $\mathbb{R}^2$ sitting in $\mathbb{R}^3$ as a horizontal plane $\{(x, y, 0)\}$ and the evolution $(\mathbb{R}^2, g_t)$ is "folding" the plane up onto itself, getting arbitrarily close to the positive z-axis $\{(0, 0, z) : z \geq 0\}$.

As the last examples suggests, besides remaining nonnegative, curvature may sometimes increase during a Cheeger deformation. Let us analyze how the curvature evolves in more precise terms. For this, it is convenient to define a reparametrization of the Grassmannian bundle Gr_2TM of planes on M, using the map C_t defined in (6.6) and computed in (6.7). Given a plane $\sigma = \text{span}\{X,Y\}$, set

$$C_t^{-1}(\sigma) := \text{span}\{C_t^{-1}X, C_t^{-1}Y\}.$$

This provides a 1-parameter family of bundle automorphisms of Gr_2TM induced by C_t^{-1}, called *Cheeger reparametrization*. As explained by Ziller [234], the crucial observation of Müter [172] is that, to analyze the evolution of sec_{g_t}, it is much more convenient to study $\text{sec}_{g_t}(C_t^{-1}(\sigma))$ instead of $\text{sec}_{g_t}(\sigma)$, cf. Remark 6.4. The following result of Müter [172, Satz 4.9] (see also [234, Cor 1.4]) summarizes how the sectional curvature of g_0-flat planes evolves for $t > 0$.

Proposition 6.6. *Let g_t be the Cheeger deformation of g_0, with $\text{sec}_{g_0} \geq 0$. Given a plane $\sigma = \text{span}\{X,Y\} \subset T_pM$ with $\text{sec}_{g_0}(\sigma) = 0$, consider the unnormalized g_t-sectional curvature of $C_t^{-1}(\sigma)$, given by*

$$k_C(t) := g_t\big(R_t(C_t^{-1}X, C_t^{-1}Y)C_t^{-1}X, C_t^{-1}Y\big)$$
$$= \big\|C_t^{-1}X \wedge C_t^{-1}Y\big\|_{g_t}^2 \, \text{sec}_{g_t}\big(C_t^{-1}(\sigma)\big),$$

where R_t is the curvature tensor of g_t. Then $k_C'(0) \geq 0$, and if the equality $k_C'(0) = 0$ holds, we have:

(i) *If $[P_0X_m, P_0Y_m] \neq 0$, then $k_C''(0) = 0$, $k_C'''(0) > 0$ and $k_C(t) > 0$ for all $t > 0$;*
(ii) *If $[P_0X_m, P_0Y_m] = 0$, then $k_C(t) = 0$ for all $t > 0$.*

In particular, if $\text{sec}_{g_0}(\sigma) = 0$ and $[P_0X_m, P_0Y_m] \neq 0$, i.e., $\text{span}\{P_0X_m, P_0Y_m\}$ has positive curvature[2] in (G,Q), then $\text{sec}_{g_t}(C_t^{-1}(\sigma)) > 0$ for all $t > 0$.

Combining the above with the Gauss-Bonnet Theorem, it follows that if σ is tangent to a totally geodesic flat torus in (M,g_0) that contains a horizontal direction, then $\text{sec}_{g_t}(C_t^{-1}(\sigma)) = 0$ for all $t \geq 0$. On the other hand, we also obtain the following important positive result:

Corollary 6.7. *Consider an isometric action of $G = SO(3)$ or $G = SU(2)$ on (M,g_0). If $\text{sec}_{g_0}(\sigma) = 0$ and the image of the projection of $\sigma \subset \mathcal{V}_p \oplus \mathcal{H}_p$ onto $\mathcal{V}_p$ is 2-dimensional, then $\text{sec}_{g_t}(C_t^{-1}(\sigma)) > 0$ for all $t > 0$.*

In fact, by Proposition 2.26 (see also Example 6.25), these Lie groups (G,Q) have $\sec > 0$ and hence $[P_0X_m, P_0Y_m] \neq 0$ whenever $X_m \neq 0$ and $Y_m \neq 0$.

---

[2] Recall Proposition 2.26 (iv).

6.2 Compact Homogeneous Spaces

In this section, we quickly introduce the basic theory of compact homogeneous spaces. For a more comprehensive introduction, see, for instance, [24, 33, 71, 179]. Let G be a compact connected Lie group and H a closed subgroup (hence H is an embedded Lie subgroup by Theorem 1.42). Consider the space of (right) cosets

$$G/H := \{gH : g \in G\}$$

and the quotient map $\rho \colon G \to G/H$, $\rho(g) = gH$. Then G/H has a unique smooth structure induced by ρ, making it a smooth manifold with $\dim G/H = \dim G - \dim H$, see Corollary 3.38. Manifolds of this form are called *homogeneous spaces*. The justification for this name comes from the natural transitive left G-action on G/H, induced by the G-action by left translations on G (see Exercise 3.40), that is,

$$\mu \colon G \times G/H \to G/H, \quad \mu(\overline{g}, gH) = \overline{g}gH. \tag{6.13}$$

We refer to the above action as the *left translation action*, cf. (1.3). We follow the usual slight abuse of notation and denote the diffeomorphism $\mu^g = \mu(g, \cdot) \colon G/H \to G/H$ simply by $g \colon G/H \to G/H$. Also, we generally assume that the above action is effective, i.e., $\mu^g = \mathrm{id}$ implies $g = e$. In some cases, however, it may be useful to relax this condition to *almost effective*, which means that the *ineffective kernel* of the action can be finite (see Definition 3.2).

Exercise 6.8. Show that the left translation action (6.13) is effective if and only if G and H have no common normal subgroups. In particular, the ineffective kernel of the action is the largest subgroup that is simultaneously normal in H and G.

Exercise 6.9. In this exercise, we show that every homogeneous space G/H comes equipped with a free (right) action by the group[3] $N(H)/H$, and is hence the total space of a *homogeneous principal $N(H)/H$-bundle*:

$$N(H)/H \to G/H \to G/N(H), \tag{6.14}$$

which already appeared in the proof of Theorem 3.95. Prove the following:

(i) The map $\mu \colon G/H \times N(H) \to G/H$, $\mu(gH, n) := gHn = gnH$ defines a smooth right action of $N(H)$ on G/H, with ineffective kernel H;
(ii) Use Remark 3.3 and Theorem 3.34 to conclude that $N(H)/H$ acts freely on G/H, making it the total space of the principal bundle (6.14).

Instead of starting from the compact Lie groups G and H to build the homogeneous space G/H, a more classical viewpoint is that a homogeneous space is

---

[3]Recall that $N(H)$ is the largest subgroup of G where H is normal, see Exercise 1.61.

a Riemannian manifold (M, g) with a transitive isometric group action. Choosing any subgroup $G \subset \text{Iso}(M, g)$ that acts transitively on M, by Proposition 3.41, we have that M is G-equivariantly diffeomorphic to G/H, where $H := G_x$ is the isotropy group of any point $x \in M$. Notice, however, that there may be many subgroups G of $\text{Iso}(M, g)$ that act transitively, hence $M \cong G/H$ is generally not a unique representation. The largest group $G = \text{Iso}(M, g)$ that acts transitively on the homogeneous space (M, g) is usually called its *full isometry group*.

Regarding examples of homogeneous spaces, note that Lie groups themselves are trivially homogeneous spaces with $H = \{e\}$. Before going any further, let us introduce a very important class of examples, which already made a quick appearance in Exercise 3.44. For the sake of brevity, in the remainder of this chapter, we use the shorthand notation $G_1 G_2$ for the product group $G_1 \times G_2$, as well as any quotient $(G_1 \times G_2)/\Gamma$ by a finite normal subgroup $\Gamma \subset G_1 \times G_2$.

Example 6.10. The following homogeneous spaces are called *Compact Rank One Symmetric Spaces*, or *CROSS* for short, since they are symmetric spaces (see Exercise 6.32) whose maximal totally geodesic flat submanifold has dimension 1, i.e., is a geodesic. They consist of spheres and projective spaces relative to the 4 real normed division algebras: $\mathbb{R}$ of reals, $\mathbb{C}$ of complex numbers, $\mathbb{H}$ of quaternions and $\mathbb{C}a$ of Cayley numbers (or octonions):

(i) $S^n = O(n+1)/O(n) = SO(n+1)/SO(n)$, the unit *sphere* in $\mathbb{R}^{n+1}$. The standard isometric action of $O(n+1)$ on $\mathbb{R}^{n+1}$ is transitive on the unit sphere, and the isotropy group $O(n)$ of a point $v \in S^n$ is formed by linear orthogonal transformations of the hyperplane $v^\perp \cong \mathbb{R}^n$;

(ii) $\mathbb{R}P^n = SO(n+1)/S(O(n)O(1))$, the *real projective space*, is the quotient of S^n by the properly discontinuous antipodal $\mathbb{Z}_2$-action;

(iii) $\mathbb{C}P^n = U(n+1)/U(n)U(1) = SU(n+1)/S(U(n)U(1))$, the *complex projective space*. The standard isometric action of $U(n+1)$ on $\mathbb{C}^{n+1}$ restricted to the unit sphere S^{2n+1} commutes with the unit multiplication $U(1)$-action (see Exercise 3.40). Thus, it descends to an isometric action on the orbit space $\mathbb{C}P^n = S^{2n+1}/U(1)$. Alternatively, consider the $U(n+1)$-action on the set of complex lines in $\mathbb{C}^{n+1}$. The isotropy group $U(n)U(1)$ of a complex line ℓ is formed by unitary transformations $U(n)$ of $\ell^\perp \cong \mathbb{C}^n$ and $U(1)$ of $\ell \cong \mathbb{C}$;

(iv) $\mathbb{H}P^n = Sp(n+1)/Sp(n)Sp(1)$ the *quaternionic projective space*. The standard isometric action of $Sp(n+1)$ on $\mathbb{H}^{n+1}$ restricted to the unit sphere S^{4n+3} commutes with the unit multiplication $Sp(1)$-action (see Exercise 3.40). Thus, it descends to an isometric action on the orbit space $\mathbb{H}P^n = S^{4n+3}/Sp(1)$. Alternatively, consider the $Sp(n+1)$-action on the set of quaternionic lines in $\mathbb{H}^{n+1}$. The isotropy group $Sp(n)Sp(1)$ of a quaternionic line ℓ is formed by unitary transformations $Sp(n)$ of $\ell^\perp \cong \mathbb{H}^n$ and $Sp(1)$ of $\ell \cong \mathbb{H}$;

(v) $\mathbb{C}aP^2 = F_4/\text{Spin}(9)$, the *Cayley plane*, or the *octonionic projective plane*. This is the homogeneous space of octonionic lines in $\mathbb{C}a^3$, and is given by the quotient of the exceptional Lie group F_4 by a subgroup isomorphic to $\text{Spin}(9)$, see [123, Thm 14.99]. For more details on its geometry, see [26, 123]. We remark that, differently from the above examples, $\mathbb{C}aP^2$ is not covered by a sphere. Also, higher dimensional analogues do not exist, due to non-associativity of the octonions.[4]

As in the case of Lie groups, one is usually interested in studying objects invariant under left translations. In particular, a metric g on G/H is G-invariant if (6.13) is an isometric action on the Riemannian manifold $(G/H, \mathrm{g})$, i.e., every $g: G/H \to G/H$ is an isometry. It is important to note that, in this case, the full isometry group $\mathrm{Iso}(G/H, \mathrm{g})$ may be strictly larger[5] than G.

Because of the presence of a transitive isometric group action, geometric properties of homogeneous manifolds are determined by information at a single point. For example, the scalar curvature of a homogeneous manifold is clearly a constant function. In order to better describe how the geometry of G/H is determined at a point, we look at the corresponding linearized objects.

Denote by $\mathfrak{g}$ and $\mathfrak{h}$ the Lie algebras of G and H, respectively. Since we are assuming G is compact, by Proposition 2.24, G admits a bi-invariant metric, that we denote by Q. Using this metric, define $\mathfrak{m}$ to be the orthogonal complement of $\mathfrak{h}$ in $\mathfrak{g}$. By the bi-invariance of Q, it follows that $\mathfrak{m}$ is invariant under the adjoint representation restricted to H, i.e., $\mathrm{Ad}(h)\mathfrak{m} \subset \mathfrak{m}$, for all $h \in H$. This representation is closely related to the geometry of G/H, as explained in the next result.

Proposition 6.11. *The tangent space $T_{eH}G/H$ is identified with the $\mathrm{Ad}(H)$-invariant complement $\mathfrak{m}$ of $\mathfrak{h}$ in $\mathfrak{g}$. Under this identification, the isotropy representation of H on $T_{eH}G/H$ is the adjoint representation $\mathrm{Ad}: H \times \mathfrak{m} \to \mathfrak{m}$.*

Proof. Consider the derivative $\mathrm{d}\rho_e: \mathfrak{g} \to T_{eH}G/H$ of the quotient map $\rho: G \to G/H$, $\rho(g) = gH$. This is a linear map given by

$$\mathrm{d}\rho_e X = \tfrac{\mathrm{d}}{\mathrm{d}t}\exp(tX)H\big|_{t=0}, \tag{6.15}$$

so its restriction to the second factor of the decomposition $\mathfrak{g} = \mathfrak{h} \oplus \mathfrak{m}$ is a linear isomorphism $\mathfrak{m} \cong T_{eH}G/H$, giving the desired identification. To make this identification H-equivariant, we need to prove that the restriction of the adjoint representation $H \ni h \mapsto \mathrm{Ad}(h) \in \mathrm{GL}(\mathfrak{m})$ corresponds to the isotropy representation $H \ni h \mapsto \mathrm{d}h_{eH} \in \mathrm{GL}(T_{eH}G/H)$. This follows at once by direct computation using (1.10), since for all $h \in H$ and $X \in \mathfrak{m}$,

---

[4]The reason why $\mathbb{C}aP^2$ can be constructed as a projectivization of $\mathbb{C}a^3$ is related to the fact that $\mathbb{C}a$ is 2-*associative*, i.e., any subalgebra generated by 2 elements is associative.

[5]For instance, one can write the sphere as $S^{4n+3} = \mathrm{Sp}(n+1)/\mathrm{Sp}(n)$, but depending on what $\mathrm{Sp}(n+1)$-homogeneous metric we consider (e.g., the round metric), the full isometry group may be strictly larger than $\mathrm{Sp}(n+1)$, see Example 6.16 and Remark 6.17.

$$\mathrm{d}\rho_e\big(\mathrm{Ad}(h)X\big) = \tfrac{\mathrm{d}}{\mathrm{d}t}\exp(t\mathrm{Ad}(h)X)H\big|_{t=0}$$

$$= \tfrac{\mathrm{d}}{\mathrm{d}t}\big(h\exp(tX)h^{-1}\big)H\big|_{t=0} \qquad (6.16)$$

$$= \mathrm{d}h_{eH}\big(\mathrm{d}\rho_e X\big). \qquad\qquad \square$$

Exercise 6.12. Use the two equivalent ways given above of writing the isotropy representation on G/H to show in two different ways that if the left translation G-action on G/H is effective, then the isotropy representation is also effective.

Hint: Interpreting the isotropy representation as:

(i) $H \ni h \mapsto \mathrm{d}h_{eH} \in \mathrm{GL}(T_{eH}G/H)$, use Proposition 2.11 to show that if $h(eH) = eH$ and $\mathrm{d}h_{eH} = \mathrm{id}$, then $h\colon G/H \to G/H$ is the identity;

(ii) $H \ni h \mapsto \mathrm{Ad}(h) \in \mathrm{GL}(\mathfrak{m})$, use Exercise 6.8 to show that if $\mathrm{Ad}(h) = \mathrm{id}$, then $h = e$.

Using the above, we can identify the moduli space of invariant metrics on a homogeneous space in terms of inner products on $\mathfrak{m}$, as follows.

Theorem 6.13. *The set of G-invariant Riemannian metrics on G/H is in 1-to-1 correspondence with the set of inner products on $\mathfrak{m}$ that are $\mathrm{Ad}(H)$-invariant.*

Proof. The restriction of any G-invariant metric g to $T_{eH}G/H$ gives an inner product invariant under the isotropy representation. By Proposition 6.11, this is an inner product on $\mathfrak{m}$ that is $\mathrm{Ad}(H)$-invariant. Conversely, given an $\mathrm{Ad}(H)$-invariant inner product $\langle\cdot,\cdot\rangle$ on $\mathfrak{m} \cong T_{eH}G/H$, define a G-invariant metric g on G/H by setting

$$\mathrm{g}(X,Y) := \big\langle \mathrm{d}L_{g^{-1}}X, \mathrm{d}L_{g^{-1}}Y \big\rangle, \quad X,Y \in T_{gH}G/H. \qquad (6.17)$$

Indeed, since $\langle\cdot,\cdot\rangle$ is $\mathrm{Ad}(H)$-invariant, by (6.16) it follows that (6.17) is independent of the choice of representative in the coset gH, and hence g is well-defined. $\square$

Remark 6.14. Provided H is connected, many of the above statements in terms of $\mathrm{Ad}(H)$-invariance have equivalent counterparts in terms of $\mathrm{ad}(H)$, obtained by differentiating the $\mathrm{Ad}(H)$-invariance equation. Namely, the complement $\mathfrak{m}$ to $\mathfrak{h}$ in $\mathfrak{g}$ is $\mathrm{Ad}(H)$-invariant if and only if $\mathfrak{m}$ is $\mathrm{ad}(H)$-invariant, that is, $[\mathfrak{h},\mathfrak{m}] \subset \mathfrak{m}$. Similarly, an inner product on $\mathfrak{m}$ is $\mathrm{Ad}(H)$-invariant if and only if $\mathrm{ad}(H)$ is skew-symmetric for this inner product, cf. Proposition 2.26 (i).

In particular, Theorem 6.13 implies that the moduli space of G-invariant metrics on G/H is a finite-dimensional open cone. To further understand the structure of this moduli space, we apply Schur's Lemma (see [56, p. 69]) to analyze the general form of an $\mathrm{Ad}(H)$-invariant inner product on $\mathfrak{m}$.

Consider the decomposition of $\mathfrak{m}$ into the direct sum of irreducible $\mathrm{Ad}(H)$-representations. Grouping the copies of irreducible summands that are equivalent into subspaces $\mathfrak{m}_i := \bigoplus_{j=1}^{n_i} \mathfrak{m}_{ij}$, we have a decomposition $\mathfrak{m} = \mathfrak{m}_0 \oplus \mathfrak{m}_1 \oplus \cdots \oplus \mathfrak{m}_l$. In other words, for each $0 \le i \le l$, any $\mathfrak{m}_{ij}$ and $\mathfrak{m}_{ik}$, with $1 \le j \le k \le n_i$, are equivalent

Ad(H)-representations, that is, there is an equivariant isomorphism $\mathfrak{m}_{ij} \cong \mathfrak{m}_{ik}$. The subspaces $\mathfrak{m}_i$ are called *isotypic components* of $\mathfrak{m}$, and $n_i \in \mathbb{N}$ is the *multiplicity* of the irreducible summand $\mathfrak{m}_{ij}$. It is usually conventioned that $\mathfrak{m}_0$ is the *trivial* isotypic component, i.e., $\mathfrak{m}_{0j}$, $1 \leq j \leq n_0$, are copies of the (1-dimensional) trivial representation. In terms of this decomposition, we have the following:

Corollary 6.15. *Let G/H be a compact homogeneous space and fix a bi-invariant metric Q on $\mathfrak{g}$. Consider the above decomposition $\mathfrak{m} = \mathfrak{m}_0 \oplus \mathfrak{m}_1 \oplus \cdots \oplus \mathfrak{m}_l$ into isotypic components. For each $X \in \mathfrak{m}$, denote by $X_i \in \mathfrak{m}_i$ its component on $\mathfrak{m}_i$, and by $X_{ij} \in \mathfrak{m}_{ij}$ its component on $\mathfrak{m}_{ij}$, so that $X_i = \sum_{j=1}^{n_i} X_{ij}$. The G-invariant metrics on G/H are those induced by inner products on $\mathfrak{m}$ that are of the form:*

$$\langle X, Y \rangle = \sum_{i=0}^{l} \langle X_i, Y_i \rangle_i, \quad where \quad \langle X_i, Y_i \rangle_i = \sum_{j,k=1}^{n_i} a_i^{jk} Q(X_{ij}, Y_{ik}) \tag{6.18}$$

and a_i^{jk} are the coefficients of a positive-definite symmetric matrix $A_i \in GL(n_i, \mathbb{R})$.

Fortunately, in many examples discussed below, the nontrivial isotypic components consist of only one irreducible summand, that is, $n_i = 1$ for all $1 \leq i \leq l$. In this case, the matrices A_i, $1 \leq i \leq l$, are simply positive real numbers a_i, and hence the inner product (6.18) is given by

$$\langle \cdot, \cdot \rangle = \langle \cdot, \cdot \rangle_0 \oplus \sum_{i=1}^{l} a_i Q|_{\mathfrak{m}_i}, \tag{6.19}$$

where $\langle \cdot, \cdot \rangle_0 = Q(A_0 \cdot, \cdot)$ is an arbitrary inner product on $\mathfrak{m}_0 \cong \mathbb{R}^{n_0}$ and $a_i > 0$.

Equipped with the above, one can study the moduli space of homogeneous metrics on G/H, provided the isotropy representation is known. For instance, let us describe all homogeneous metrics on spheres, following the classification of Ziller [235]. All these metrics are of the form (6.19).

Example 6.16 (Homogeneous metrics on spheres). The classification of transitive Lie group actions on spheres was obtained by Borel [47] and Montgomery and Samelson [167]. We list in Table 6.1 all possible transitive G-actions[6] on S^n, with the corresponding isotropy H and the decomposition of $\mathfrak{m}$ into isotypic components.

---

[6]Here, we assume G is connected. This entails no loss of generality, since the restriction to the identity component of a continuous transitive group action on a connected space is still transitive.

Table 6.1 Transitive actions on spheres, G connected

	G	H	$\dim G/H$	Isotropy repres.
(i)	$SO(n+1)$	$SO(n)$	n	Irreducible
(ii)	$SU(n+1)$	$SU(n)$	$2n+1$	$m = m_0 \oplus m_1$
(iii)	$U(n+1)$	$U(n)$	$2n+1$	$m = m_0 \oplus m_1$
(iv)	$Sp(n+1)$	$Sp(n)$	$4n+3$	$m = m_0 \oplus m_1$
(v)	$Sp(n+1)Sp(1)$	$Sp(n)Sp(1)$	$4n+3$	$m = m_1 \oplus m_2$
(vi)	$Sp(n+1)U(1)$	$Sp(n)U(1)$	$4n+3$	$m = m_0 \oplus m_1 \oplus m_2$
(vii)	$Spin(9)$	$Spin(7)$	15	$m = m_1 \oplus m_2$
(viii)	$Spin(7)$	G_2	7	Irreducible
(ix)	G_2	$SU(3)$	6	Irreducible

All of the above representations satisfy $n_i = 1$, $1 \le i \le l$, that is, each isotypic component consists of a unique irreducible summand. Carefully analyzing each of the above cases,[7] it follows from Corollary 6.15, see also (6.19), that the moduli space of homogeneous metrics on spheres is formed by:

(i) A 1-parameter family of $U(n+1)$-homogeneous metrics on S^{2n+1};
(ii) A 3-parameter family of $Sp(n+1)$-homogeneous metrics on S^{4n+3};
(iii) A 1-parameter family of $Spin(9)$-homogeneous metrics on S^{15}.

Any homogeneous metric on S^n must be, up to rescaling, isometric to one of the above. Note that the round metric is the only one at which all families intersect.

The geometric interpretation of these families is in terms of the Hopf bundles:

$$S^1 \to S^{2n+1} \to \mathbb{C}P^n, \quad S^3 \to S^{4n+3} \to \mathbb{H}P^n, \quad S^7 \to S^{15} \to S^8\left(\tfrac{1}{2}\right). \tag{6.20}$$

Namely, each family of homogeneous metrics above coincides with the family obtained by rescaling the unit round metric on the total space of the corresponding Hopf bundle in the vertical directions. On each of these total spaces, consider the metric

$$g_\lambda := g|_{\mathscr{H}} \oplus \lambda\, g|_{\mathscr{V}}, \quad \lambda > 0, \tag{6.21}$$

---

[7]Let us give a brief account of this analysis, see Ziller [235, Sec. 1] for details. On cases (i), (viii) and (ix), since the H-action on m is irreducible, the unique G-homogeneous metric is the round metric. On cases (ii) and (iii), the H-action on m_1 is the standard $U(n)$-action on $\mathbb{C}^n$. The metrics obtained in both cases are the same, yielding a family of $U(n+1)$-homogeneous metrics on S^{2n+1} that depends on two parameters. On case (iv), the H-action on m_1 is the standard $Sp(n)$-action on $\mathbb{H}^n$. This gives a family of $Sp(n+1)$-homogeneous metrics on S^{4n+3} that depends on seven parameters, six of which determine the restriction $\langle \cdot, \cdot \rangle_0$ to m_0. By appropriately diagonalizing $A_0 \in GL(3, \mathbb{R})$, this family can be further reduced to one that depends on only four parameters. Cases (v) and (vi) are contained in case (iv). Finally, on case (vii), the H-action on the 7-dimensional space m_1 and on the 8-dimensional space m_2 are the only irreducible representations of $Spin(7)$ in those dimensions. This gives a family of $Spin(9)$-homogeneous metrics on S^{15} that depends on two parameters. All the above can be normalized so that $a_1 = 1$, reducing by one the number of parameters of each family.

where $\mathcal{H}$ and $\mathcal{V}$ denote horizontal and vertical components, and $g_1 = g_{\mathcal{H}} \oplus g_{\mathcal{V}}$ is the unit round metric.[8] The metrics g_λ account for the entire 1-parameter families of homogeneous metrics in cases (i) and (iii) above, and for a subfamily in case (ii). The general homogeneous metrics in case (ii) can be described as follows. Identify the vertical space $\mathcal{V}$ of S^{4n+3}, i.e., the tangent space to the fiber S^3, with $(\mathbb{R}^3, \times) \cong \mathfrak{su}(2)$, recall Exercises 1.9 and 1.55. Use 3 positive parameters λ_1, λ_2 and λ_3, to respectively rescale the 3 elements of the canonical orthonormal basis of $(\mathbb{R}^3, \times)$, defining a left-invariant metric on the fiber $S^3 \cong SU(2)$. This metric on the vertical directions and the round metric on the horizontal directions determines a homogeneous metric $g_{(\lambda_1,\lambda_2,\lambda_3)}$ on the total space S^{4n+3}. This 3-parameter family is precisely the family in (ii). Note that if $\lambda_1 = \lambda_2 = \lambda_3 = \lambda$, then $g_{(\lambda_1,\lambda_2,\lambda_3)} = g_\lambda$.

Remark 6.17. Let us analyze the full isometry group of the $Sp(n+1)$-homogeneous space $\left(S^{4n+3}, g_{(\lambda_1,\lambda_2,\lambda_3)}\right)$, to illustrate how a G-homogeneous space G/H might have full isometry group *strictly larger* than G. When two of the parameters λ_i are equal, rotations in the corresponding vertical plane are isometries, so the full isometry group is $Sp(n+1)U(1)$. Similarly, when all three parameters λ_i are equal, the full isometry group is $Sp(n+1)Sp(1)$. These respectively correspond to cases (vi) and (v) in Table 6.1. Finally, if the λ_i are all equal to 1, then $g_{(1,1,1)} = g_1$ is the round metric, which has the largest full isometry group $O(4n+4)$.

Remark 6.18. For $0 < \lambda \leq 1$, the above family of homogeneous metrics g_λ on the total space of the Hopf bundles $S^1 \to S^{2n+1} \to \mathbb{C}P^n$ and $S^3 \to S^{4n+3} \to \mathbb{H}P^n$ can be interpreted as a Cheeger deformation of the round metric g_1. In fact, these are principal bundles, and $S^1 \cong U(1)$, respectively $S^3 \cong SU(2)$, acts isometrically on the round sphere S^{2n+1}, respectively S^{4n+3}, with Hopf fibers as orbits (see Example 3.30). An algebraic description of these actions is given in Exercise 6.19.

The Cheeger deformation of the round metric g_1 is the metric $g_{\lambda(t)}$ in (6.21) with $\lambda(t) = \frac{1}{1+t}, t \geq 0$. In other words, the Cheeger deformation shrinks the Hopf fibers uniformly, and the corresponding spheres converge (in Gromov-Hausdorff sense) to $\mathbb{C}P^n$ (respectively $\mathbb{H}P^n$) as $t \nearrow +\infty$, which corresponds to $\lambda(t) \searrow 0$. In particular, from Proposition 6.1 and Corollary 6.7, all such metrics have sec > 0. On the other hand, the invariant metrics obtained by *dilating* the Hopf fibers, i.e., g_λ for $\lambda > 1$, are not Cheeger deformations of g_1; and if the dilating factor is $\lambda \geq 4/3$ these manifolds actually fail to have sec > 0. A complete description of homogeneous metrics with sec > 0 on spheres is given in Theorem 6.31.

We also stress that no analogous statement holds for the Hopf bundle $S^7 \to S^{15} \to S^8\left(\frac{1}{2}\right)$, which is not a homogeneous bundle. In fact, there are no isometric group actions on S^{15} whose orbits are the 7-dimensional Hopf fibers, see [120].

Exercise 6.19. Let $H \subset K \subset G$ be compact Lie groups, and consider the *homogeneous bundle* $K/H \to G/H \to G/K$, whose projection map $\pi \colon G/H \to G/K$ is

---

[8]Spheres equipped with the metric g_λ are known as *Berger spheres*; though, historically, this term was mostly used referring to the case of (S^3, g_λ) first studied by Berger.

given by $\pi(gH) = gK$. Show that if H is a normal subgroup of K, then the right K-action on G/H given by $gH \cdot k := gHk = gkH$ is well-defined and its orbits are the fibers of π. Notice that $K \subset N(H)$ and compare with (6.14) in Exercise 6.9.

The result in the above exercise applies to the homogeneous Hopf bundles in Remark 6.18, where $H \subset K \subset G$ are respectively $U(n) \subset U(n)U(1) \subset U(n+1)$ and $Sp(n) \subset Sp(n)Sp(1) \subset Sp(n+1)$, see Example 6.10.

Remark 6.20. Besides the round metric, there are other homogeneous Einstein metrics on spheres. Namely, Jensen [134] found that g_λ with $\lambda = \frac{1}{2n+3}$ is an $Sp(n+1)$-invariant Einstein metric on S^{4n+3}, and Bourguignon and Karcher [52] found that g_λ with $\lambda = \frac{3}{11}$ is a $Spin(9)$-invariant Einstein metric on S^{15}. Ziller [235] proved that these are in fact all homogeneous Einstein metrics on spheres.

Note that, in order to verify that a G-invariant metric g on a homogeneous space G/H is Einstein, it clearly suffices to verify that $Ric = \lambda g$ at a single point, for instance eH. This viewpoint reduces the Einstein equation, which is a partial differential equation on a manifold, to an algebraic system on $\mathfrak{m}$, see [44, 225, 226]. This approach also allows to study homogeneous Ricci solitons, and the Ricci flow of homogeneous metrics, see [147, 148, 150].

Despite not necessarily being Einstein, homogeneous metrics always have constant scalar curvature, since this is a function invariant under isometries. However, there may be constant scalar curvature metrics on homogeneous spaces G/H that are *not G-invariant*. Combining classical variational bifurcation techniques and the above characterizations of the homogeneous metrics on spheres, Bettiol and Piccione [37] recently obtained new (non-homogeneous) constant scalar curvature metrics on S^{4n+3}, $n \geq 1$, in the conformal classes of g_λ, for infinitely many $\lambda \in (0,1)$, that accumulate at 0. This was later generalized in [38] to the homogeneous bundles described in Exercise 6.19 for which K/H has positive scalar curvature.

We now compute the moduli space of invariant metrics on the other CROSS, following Ziller [235, Sec. 3].

Example 6.21 (Homogeneous metrics on projective spaces). Analogously to Table 6.1, we list in Table 6.2 all transitive actions by connected Lie groups on the remaining CROSS, obtained by Oniscik [178, p. 168], see also [179].

Table 6.2 Transitive actions on projective spaces, G connected

	G	H	G/H	Isotropy repres.
(i)	$SU(n+1)$	$S(U(n)U(1))$	$\mathbb{C}P^n$	Irreducible
(ii)	$Sp(n+1)$	$Sp(n)U(1)$	$\mathbb{C}P^{2n+1}$	$\mathfrak{m} = \mathfrak{m}_1 \oplus \mathfrak{m}_2$
(iii)	$Sp(n+1)$	$Sp(n)Sp(1)$	$\mathbb{H}P^n$	Irreducible
(iv)	F_4	$Spin(9)$	$\mathbb{C}aP^2$	Irreducible

In all but one case in Table 6.2, the isotropy representation is irreducible; so (up to rescaling) there exists a unique G-invariant metric on the corresponding homogeneous space G/H. This is the Fubini-Study metric g_{FS}, which, in the cases of $\mathbb{C}P^n$ and $\mathbb{H}P^n$ is the metric that turns the Hopf bundles $S^{2n+1} \to \mathbb{C}P^n$ and

$S^{4n+3} \to \mathbb{H}P^n$ into Riemannian submersions, where the corresponding spheres are endowed with the round metric.[9] Since the isotropy representation is irreducible, it also follows that these metrics are Einstein, since there is (up to rescaling) only one invariant symmetric $(0,2)$-tensor on these manifolds, by Schur's Lemma.

In the remaining case (ii), the isotropy representation has two inequivalent irreducible factors $\mathfrak{m}_1$ and $\mathfrak{m}_2$, of dimensions 2 and $4n$ respectively. Consequently, there is (up to normalization) a 1-parameter family of $Sp(n+1)$-invariant metrics on $\mathbb{C}P^{2n+1}$. These can be interpreted, similarly to the case (6.21) of spheres, as a deformation $g_{FS}|_{\mathscr{H}} \oplus \lambda\, g_{FS}|_{\mathscr{V}}$ of the Fubini-Study metric g_{FS}, rescaling by λ the vertical directions of the homogeneous bundle $\mathbb{C}P^1 \to \mathbb{C}P^{2n+1} \to \mathbb{H}P^n$.

Remark 6.22. Ziller [235] found that $g_{FS}|_{\mathscr{H}} \oplus \lambda\, g_{FS}|_{\mathscr{V}}$ with $\lambda = \frac{1}{n+1}$ is an $Sp(n+1)$-invariant Einstein metric on $\mathbb{C}P^{2n+1}$, cf. Remark 6.20.

Among all homogeneous metrics on G/H, there are special metrics which are induced by a bi-invariant metric Q on G. These metrics correspond to the inner product $Q|_{\mathfrak{m}}$ via Corollary 6.15, that is, to the case in which $A_i \in GL(n_i, \mathbb{R})$, $1 \leq i \leq l$, in (6.18) are identity matrices. Metrics of this form are called *normal homogeneous metrics*. If g is one such metric on G/H, then the projection $\rho : (G,Q) \to (G/H, g)$ is a Riemannian submersion. In particular, $(G/H, g)$ has sec ≥ 0 by the Gray-O'Neill formula (2.14), since (G,Q) has sec ≥ 0 by Proposition 2.26.

The following result is an interesting application of Cheeger deformations, due to Schwachhöfer and Tapp [192, Prop 1.1], regarding the structure of the moduli space of invariant metrics with sec ≥ 0 on a homogeneous space.

Theorem 6.23. *The moduli space of invariant metrics on a compact homogeneous space G/H with sec ≥ 0 is path-connected.*

Proof. Let $g_* := Q|_{\mathfrak{m}}$ be a normal homogeneous metric. Given an invariant metric g with $\sec_g \geq 0$, consider the path of metrics $(1+t)g_t$ obtained by rescaling the Cheeger deformation of g by $(1+t)$. From (6.7), using the notation of Proposition 6.3, we have

$$(1+t)g_t(X,Y) = Q\big((1+t)P_t(X_{\mathfrak{m}}), Y_{\mathfrak{m}}\big) = Q\big((1+t)P_0\,(\mathrm{id}+tP_0)^{-1}X_{\mathfrak{m}}, Y_{\mathfrak{m}}\big).$$

As $t \nearrow +\infty$, the above clearly converges to $Q(X_{\mathfrak{m}}, Y_{\mathfrak{m}})$, proving that $(1+t)g_t$ converges to g_* as $t \nearrow +\infty$. Evidently, $(1+t)g_t$ converges to g as $t \searrow 0$. By Proposition 6.1, the entire path $(1+t)g_t$ is through invariant metrics with sec ≥ 0.[10] $\qquad\square$

---

[9]Note that, as no sphere fibers over the Cayley plane $\mathbb{C}aP^2$, there is no analogous interpretation for the Fubini-Study metric in this case.

[10]Besides proving the above statement, this proof indicates that the moduli space of homogeneous metrics with sec ≥ 0 is, in some sense, *star-shaped* with respect to normal homogeneous metrics.

Example 6.24. Some normal homogeneous metrics on spheres (see Example 6.16) *do not coincide* with the round metric. In fact, apart from $S^3 \cong \mathrm{SU}(2)$, the round metric is only normal homogeneous in the cases (i), (viii) and (ix) in Table 6.1, for which there is a unique invariant metric up to rescaling. In all other cases, the normal homogeneous metric is of the form g_λ for some $0 < \lambda < 1$, see (6.21). More precisely, up to rescaling, the normal homogeneous metric g_λ on $S^{2n+1} = \mathrm{SU}(n+1)/\mathrm{SU}(n)$ corresponds to $\lambda = \frac{n+1}{2n}$, on $S^{4n+3} = \mathrm{Sp}(n+1)/\mathrm{Sp}(n)$ corresponds to $\lambda = \frac{1}{2}$ and on $S^{15} = \mathrm{Spin}(9)/\mathrm{Spin}(7)$ corresponds to $\lambda = \frac{1}{4}$. For details, see [116, Lemma 2.4].

Example 6.25. Compact Lie groups (G, Q) with bi-invariant metrics are not only normal homogeneous spaces, but also *symmetric spaces*, see Theorem 2.30 and Remark 4.12. From Proposition 2.26, (G, Q) can only have sec > 0 if G has rank one. Thus, by Theorem 4.6, the only compact Lie groups that admit a bi-invariant metric with sec > 0 are $\mathrm{SU}(2)$ and $\mathrm{SO}(3)$.

We now present the curvature formula for a general G-invariant metric on a homogeneous space G/H, in a slightly different way from usual textbooks in the area, e.g., [24, 33, 71]. Namely, we use a formalism introduced by Püttmann [187] that is very convenient for later purposes. In what follows, we repeatedly use Theorem 6.13 to write G-invariant metrics on G/H as $\mathrm{Ad}(H)$-invariant inner products on $\mathfrak{m}$. Also, we fix a bi-invariant metric Q on G and use the corresponding normal homogeneous metric on G/H as a reference metric to write general G-invariant metrics. More precisely, given a G-invariant metric g on G/H, there exists a unique Q-symmetric $\mathrm{Ad}(H)$-equivariant automorphism $P \colon \mathfrak{m} \to \mathfrak{m}$ such that

$$\mathrm{g}(X, Y) = Q(PX, Y), \quad \text{for all } X, Y \in \mathfrak{m}. \tag{6.22}$$

We define the auxiliary symmetric and skew-symmetric bilinear maps

$$B_\pm(X, Y) := \tfrac{1}{2}\big([X, PY] \mp [PX, Y]\big). \tag{6.23}$$

With these definitions, using the Gray-O'Neill formula (2.14) one can prove the following, cf. [187, Lemma 3.6], [33, Chap. 7].

Proposition 6.26. *The curvature tensor of the invariant metric* $\mathrm{g}(\cdot, \cdot) = Q(P\cdot, \cdot)$ *on* G/H *is given by*

$$
\begin{aligned}
R(X, Y, Z, W) = {} & \tfrac{1}{2}\big(Q(B_-(X, Y), [Z, W]) + Q([X, Y], B_-(Z, W))\big) \\
& + \tfrac{1}{4}\big(Q([X, W], P[Y, Z]_\mathfrak{m}) - Q([X, Z], P[Y, W]_\mathfrak{m}) \\
& - 2Q([X, Y], P[Z, W]_\mathfrak{m})\big) \\
& + Q(B_+(X, W), P^{-1}B_+(Y, Z)) \\
& - Q(B_+(X, Z), P^{-1}B_+(Y, W)),
\end{aligned}
$$

where $X_\mathfrak{m}$ *denotes the component in* $\mathfrak{m}$ *of an element* $X \in \mathfrak{g}$.

Exercise 6.27. The Lie group (G,Q) endowed with a bi-invariant metric is a particular case of an invariant metric on a homogeneous space, where $H = \{e\}$ and $P = \mathrm{id}$. Compute $B_\pm$ in this case and substitute in the above formula to recover formula (iii) in Proposition 2.26.

Exercise 6.28 ($\star$). Note that when $P = \mathrm{id}$, the metric g on G/H is normal homogeneous. For one such metric:

(i) Use Proposition 6.26 to compute $\sec(X,Y) = R(X,Y,X,Y)$, where $X,Y \in \mathfrak{m}$ are orthonormal, obtaining

$$\sec(X,Y) = \tfrac{1}{4}\big\|[X,Y]\big\|^2 + \tfrac{3}{4}\big\|[X,Y]_\flat\big\|^2, \tag{6.24}$$

and verifying the above claim that normal homogeneous metrics have $\sec \geq 0$;

(ii) Characterize the vectors X such that $\sec(X,Y) = 0$ for any orthogonal vector Y;

(iii) Use the above to show that $\mathrm{Ric}(X,X) \geq 0$ for all $X \in \mathfrak{m}$ and $\mathrm{Ric}(X,X) = 0$ only if $X \in Z(\mathfrak{g})$, see (1.15).

Remark 6.29. Recall that, by the Bonnet-Myers Theorem 2.19, if a compact Riemannian manifold (M,g) has $\mathrm{Ric} > 0$, then its fundamental group $\pi_1(M)$ is finite. The converse statement holds for compact homogeneous spaces, that is, if G/H is a compact homogeneous space with $\pi_1(G/H)$ finite, then G/H admits a Riemannian metric with $\mathrm{Ric} > 0$. Indeed, any normal homogeneous metric on G/H has $\mathrm{Ric} > 0$ by Exercise 6.28 (iii), since it is possible to prove that $Z(\mathfrak{g}) \cap \mathfrak{m} = \{0\}$ provided $\pi_1(G/H)$ is finite, see [28].

It is sometimes possible to increase curvature from $\sec \geq 0$ to $\sec > 0$ along the submersion $\rho \colon G \to G/H$. We stress this just happens in *very special cases* (only in dimensions 6, 7, 12, 13 and 24), and it is typically achieved with a left-invariant metric on G which is not bi-invariant, but rather a deformation[11] of a bi-invariant metric. The classification of simply-connected closed homogeneous manifolds with an invariant metric of positive sectional curvature was obtained through the work of Berger [30], Wallach [221], Aloff-Wallach [21] and Berard-Bergery [27], see also [232].

Theorem 6.30 (Homogeneous classification of $\sec > 0$). *Apart from the CROSS (see Example 6.10), the only simply-connected closed manifolds to admit a homogeneous metric with $\sec > 0$ are:*

- *Aloff-Wallach spaces $W_{k,l}^7 = \mathrm{SU}(3)/S_{k,l}^1$, $\gcd(k,l) = 1$, $kl(k+l) \neq 0$, including $W_{1,1}^7 = \mathrm{SU}(3)\mathrm{SO}(3)/\mathrm{U}(2)$;*
- *Berger spaces $B^7 = \mathrm{SO}(5)/\mathrm{SO}(3)$ and $B^{13} = \mathrm{SU}(5)/S^1\mathrm{Sp}(2)$;*

---

[11]More precisely, a Cheeger deformation of a bi-invariant metric on G with respect to the action of an intermediate subgroup $H \subset K \subset G$.

- *Wallach flag manifolds* $W^6 = SU(3)/T^2$, $W^{12} = Sp(3)/Sp(1)Sp(1)Sp(1)$ *and* $W^{24} = F_4/Spin(8)$.

For details on the above homogeneous spaces, we refer the reader to Ziller [236] and Wilking and Ziller [232].

After knowing which homogeneous spaces admit invariant metrics of positive curvature, a natural question is to study the moduli space of such invariant metrics, and optimal pinching constants. Recall that the *pinching constant* of a Riemannian manifold (M, g) is the ratio between extremal values of $\sec_g$ at the same point:

$$\delta_g = \frac{\min \sec_g}{\max \sec_g} \leq 1$$

It is a classical fact in Riemannian geometry that if $\delta_g = 1$, then (M, g) is isometric to a spherical space form, that is, a quotient of a round sphere. A classical result in modern Riemannian geometry is the Topological Sphere Theorem, which states that a complete simply-connected manifold (M, g) with $\delta_g > \frac{1}{4}$ must be homeomorphic to a sphere. This deep result was proved by Berger [29] and Klingenberg [142, 143], elaborating on a novel idea of Rauch [190]. An exciting recent development was the proof of the Differentiable Sphere Theorem by Brendle and Schoen [55], which improves the conclusion from *homeomorphism* to *diffeomorphism*. In part, this was made possible by the remarkable work of Böhm and Wilking [45], who proved that the Ricci flow evolves a metric with positive-definite curvature operator to a limit with constant sectional curvature. Notice that for a CROSS other than S^n (recall Example 6.10), such as $\mathbb{C}P^n$ and $\mathbb{H}P^n$, the pinching of the Fubini-Study metric is $\delta_{g_{FS}} = \frac{1}{4}$, showing that the *quarter pinched* condition $\delta_g > \frac{1}{4}$ is sharp in even dimensions.

The *optimal pinching constant* for a homogeneous space is the largest possible pinching δ_{max} among invariant metrics, providing a notion of the *best* invariant metric on such space. For homogeneous spaces that are not symmetric, computing the exact value of δ_{max} is an extremely difficult task. This was achieved by Valiev [215] on the flag manifolds W^6, W^{12} and W^{24}, for which $\delta_{max} = \frac{1}{64}$, by Eliasson [80] on the Berger space B^7, for which $\delta_{max} = \frac{1}{37}$, and by Püttmann [187] on the Berger space B^{13} and the Aloff-Wallach space $W_{1,1}^7$, for which $\delta_{max} = \frac{1}{37}$. Finally, Püttmann [187] also proved that $\delta_{max} \leq \frac{1}{37}$ for all remaining Aloff-Wallach spaces $W_{k,l}^7$.

Consider now the problem of determining the moduli space of positively curved metrics on a given homogeneous space. A complete answer in the case of spheres (recall Example 6.16), together with their pinching constants, was recently obtained by Verdiani and Ziller [219].

Theorem 6.31. *Denote by g_λ the homogeneous metric* (6.21) *obtained by scaling the round metric on each Hopf bundle* (6.20) *by λ in the vertical direction. Then g_λ has $\sec > 0$ if and only if $0 < \lambda < 4/3$. Denote by $g_{(\lambda_1, \lambda_2, \lambda_3)}$ the 3-parameter family of homogeneous metric on S^{4n+3}, and set*

$$V_i := (\lambda_j^2 + \lambda_k^2 - 3\lambda_i^2 + 2\lambda_i\lambda_j + 2\lambda_i\lambda_k - 2\lambda_j\lambda_k)/\lambda_i \quad \text{and} \quad H_i := 4 - 3\lambda_i,$$

where (i, j, k) denotes a cyclic permutation of $(1, 2, 3)$. Then $g_{(\lambda_1, \lambda_2, \lambda_3)}$ has sec > 0 if and only if

$$V_i > 0, \quad H_i > 0, \quad \text{and} \quad 3|\lambda_j\lambda_k - \lambda_j - \lambda_k + \lambda_i| < \lambda_j\lambda_k + \sqrt{H_iV_i}.$$

We conclude this section with an exercise that provides a glimpse of the theory of *symmetric spaces*, for which we recommend as further references [71, 126, 129].

Exercise 6.32 (⋆). By definition, (M, g) is a symmetric space if for all $p \in M$ there exists an isometry $I^p : M \to M$ that fixes p and reverses geodesics through p, i.e., $I^p(p) = p$ and $d(I^p)_p = -\text{id}$, see Exercise 2.11 (and recall Theorem 2.30).

(i) Use the above definition to verify that a symmetric space is complete;
(ii) Use the Hopf-Rinow Theorem 2.9 to prove that a symmetric space (M, g) is a homogeneous space G/H, where $G = \text{Iso}(M, g)$ is its isometry group and $H := G_p$ is the isotropy of a point $p \in M$;
(iii) If $M = G/H$ is a symmetric space as above, show that $g \mapsto I^p g I^p$ defines an involution of G, whose fixed point set is a closed subgroup with identity component equal to the identity component of H;
(iv) Use this involution to show that the Lie algebra $\mathfrak{g}$ of G splits as $\mathfrak{g} = \mathfrak{h} \oplus \mathfrak{m}$, where $\mathfrak{h}$ is the Lie algebra of H, $[\mathfrak{h}, \mathfrak{m}] \subset \mathfrak{m}$ and $[\mathfrak{m}, \mathfrak{m}] \subset \mathfrak{h}$, cf. Proposition 6.11.

A pair of Lie groups (G, H) as above is called a *symmetric pair*.

6.3 Cohomogeneity One Actions

In this section, we describe the basic theory of cohomogeneity one manifolds, exploring both their topological and geometric aspects. Although this material is available in research papers and is well-known among specialists, to our knowledge, it has not yet appeared in textbooks with a more didactic approach. A sample of recent papers for further study, whose main results are related to the material presented in this section, is [34, 39, 41–43, 75, 83, 91, 111, 112, 114–116].

A connected Riemannian manifold (M, g) is said to have *cohomogeneity one* if it supports an isometric action of a compact Lie group G with at least one orbit of codimension 1. This means that the orbit space M/G is one-dimensional, and since G can be assumed closed in the full isometry group of (M, g), we must have, up to rescaling, one of:

(i) $M/G = \mathbb{R}$, the real line;
(ii) $M/G = \mathbb{R}_+ = [0, +\infty)$, a half-line;
(iii) $M/G = S^1$, the circle;
(iv) $M/G = [-1, 1]$, an interval.

Let us analyze the orbit structure of M in each of the above cases, see Definition 3.84, by studying the fibers of the quotient map $\pi \colon M \to M/G$. Interior points of M/G correspond to principal orbits, and boundary points correspond to nonprincipal orbits, see Remark 3.106. In cases (i) and (iii) above, all G-orbits are principal, and M is a fiber bundle over M/G with fiber the principal orbit. In case (ii), there is one nonprincipal orbit, namely $G(x_0) = \pi^{-1}(0)$, so $M \cong G \times_{G_{x_0}} D$ is equivariantly diffeomorphic to the tubular neighborhood of $G(x_0)$, by the Tubular Neighborhood Theorem 3.57.[12] Finally, in case (iv), there are two nonprincipal orbits $\pi^{-1}(\pm 1)$. Topologically, the latter is the most interesting case, since it is not a fiber bundle over a simpler manifold in any obvious way.

We now specialize to this case $M/G = [-1, 1]$ to give a more in-depth analysis. Choose some $p_0 \in \pi^{-1}(0)$ and let $\gamma \colon [-1, 1] \to M$ be the unique minimal horizontal geodesic with $\gamma(0) = p_0$. Then γ can be extended to $\gamma \colon \mathbb{R} \to M$, and its image $\gamma(\mathbb{R})$ is either an embedded circle or a one-to-one immersed line; in particular, it does not have self-intersections [1, Thm 6.1]. Moreover, since it is horizontal, $\gamma(\mathbb{R})$ intersects all orbits orthogonally. This implies that the G-action on M is polar,[13] see Definition 4.8. Denote by $H := G_{\gamma(0)}$ the isotropy group at $\gamma(0)$, which is equal to the isotropy groups $G_{\gamma(t)}$ for all $t \not\equiv 1 \mod 2\mathbb{Z}$ (Exercise 6.35); and by $K_\pm := G_{\gamma(\pm 1)}$ the isotropy groups at $p_\pm := \gamma(\pm 1)$, respectively. Since $G/H = \pi^{-1}(t)$, $t \in (-1, 1)$, is a principal orbit, we call H the *principal isotropy group*, while each $K_\pm$ is called *singular* or *exceptional isotropy group*, according to $G/K_\pm = \pi^{-1}(\pm 1)$ being a singular or exceptional orbit. Denote by $D_\pm$ a normal slice to the nonprincipal orbit $G/K_\pm$ at $p_\pm$. We are now ready to describe a decomposition of such a cohomogeneity one manifold as union of two disk bundles.

Proposition 6.33. *There exists an equivariant diffeomorphism*

$$M \cong (G \times_{K_-} D_-) \cup_{G/H} (G \times_{K_+} D_+), \qquad (6.25)$$

i.e., M is the union of tubular neighborhoods of the nonprincipal orbits $G/K_\pm$, glued along their common boundary G/H.

Proof. According to the Tubular Neighborhood Theorem 3.57, $\pi^{-1}([-1, 0])$ and $\pi^{-1}([0, 1])$ are equivariantly diffeomorphic to the tubular neighborhoods

$$\mathrm{Tub}\big(G(p_\pm)\big) = G \times_{K_\pm} D_\pm$$

of the singular orbits $G/K_\pm$, respectively. The common boundary of each of these open subsets is the principal orbit $\pi^{-1}(0) = G/H$. Thus, the equivariant decomposition $M \cong \pi^{-1}([-1, 0]) \cup_{\pi^{-1}(0)} \pi^{-1}([0, 1])$ is precisely (6.25). $\square$

---

[12]Slices in this section are generally denoted by D, instead of S as in Chap. 3.

[13]Note that any cohomogeneity one action is polar, regardless of M/G being a closed interval.

Remark 6.34. The notion of cohomogeneity one action can be extended to the framework of Alexandrov spaces. In this context, analogous structure results to Proposition 6.33 have been proved by Galaz-Garcia and Searle [91].

Exercise 6.35. From Exercise 3.77, we know that the slice representation at $\gamma(0)$ is trivial. Use that the principal isotropy group $H = G_{\gamma(0)}$ preserves the horizontal geodesic γ pointwise to verify the claim that $G_{\gamma(t)} = H$ for all $t \in (-1, 1)$. Note that this also follows immediately from Kleiner's Lemma 3.70.

It is often more natural to consider a closed interval orbit space $M/G = [a, b]$, different from $[-1, 1]$, for geometric reasons. We discuss the abstract theory using $[-1, 1]$ as a standard rescaling, but in many examples the interval will be different, as in the following:

Exercise 6.36. As a first (somewhat trivial) example of cohomogeneity one manifold, consider the unit round sphere $S^2 \subset \mathbb{R}^3$ with an action of S^1 by rotation about the z-axis. Identify the quotient map with the distance function from the North pole $\pi \colon S^2 \to [0, \pi]$. Describe the isotropy groups, finding the corresponding principal and singular orbits. Write explicitly the decomposition (6.25). A generalization of this example is given in Example 6.47.

The normal slices $D_\pm$ are normal disks at $p_\pm$, and their boundary is called the *normal sphere* $S^\perp_\pm = \partial D_\pm$ to the singular orbit $G(p_\pm)$. We write

$$\ell_\pm := \dim S^\perp_\pm = \operatorname{codim} G/K_\pm - 1.$$

The singular isotropy group $K_\pm$ acts on the slice $D_\pm$ via the slice representation, see Definition 3.72. The restriction of this $K_\pm$-representation to $S^\perp_\pm$ is transitive, so the normal sphere is precisely the (possibly ineffective) homogeneous space $S^\perp_\pm = K_\pm/H$. This allows us to define the *group diagram* of this cohomogeneity one manifold:

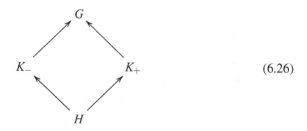

$$(6.26)$$

where the arrows denote the natural inclusions. We also denote the above diagram by $H \subset \{K_-, K_+\} \subset G$. Notice that such a group diagram determines two homogeneous bundles $K_\pm/H \to G/H \to G/K_\pm$, that have the principal orbit G/H as the total space of a sphere bundle over each singular orbit $G/K_\pm$.

Conversely, given compact Lie groups $H \subset \{K_-, K_+\} \subset G$ such that $K_\pm/H$ are spheres,[14] there exists a cohomogeneity one G-manifold whose group diagram is (6.26). Namely, define M using (6.25). In this way, the group diagram determines a cohomogeneity one manifold, up to certain ambiguities (see [114, p. 44]).

Proposition 6.37. *Two group diagrams* $H \subset \{K_-, K_+\} \subset G$ *and* $\widetilde{H} \subset \{\widetilde{K}_-, \widetilde{K}_+\} \subset G$ *determine the same cohomogeneity one manifold (up to equivariant diffeomorphism) if and only if, up to switching K_- and K_+, there exist a* $\in N(H)_0$ *and* $b \in G$ *such that* $K_- = b \widetilde{K}_- b^{-1}$, $H = b \widetilde{H} b^{-1}$ *and* $K_+ = ab \widetilde{K}_+ b^{-1} a^{-1}$.

Remark 6.38. The above results on how a cohomogeneity one manifold is determined by its group diagram illustrate a particular case of Theorem 5.42 regarding reconstructing polar manifolds from the polar data, due to Grove and Ziller [118]. Cohomogeneity one manifolds are polar, and its polar data is precisely its group diagram.

Using the decomposition (6.25), we can deduce important relations between the orbit structure and the topology of cohomogeneity one manifolds. For instance, it follows that the inclusion $G/K_\pm \hookrightarrow M$ is $\ell_\mp$-connected and $G/H \hookrightarrow M$ is $\min\{\ell_-, \ell_+\}$-connected.[15] One can also use this decomposition to apply the Mayer-Vietoris exact sequence, relating the topology of the orbits with that of M. We now collect some immediate consequences of the above observations, see [2, 114, 116].

Proposition 6.39. *Let M be a cohomogeneity one manifold with $M/G = [-1, 1]$. Then the following hold:*

  (i) *If both nonprincipal orbits $G/K_\pm$ are exceptional, i.e., $\ell_\pm = 0$, then M has infinite fundamental group;*
 (ii) *If M is simply-connected, then there are no exceptional orbits, i.e., $\ell_\pm \geq 1$;*
(iii) *If $\ell_- \geq 2$ and $\ell_+ \geq 1$, then K_+ is connected. If both $\ell_\pm \geq 2$, then all isotropy groups H and $K_\pm$ are connected;*
(iv) *The Euler characteristic of M is $\chi(M) = \chi(G/K_-) + \chi(G/K_+) - \chi(G/H)$.*

Remark 6.40. Item (ii) of the above result is actually true for any polar manifolds, see Corollary 5.35. In the particular case of cohomogeneity one, however, the proof is elementary from $G/K_\pm \hookrightarrow M$ being $\ell_\mp$-connected. For an example of a cohomogeneity one action with an exceptional orbit, consider $\mathbb{R}P^n$ with a rotation action of $SO(n)$, see Exercise 3.87.

We now turn to the question of determining the moduli space of invariant metrics on a cohomogeneity one manifold. An invariant smooth metric g on M is clearly determined by its restriction to the open dense set M_{princ} of regular points (recall the Principal Orbit Theorem 3.82), which in the cohomogeneity one case is $M_{\mathrm{princ}} = M \setminus$

---

[14]The possible pairs $(K_\pm, H)$ with $K_\pm$ connected for which $K_\pm/H$ is a sphere are listed in Table 6.1.

[15]Recall that a map $f: X \to Y$ is ℓ-connected if the induced homomorphisms $f_i: \pi_i(X) \to \pi_i(Y)$ is an isomorphism for $i < \ell$ and onto for $i = \ell$.

$\{G/K_\pm\}$. The choice of the horizontal geodesic $\gamma(t)$ on M determines an equivariant diffeomorphism $M_{\text{princ}} \cong (-1,1) \times G/H$, for which $\gamma(t) = (t, eH)$. Making this our standard choice of (global) coordinates on the regular part of M, an invariant smooth metric g can be written as

$$g = dt^2 + g_t, \quad t \in (-1,1), \tag{6.27}$$

where g_t is a smooth 1-parameter family of homogeneous metrics on G/H, satisfying appropriate smoothness conditions as $t \to \pm 1$. As usual, denote by $\mathfrak{h}$ and $\mathfrak{g}$ the Lie algebras of H and G. Using a choice of background bi-invariant metric Q on G, let $\mathfrak{m}$ be the orthogonal complement of $\mathfrak{h}$ in $\mathfrak{g}$. Then for each $t \in (-1,1)$, the tangent space $T_{\gamma(t)}M$ splits (orthogonally) as the direct sum of the normal space to the orbit $G(\gamma(t))$, spanned by $\dot{\gamma}(t) = \frac{\partial}{\partial t}$, and $T_{\gamma(t)}G(\gamma(t))$. We identify the latter with $T_{eH}G/H$ via action fields, as in (3.11), and hence with $\mathfrak{m}$, by Proposition 6.11. Gathering all of the above, we proved the following:

Proposition 6.41. *Let M be a cohomogeneity one manifold with $M/G = [-1,1]$ and $\gamma \colon [-1,1] \to M$ be a minimizing horizontal geodesic between the nonprincipal orbits. Every G-invariant metric g on M for which γ is a horizontal geodesic is of the form (6.27), where g_t is a 1-parameter family of G-invariant metrics on G/H, i.e., $\mathrm{Ad}(H)$-invariant inner products on $\mathfrak{m}$.*

By the above, one can find a smooth 1-parameter family of Q-symmetric $\mathrm{Ad}(H)$-equivariant automorphisms $P_t \colon \mathfrak{m} \to \mathfrak{m}$, such that

$$g\big(X^*_{\gamma(t)}, Y^*_{\gamma(t)}\big) = g_t(X,Y) = Q(P_t X, Y), \quad \text{for all } X, Y \in \mathfrak{m},\, t \in (-1,1).$$

The family P_t encodes the homogeneous metrics g_t and also contains information about the extrinsic geometry of principal orbits. For instance, we now compute the shape operator of principal orbits in terms of P_t, following Grove and Ziller [116].

Proposition 6.42. *For each $t \in (-1,1)$, the shape operator $\mathscr{S}_t$ of the hypersurface $G(\gamma(t))$ at $\gamma(t)$ is identified with*

$$\mathscr{S}_t \colon \mathfrak{m} \to \mathfrak{m}, \quad \mathscr{S}_t = -\tfrac{1}{2} P_t^{-1} P_t', \tag{6.28}$$

where $P_t' \colon \mathfrak{m} \to \mathfrak{m}$ is the first derivative $P_t'(X) = \frac{d}{ds} P_s(X)\big|_{s=t}$.

Proof. To simplify notation, we omit t whenever possible. Let $X \in \mathfrak{m}$ and consider its action field X^* on M. It is easy to see that, up to a change in coordinates that locally straightens the hypersurface $G(\gamma(t))$, the normal vector $\dot{\gamma}(t)$ is locally constant along $G(\gamma(t))$, hence has zero derivative in the direction of X^* in this coordinate system. This means that $[X^*, \dot{\gamma}] = 0$, hence $\nabla_{X^*} \dot{\gamma} = \nabla_{\dot{\gamma}} X^*$. In particular, using that X^* and Y^* are Killing vector fields (see Proposition 2.14), we have

$$g(\nabla_{\dot{\gamma}} X^*, Y^*) = g(\nabla_{X^*} \dot{\gamma}, Y^*) = X^* \left(g(\dot{\gamma}, Y^*) \right) - g(\dot{\gamma}, \nabla_{X^*} Y^*)$$
$$= g(X^*, \nabla_{\dot{\gamma}} Y^*).$$

Thus, we conclude the proof with

$$g_t(\mathscr{S}_t(X), Y) = -g(\nabla_{X^*} \dot{\gamma}, Y^*)$$
$$= -g(\nabla_{\dot{\gamma}} X^*, Y^*)$$
$$= -\tfrac{1}{2} \tfrac{d}{dt} g(X^*, Y^*)$$
$$= -\tfrac{1}{2} \tfrac{d}{dt} Q(P_t X, Y)$$
$$= -\tfrac{1}{2} Q(P'_t X, Y)$$
$$= -\tfrac{1}{2} g_t \left(P_t^{-1} P'_t X, Y \right).$$

$\square$

Remark 6.43. Let us discuss some aspects of submanifold geometry of orbits of a cohomogeneity one manifold. By G-equivariance of the mean curvature vector $\mathbf{H}$, every principal orbit $G(\gamma(t))$ is a *constant mean curvature* hypersurface, i.e., $\|\mathbf{H}\| = H_t$ is constant on $G(\gamma(t))$, for each t. This is equivalent to saying that the sum of eigenvalues of the shape operator $\mathscr{S}_t = -\tfrac{1}{2} P_t^{-1} P'_t$ at $\gamma(t)$ is the same in any other point of $G(\gamma(t))$, cf. Proposition 3.78 (iii).

Consider the mean curvature $\|\mathbf{H}\|$ of $G(\gamma(t))$ as a function of $t \in (-1, 1)$. It can be proved that its global minimum value is zero (corresponding to a principal orbit of critical volume, which must hence be minimal), and $\|\mathbf{H}\|$ diverges to $+\infty$ as $t \to \pm 1$, provided that the nonprincipal orbits are singular. As for the singular orbits themselves, they are isolated orbits of their type, hence a stratum of the stratification of M by orbit types (see Theorem 3.102). In particular, singular orbits are minimal submanifolds, see Remark 3.105. Furthermore, singular orbits are sometimes forced to be totally geodesic, e.g., due to the isotropy representations, since the second fundamental form is equivariant and the space of such objects may be trivial. Also, if the dimension of a singular orbit is $< \tfrac{1}{2}(\dim M - 1)$, then it must be totally geodesic.

Using these properties and equivariant bifurcation techniques, Bettiol and Piccione [39] recently proved the existence of infinitely many new constant mean curvature hypersurfaces that are diffeomorphic (but not isometric) to principal orbits on some cohomogeneity one manifolds, generalizing *Delaunay surfaces* on spheres. In the case of S^3, these are classical Delaunay surfaces, which are *nonround tori* (with isometry group $\mathbb{Z}_m \times S^1$) that bifurcate from the family (6.33) in Example 6.48.

Remark 6.44. Some of the above submanifold geometry properties of orbits of a cohomogeneity one action were proved to remain true for general isoparametric foliations by Ge and Tang [95, 96], see Remark 5.52.

We now describe a particularly useful type of cohomogeneity one metric, that Grove and Ziller [116] call *diagonal*. Suppose $\mathfrak{m} = \mathfrak{m}_1 \oplus \cdots \oplus \mathfrak{m}_l$ is a decomposition in mutually orthogonal $\mathrm{Ad}(H)$-invariant subspaces (that are not necessarily $\mathrm{Ad}(H)$-irreducible). Consider, along $\gamma(t)$, the metric

$$g_t := \sum_{i=1}^{l} f_i(t)^2 Q|_{\mathfrak{m}_i}, \tag{6.29}$$

where $f_i \colon (-1,1) \to \mathbb{R}$ are positive functions satisfying appropriate smoothness conditions as $t \to \pm 1$, and Q is a bi-invariant metric on G. The resulting invariant metric $g = dt^2 + g_t$ is called a *diagonal* cohomogeneity one metric. The following exercise justifies this terminology.

Exercise 6.45. Verify that the family of automorphisms $P_t \colon \mathfrak{m} \to \mathfrak{m}$ and the shape operator $\mathscr{S}_t$ of principal orbits with respect to a diagonal metric as above are:

$$P_t = \begin{pmatrix} f_1^2 & & \\ & \ddots & \\ & & f_l^2 \end{pmatrix} \quad \text{and} \quad \mathscr{S}_t = - \begin{pmatrix} \frac{f_1'}{f_1} & & \\ & \ddots & \\ & & \frac{f_l'}{f_l} \end{pmatrix}. \tag{6.30}$$

Exercise 6.46. Let $g = dt^2 + g_t$ be a diagonal cohomogeneity one metric and consider its Cheeger deformation $g_s = dt^2 + g_{s,t}$ with respect to the cohomogeneity one action. Here, s denotes the parameter of the Cheeger deformation. Verify that g_s, $s \geq 0$, are also diagonal metrics, and use Proposition 6.3 to check that $g_{s,t}(X,Y) = Q(P_{s,t}X,Y)$, where the Q-symmetric automorphism $P_{s,t} \colon \mathfrak{m} \to \mathfrak{m}$ is

$$P_{s,t} = \begin{pmatrix} \dfrac{f_1(t)^2}{1+sf_1(t)^2} & & \\ & \ddots & \\ & & \dfrac{f_l(t)^2}{1+sf_l(t)^2} \end{pmatrix}. \tag{6.31}$$

Note that as $s \to +\infty$, for each t, $g_{s,t} \to 0$, and hence (M,g_s) converges in Gromov-Hausdorff sense to its orbit space M/G.

Before proceeding with abstract cohomogeneity one manifolds, let us examine in details a few examples where many of the above results can be seen more explicitly.

Example 6.47 (Surfaces of revolution). Surfaces of revolution M in $\mathbb{R}^3$ can be seen as a very simple type of diagonal cohomogeneity one manifold, where the group acting is S^1. Consider the embedding

$$F\colon [0,R] \times [0,2\pi] \to \mathbb{R}^3, \quad F(r,\theta) = \big(\phi(r)\cos\theta, \phi(r)\sin\theta, \psi(r)\big),$$

where $\phi\colon [0,R] \to \mathbb{R}$ and $\psi\colon [0,R] \to \mathbb{R}$ are smooth functions, such that ϕ is positive and only vanishes[16] at the endpoints. For simplicity, we assume that the profile curve $F(r,0) = (\phi(r),0,\psi(r))$ is parametrized with unit speed, i.e., $(\phi')^2 + (\psi')^2 = 1$. The manifold $M = F([0,R] \times [0,2\pi])$ is a surface of revolution in $\mathbb{R}^3$, diffeomorphic to S^2. It has a cohomogeneity one action of S^1 by rotating the parameter θ, i.e., rotations around the z-axis. The quotient map is given by the distance function to $F(0,\theta)$ on M,

$$\pi\colon M \subset \mathbb{R}^3 \to [0,R], \quad \pi(F(r,\theta)) = r.$$

The singular orbits are fixed points $F(0,\theta)$ and $F(R,\theta)$, and principal orbits are circles $\{F(r,\theta) : \theta \in [0,2\pi]\}$. In particular, the horizontal[17] geodesic $\gamma\colon [0,R] \to M$ can be taken as the profile curve $\gamma(r) := F(r,0) = (\phi(r),0,\psi(r))$. The disk bundles in (6.25) are simply disks that, glued along their common circle boundary (a principal orbit), give the usual topological decomposition of $M \cong S^2$.

The metric induced by F on M is easily computed to be

$$g = \big((\phi')^2 + (\psi')^2\big)\mathrm{d}r^2 + \phi^2\mathrm{d}\theta^2 = \mathrm{d}r^2 + \phi^2\mathrm{d}\theta^2, \quad r \in (0,R), \tag{6.32}$$

i.e., $g = \mathrm{d}r^2 + g_r$, where g_r is a round metric on S^1 with length $\phi(r)$. This is trivially a diagonal metric, with $P_r = \phi(r)^2$, see Exercise 6.45. This generalizes the example of the round unit sphere in Exercise 6.36, where $R = \pi$, $\phi(r) = \sin r$ and $\psi(r) = \cos r$.

Example 6.48. Let $S^3 = \{(z,w) \in \mathbb{C}^2 : |z|^2 + |w|^2 = 1\}$ be the round sphere, and consider the action of the torus $G = T^2 = \{(e^{i\theta_1}, e^{i\theta_2}) : \theta_j \in \mathbb{R}\}$ by

$$\big(e^{i\theta_1}, e^{i\theta_2}\big) \cdot (z,w) := \big(e^{i\theta_1}z, e^{i\theta_2}w\big).$$

The closed geodesics $\{(z,0) \in S^3\}$ and $\{(0,w) \in S^3\}$ are singular orbits of this T^2-action, corresponding to a circle isotropy subgroup. The remaining orbits are principal orbits that have trivial isotropy group, and are hence diffeomorphic to T^2. Let $\gamma(t) = (\cos t, \sin t)$, $0 \le t \le \frac{\pi}{2}$, be a horizontal geodesic joining the singular orbits. Then the orbits of $G = T^2$ are

$$G(\gamma(t)) = \{(z,w) \in S^3 : |z| = \cos t, |w| = \sin t\}, \quad 0 \le t \le \tfrac{\pi}{2}. \tag{6.33}$$

---

[16] In order for $M = F([0,R] \times [0,2\pi])$ to be a smooth closed submanifold of $\mathbb{R}^3$, some conditions on the derivatives of ϕ and ψ at the endpoints are needed. This illustrates the fact that the metrics g_r in (6.27) must satisfy appropriate conditions at the corresponding endpoints $t \to \pm 1$.

[17] This is a *horizontal* curve in the sense that it is orthogonal to all circle orbits. However, in $\mathbb{R}^3$, it sits in the plane $y = 0$, while the circle orbits are in planes $z = const$.

The group diagram of this cohomogeneity one action is

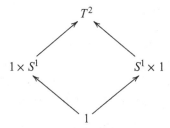

Notice that the decomposition (6.25) is the usual decomposition of S^3 as two solid tori glued along their boundary. We can write the round metric g on S^3 in the form (6.27), obtaining

$$g = dt^2 + g_t = dt^2 + \cos^2 t \, dx^2 + \sin^2 t \, dy^2, \qquad (6.34)$$

where $z = e^{ix} \cos t$ and $w = e^{iy} \sin t$. In other words, $\frac{\partial}{\partial x}$ and $\frac{\partial}{\partial y}$ are the vertical vector fields along $\gamma(t)$, that span the tangent space to (6.33). Thus, (6.34) is a diagonal metric with

$$f_1(t) = \cos t \quad \text{and} \quad f_2(t) = \sin t.$$

In particular, the shape operator and hence the mean curvature of (6.33) can be computed as in Exercise 6.45, and the result is

$$H_t = \tan t - \cot t, \quad 0 < t < \tfrac{\pi}{2}. \qquad (6.35)$$

Thus, we have a minimal torus when $t = \frac{\pi}{4}$, corresponding to the principal orbit of largest volume; and the constant mean curvature (6.35) of the tori $G(\gamma(t))$ blows up as $t \to 0$ and $t \to \frac{\pi}{2}$, cf. Remark 6.43. Notice that this partition of S^3 is precisely the singular Riemannian foliation described in Exercise 5.54, in the particular case $k = 1$ and $n = 2$, see also Remark 5.55.

Remark 6.49. The torus $G(\gamma(\frac{\pi}{4})) = f^{-1}\left(\frac{1}{2}\right)$ is called the (minimal) Clifford torus in S^3. As a curiosity, Brendle [54] recently gave a proof of the celebrated *Lawson conjecture*, which asserts that the Clifford torus is the unique embedded minimal torus in S^3, up to isometries.

Exercise 6.50 ($\star$). The goal of this exercise is to generalize the above Example 6.48 to S^{n+1}, $n \geq 2$. Use the direct sum decomposition $\mathbb{R}^{n+2} = \mathbb{R}^{k+1} \oplus \mathbb{R}^{n-k+1}$ to find an action of $G = \mathrm{SO}(k+1) \times \mathrm{SO}(n-k+1)$ on S^{n+1} whose orbits are the level sets of the isoparametric function $f \colon S^{n+1} \to \mathbb{R}$ from Exercise 5.54. Verify that the round metric on S^{n+1} is a diagonal cohomogeneity one metric with respect to this action, with

$$f_1(t) = \cdots = f_k(t) = \cos t \quad \text{and} \quad f_{k+1}(t) = \cdots = f_n(t) = \sin t.$$

Using Exercise 6.45, compute the principal curvatures of the principal orbits and see that they match the result in Exercise 5.54 (iii) and Remark 5.55; in particular, prove that their mean curvature is

$$H_t = k \tan t - (n-k) \cot t, \quad 0 < t < \tfrac{\pi}{2}.$$

Note that when $t = \arctan \sqrt{\frac{k}{n-k}}$, we get a minimal $S^k \times S^{n-k}$ in S^{n+1}; and as $t \to 0$ and $t \to \frac{\pi}{2}$, the mean curvature blows up, cf. Remark 6.43.

Remark 6.51. Analogously to Exercise 6.50, the singular Riemannian foliation of hyperbolic space $H^{n+1} \subset (\mathbb{R}^{n+2}, g_L)$ described in Exercise 5.56 can also be seen as a partition into orbits of a cohomogeneity one G-action, via the direct sum decomposition $\mathbb{R}^{n+2} = \mathbb{R}^{k+1} \oplus \mathbb{R}^{n-k+1}$, with $G = \mathrm{SO}(k,1) \times \mathrm{SO}(n-k+1)$. Here, $\mathrm{SO}(k,1)$ is the *Lorentz group*, which is the (identity component of the) group of isometries of the Lorentz space $(\mathbb{R}^{k+1}, g_L)$ that fix the origin, i.e., the isometry group of H^k.

Example 6.52. Let $\ell \subset \mathbb{C}^{n+2}$ be a complex line, and consider its orthogonal complement $\ell^\perp \cong \mathbb{C}^{n+1}$. There is a natural linear action of $\mathrm{SU}(n+1)$ on $\mathbb{C}^{n+2} = \ell \oplus \ell^\perp$ that fixes ℓ and acts by multiplication on $\ell^\perp$. This $\mathrm{SU}(n+1)$-action on $\mathbb{C}^{n+2}$ commutes with the circle action by complex multiplication whose quotient is $\mathbb{C}P^{n+1}$. As a result, there is an induced $\mathrm{SU}(n+1)$-action on $\mathbb{C}P^{n+1}$. The image $[\ell] \in \mathbb{C}P^{n+1}$ is clearly a fixed point of this action, and it is easy to see that the orbit space is isometric to $[0, \pi/2]$. More precisely, consider a geodesic arc in the unit sphere of $\mathbb{C}^{n+2}$ that joins ℓ to $\ell^\perp$. The image of this geodesic in $\mathbb{C}P^{n+1}$ is a geodesic γ that intersects all orbits perpendicularly. Moreover, the quotient map $\pi \colon \mathbb{C}P^{n+1} \to [0, \pi/2]$ can be identified with the distance function from $[\ell]$. It is easy to check that the group diagram for this action is:

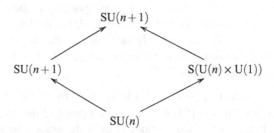

cf. Exercise 3.44, so the normal spheres to the singular orbits have dimensions $\ell_- = 2n+1$ and $\ell_+ = 1$. The singular orbits are the fixed point $\{[\ell]\} = \pi^{-1}(0)$ and its cut locus $\mathbb{C}P^n = \pi^{-1}(\pi/2)$. Principal orbits are geodesic spheres $S^{2n+1} = \pi^{-1}(t)$ of radius $t \in (0, \pi/2)$ centered at $[\ell]$. Notice that the decomposition (6.25) is homotopically the cell decomposition of $\mathbb{C}P^{n+1}$ as $\mathbb{C}P^n$ with a $2n$-cell attached to it via the Hopf map.

Using that the quotient map $S^{2n+3} \to \mathbb{C}P^{n+1}$ is a Riemannian submersion from the round sphere to $(\mathbb{C}P^{n+1}, g_{FS})$, we can write the Fubini-Study metric as a cohomogeneity one metric (6.27), obtaining:

$$g_{FS} = dt^2 + g_t = dt^2 + \sin^2 t \left(g|_{\mathscr{H}} + \cos^2 t \, g|_{\mathscr{V}} \right). \tag{6.36}$$

This means that the principal orbits (S^{2n+1}, g_t) are isometric to Berger spheres, obtained by rescaling by $\sin^2 t$ the $SU(n+1)$-invariant metric g_λ with $\lambda = \cos^2 t$, from Example 6.16. In order to write g_t as a diagonal metric in the form (6.29), we consider the bi-invariant metric Q on $SU(n+1)$, recall Exercise 2.25 and Proposition 2.48. There is a Q-orthogonal decomposition $\mathfrak{m} = \mathfrak{m}_1 \oplus \mathfrak{m}_2$ in $\mathrm{Ad}(SU(n))$-invariant subspaces $\mathfrak{m}_i$ of dimensions $2n$ and 1 respectively, corresponding to the horizontal $\mathscr{H}$ and vertical $\mathscr{V}$ directions of the Hopf bundle. These $\mathrm{Ad}(SU(n))$-subrepresentations are irreducible, since $SU(n)$ acts by multiplication on $\mathfrak{m}_1$ and trivially on $\mathfrak{m}_2$. From (6.36) and Example 6.24, we have that the corresponding functions f_i are

$$f_1(t) = \sin t \quad \text{and} \quad f_2(t) = \sqrt{\tfrac{2n}{n+1}} \sin t \cos t. \tag{6.37}$$

This gives a full description of the intrinsic geometry of the principal orbits.

As to their extrinsic geometry, we first note that (S^{2n+1}, g_t) are the distance spheres in $(\mathbb{C}P^{n+1}, g_{FS})$, of radius t centered at the fixed point $[\ell]$. We can compute their shape operator using Exercise 6.45 and (6.37), obtaining

$$\mathscr{S}_t = -\begin{pmatrix} \cot t & & & \\ & \ddots & & \\ & & \cot t & \\ & & & 2\cot 2t \end{pmatrix}. \tag{6.38}$$

In particular, the mean curvature of the principal orbit (S^{2n+1}, g_t) is

$$H_t = 2\cot 2t + 2n \cot t, \tag{6.39}$$

so, we have a minimal hypersurface when $t = \arctan \sqrt{2n+1}$, corresponding to the principal orbit of largest volume, and (6.39) blows up as $t \to 0$ and $t \to \pi/2$, cf. Remark 6.43.

Exercise 6.53. Verify formula (iv) for Euler characteristics in Proposition 6.39 in the case of the rotation action on the sphere (Exercise 6.36) and in the case of the $SU(n+1)$-action on $\mathbb{C}P^{n+1}$ (Example 6.52). Recall that $\chi(\mathbb{C}P^n) = n+1$ and $\chi(S^n) = 2$ if n is even and 0 if n is odd, see Example A.32.

Exercise 6.54 ($\star$). Describe a quaternionic version of the cohomogeneity one action in Example 6.52, using a quaternionic line ℓ in $\mathbb{H}^{n+2}$ and considering the action of $Sp(n+1)$ that fixes it. Identify the orbits of the induced $Sp(n+1)$-action on $\mathbb{H}P^{n+1}$, write the group diagram of this action, and study the intrinsic and extrinsic geometry of the principal orbits.

Exercise 6.55 ($\star$). Consider the family of cohomogeneity one manifolds M_k, $k \in \mathbb{N}$, defined by the group diagrams $\mathbb{Z}_k \subset \{S^1, S^1\} \subset S^3$, where both embeddings of $K_{\pm} = S^1$ inside $G = S^3$ are given by the maximal torus (4.1).

  (i) Describe an equivariant diffeomorphism between M_k and the associated bundle $S^3 \times_{S^1} S^2$ to the Hopf bundle $S^1 \to S^3 \to S^2$, where the S^1-action on S^2 is by rotations of speed k, studied in Example 3.55;

  (ii) Conclude that $S^2 \times S^2$ and $\mathbb{C}P^2 \# \overline{\mathbb{C}P}^2$ admit cohomogeneity one G-actions, and verify that the G-action on $S^2 \times S^2$ descends to an $SO(3)$-action;

  (iii) Find cohomogeneity one metrics on $S^2 \times S^2$ and $\mathbb{C}P^2 \# \overline{\mathbb{C}P}^2$ with $\sec \geq 0$, and study the intrinsic and extrinsic geometry of the principal orbits.

Hint: For (i), verify that a curve joining the North and South poles on S^2 defines a curve on $S^3 \times S^2$ that intersects each orbit of the diagonal S^1-action, and use (6.25). Item (ii) follows from Example 3.55 by direct inspection. For (iii), consider the product of round metrics on $S^3 \times S^2$ and use the Gray-O'Neill formula (2.14), see Remark 6.80 for more details.

Example 6.56. Consider the *Brieskhorn variety* $M_d^{2n-1} \subset \mathbb{C}^{n+1}$ defined by

$$\begin{cases} z_0^d + z_1^2 + \cdots + z_n^2 = 0, \\ |z_0|^2 + |z_1|^2 + \cdots + |z_n|^2 = 1. \end{cases}$$

When n and d are odd, M_d^{2n-1} is homeomorphic to the sphere S^{2n-1}. However, if $2n - 1 \equiv 1 \mod 8$, then M_d^{2n-1} is *not diffeomorphic* to S^{2n-1}. Such manifolds are exotic spheres called *Kervaire spheres*.

According to [111], it was first observed by Calabi in dimension 5 and later by Hsiang and Hsiang [130] that M_d^{2n-1} carries a cohomogeneity one action by $SO(2)SO(n)$, given by $(e^{i\theta}, A) \cdot (z_0, \ldots, z_n) = (e^{2i\theta}z_0, e^{id\theta}A(z_1, \ldots, z_n)^t)$. The principal isotropy of this action is

$$H = \mathbb{Z}_2 \times SO(n-2) = \begin{cases} (\varepsilon, \mathrm{diag}(\varepsilon, \varepsilon, A)) & \text{if } d \text{ is odd} \\ (\varepsilon, \mathrm{diag}(1, 1, A)) & \text{if } d \text{ is even,} \end{cases} \qquad (6.40)$$

where $\varepsilon = \pm 1$ and $A \in SO(n-2)$. Thus, principal orbits are diffeomorphic to $S^1 \times T_1 S^{n-1}$, where $T_1 S^{n-1}$ denotes the unit tangent bundle of S^{n-1}. The singular isotropy groups are

$$K_- = SO(2)SO(n-2) = (e^{-i\theta}, \mathrm{diag}(R(d\theta), A)), \qquad (6.41)$$

where $R(\theta)$ is a counterclockwise rotation of angle θ, and

$$K_+ = \begin{cases} O(n-1) = (\det B, \operatorname{diag}(\det B, B)) & \text{if } d \text{ is odd} \\ \mathbb{Z}_2 \times SO(n-1) = (\varepsilon, \operatorname{diag}(1, B')) & \text{if } d \text{ is even,} \end{cases} \tag{6.42}$$

where $\varepsilon = \pm 1$, $B \in O(n-1)$ and $B' \in SO(n-1)$.

Exercise 6.57 ($\star$). The goal of this exercise is to prove a result of Back and Hsiang [25, Thm VI.3], showing that the Brieskorn variety M_d^{2n-1} with the above cohomogeneity one action does not admit an invariant metric with sec > 0, provided $n \geq 4$ and $d \geq 3$. For an improvement of this result to sec ≥ 0, see Remark 6.74.

The proof is by contradiction, verifying that one of the singular orbits must be totally geodesic and hence would have sec > 0 intrinsically, but this is impossible since it does not appear in the classification (Theorem 6.30). More precisely:

(i) Show that the cyclic subgroup of G generated by $(e^{-i\pi/d}, -I)$ for n even, and by $(e^{-i\pi/d}, I)$ for n odd is a normal subgroup in both G and K_-. This is the largest normal subgroup common to G and K_- hence the ineffective kernel of the G-action on G/K_-;

(ii) Use that $e^{i\pi} \in H$ to show that the action of $e^{i\theta} \in SO(2) \subset K_-$ on the normal slice D_- is given by a rotation of angle 2θ;

(iii) Conclude that the singular orbit G/K_- is the fixed point set of a group of isometries, hence a totally geodesic submanifold by Proposition 3.93. Now, use Theorem 6.30 to obtain a contradiction if M_d^{2n-1} had an invariant metric with sec > 0.

Given that cohomogeneity one actions are polar with section $\gamma(\mathbb{R})$, an important object of study is the corresponding Weyl group (also called polar group), i.e., the group W such that $M/G = \gamma(\mathbb{R})/W$, see Remark 4.43. In our case, the *cohomogeneity one Weyl group* W is the stabilizer of the normal geodesic γ modulo its kernel H. More precisely, W is the dihedral subgroup of $N(H)/H$ generated by the unique involutions $w_\pm \in (N(H) \cap K_\pm)/H$. Since $\gamma(\mathbb{R})/W = M/G = [-1, 1]$, the Weyl group is finite if and only if γ is a closed geodesic, in which case $|W|$ is precisely twice the number of points where γ intersects the singular orbits, cf. Remark 4.40. It also follows that the nonprincipal isotropy groups along $\gamma(\mathbb{R})$ are conjugate to one of $K_\pm$ via an element of W. Although we do not carry out a longer discussion of the Weyl group in cohomogeneity one here, it is a fundamental tool in classification results such as Theorem 6.66.

We now study the curvature of cohomogeneity one manifolds. Any plane tangent to a cohomogeneity one manifold at $\gamma(t)$ is spanned by orthonormal vectors of the form X and $Y + \alpha \dot\gamma(t)$, where X and Y are vertical vectors, i.e., tangent to the hypersurface $G(\gamma(t))$. The sectional curvature of such a plane can be written as

$$\sec(X, Y + \alpha\dot\gamma(t)) = R(X, Y, X, Y) + 2\alpha R(X, Y, X, \dot\gamma(t))$$
$$+ \alpha^2 R(X, \dot\gamma(t), X, \dot\gamma(t)).$$

By G-invariance, any plane tangent to M at a regular point can be moved to a plane of the above form via an isometry in G. Thus, computing the sectional curvatures of these planes tangent at $\gamma(t)$ is sufficient to compute the sectional curvature of all planes at regular points. Using Proposition 6.26 and the Gauss equation (2.7), one can compute each of the terms in the right-hand side of the above formula. We follow the same notation as in the previous section, where $B_\pm \colon \mathfrak{m} \times \mathfrak{m} \to \mathfrak{g}$ is defined as (6.23), but is now t-dependent, just like $P \colon \mathfrak{m} \to \mathfrak{m}$. For convenience, we drop the subindex t. The outcome of this computation is as follows, see [116, Cor 1.10].

Proposition 6.58. *The sectional curvatures of a cohomogeneity one metric* g *at* $\gamma(t)$, $t \in (-1,1)$, *are determined by*

$$R(X,Y,X,Y) = Q(B_-(X,Y),[X,Y]) - \tfrac{3}{4}Q(P[X,Y]_\mathfrak{m},[X,Y]_\mathfrak{m})$$
$$+ Q(B_+(X,Y),P^{-1}B_+(X,Y))$$
$$- Q(B_+(X,X),P^{-1}B_+(Y,Y))$$
$$+ \tfrac{1}{4}Q(P'X,Y)^2 - \tfrac{1}{4}Q(P'X,X)Q(P'Y,Y)$$

$$R(X,Y,\dot{\gamma}(t),Y) = \tfrac{3}{4}Q([X,Y],P'Y)$$
$$- \tfrac{1}{2}Q(P'X,P^{-1}B_+(Y,Y))$$
$$+ \tfrac{1}{2}Q(P'Y,P^{-1}B_+(X,Y))$$

$$R(X,\dot{\gamma}(t),X,\dot{\gamma}(t)) = Q\big(\big(-\tfrac{1}{2}P'' + \tfrac{1}{4}P'P^{-1}P'\big)X,X\big).$$

Exercise 6.59 (⋆). Using Exercise 6.45 and Proposition 6.58, compute the sectional curvatures of a diagonal cohomogeneity one metric, i.e., with g_t as in (6.29). The answers can be found in [116, Cor 1.13].

Hint: It suffices to compute $\sec(X,\dot{\gamma}(t))$ for $X \in \mathfrak{m}_i$ and $\sec(X,Y)$ for $X,Y \in \mathfrak{m}_i$ or $X \in \mathfrak{m}_i$ and $Y \in \mathfrak{m}_j$. To simplify the computations, assume X and Y are Q-orthogonal and use that $Q([\mathfrak{m}_i,\mathfrak{m}_j],\mathfrak{h}) = 0$ by skew-symmetry of $\mathrm{ad}(X)$, see Proposition 2.26 (i).

Example 6.60. Let $M \subset \mathbb{R}^3$ be a surface of revolution as in Example 6.47, with the cohomogeneity one metric g given by (6.32). There is a unique plane tangent at each point of M, and its sectional curvature coincides with the Gaussian curvature of M, see Remark 2.21. Recall that $P_r = \phi(r)^2$, so if $X \in T_{\gamma(r)}M$ is a vector normal to $\dot{\gamma}(r)$ with $Q(X,X) = 1$, then by Proposition 6.58, $R(X,\dot{\gamma}(r),X,\dot{\gamma}(r)) = -\phi\phi''$. Thus, the curvature of (M,g) at $\gamma(r)$ is given by

$$\sec\left(X,\dot{\gamma}(r)\right) = -\frac{\phi''(r)}{\phi(r)}, \tag{6.43}$$

since $g(X,X) = \phi^2$. By the rotational symmetry, all points $F(r,\theta) \in M$ have the same curvature (6.43) as $F(r,0)$.

Exercise 6.61. Consider a surface of revolution (M,g) as in Examples 6.47 and 6.60. Integrating (6.43), verify that the total curvature of (M,g) is

$$\int_M \sec(T_pM)\,\mathrm{vol}_g = 2\pi\big(\phi'(0) - \phi'(R)\big). \tag{6.44}$$

Since $\chi(M) = \chi(S^2) = 2$, conclude that the *smoothness conditions* $\phi'(0) = 1$ and $\phi'(R) = -1$ are necessary for the Gauss-Bonnet Theorem to hold in its usual smooth formulation.

Hint: The volume form of M is $\mathrm{vol}_g = \phi(r)\,\mathrm{d}r\,\mathrm{d}\theta$.

Exercise 6.62 ($\star$). Use Exercise 6.46 and Examples 6.47 and 6.60 to compute the sectional curvatures of the Cheeger deformation of the round sphere (S^2,g) and the flat plane $(\mathbb{R}^2,g)$ with respect to the rotation action by S^1, verifying that

$$\sec_{g_t}\big(T_{\gamma(r)}\mathbb{R}^2\big) = \frac{3t}{(1+tr^2)^2}, \quad \text{and} \quad \sec_{g_t}\big(T_{\gamma(r)}S^2\big) = \frac{1+2t+t\cos(2r)}{(1+t\sin^2 r)^2},$$

where $\gamma(r)$ is a radial unit speed geodesic in this surface of revolution, and $\gamma(0)$ is a fixed point. Use the above to verify the claims in Example 6.5.

As mentioned in the previous section, the converse statement to the Bonnet-Myers Theorem 2.19 holds for compact homogeneous spaces, see Remark 6.29. The fact that such converse statement also holds for compact cohomogeneity one manifolds was proved by Grove and Ziller [116].

Theorem 6.63. *A compact cohomogeneity one manifold M admits an invariant metric with* $\mathrm{Ric} > 0$ *if and only if* $\pi_1(M)$ *is finite.*

We conclude this section mentioning an important principle that allows to find applications of cohomogeneity one actions in Geometric Analysis, based on the fact that many geometrically motivated partial differential equations (PDEs) on manifolds are invariant under isometries. Any such PDE on a manifold (M,g) with a cohomogeneity one G-action reduces to a differential equation on the 1-dimensional orbit space M/G, which is hence an ordinary differential equation (ODE), cf. Remark 6.20. This means that cohomogeneity one is a useful Ansatz to simplify the search for special solutions to PDEs with many symmetries. This approach has been extensively used to study the Einstein equation $\mathrm{Ric} = \lambda g$, see [41–43, 74, 83], and some of its generalizations, such as Ricci solitons, see [73, 75].

6.4 Positive and Nonnegative Curvature via Symmetries

In this section, we provide a survey on manifolds with $\sec > 0$ and $\sec \geq 0$ that have many symmetries, in the sense of admitting an isometric action with low cohomogeneity. We first give an overview of the main results in the area, and then discuss some recent results in Bettiol [34] and Bettiol and Mendes [35, 36]. The former is a construction related to the Hopf Problem, where cohomogeneity one and conformal techniques are used to produce metrics on $S^2 \times S^2$ satisfying an intermediate curvature condition between Ric > 0 and sec > 0, called sec$^{0+} > 0$. The latter are related to an intermediate curvature condition between sec > 0 and positive-definiteness of the curvature operator, known as *strongly positive curvature*.

The study of closed manifolds with sec > 0 is one of the most challenging areas in Riemannian geometry. Despite being a classical subject, surprisingly little is understood about this class. In addition, some of the oldest open problems in global Riemannian geometry regard manifolds with sec > 0, such as the celebrated:

Hopf Problem 6.64. *Does $S^2 \times S^2$ admit a Riemannian metric with* sec > 0?

On the one hand, there are very few known topological obstructions to sec > 0. On the other hand, it is notoriously difficult to construct examples of closed manifolds with sec > 0 different from the compact rank one symmetric spaces (CROSS) discussed in Example 6.10. Despite these difficulties, many advances in the area were achieved through the use of symmetries.

Since new examples may be discovered in the process of classifying manifolds with sec > 0 with certain symmetries, classification results such as Theorem 6.30 play a very important role. Symmetries can also be used as an aid to settle open problems under stronger hypotheses, providing evidence for the general case. The archetypical example is the following result of Hsiang and Kleiner [131], which settles the above Hopf Problem 6.64 under the assumption of continuous symmetry.

Theorem 6.65. *Let M^4 be an orientable compact Riemannian manifold with* sec > 0. *If M has a nontrivial isometric S^1-action, then M is homeomorphic to S^4 or $\mathbb{C}P^2$. In particular, if $S^2 \times S^2$ has a metric with* sec > 0, *then its isometry group is finite.*

Motivated by the above ideas, Grove [107] formulated a classification program for manifolds with sec > 0 or sec ≥ 0 and many symmetries, which led to several interesting developments in the last decade. A key feature in this program is the ambiguity in what it means for a Riemannian manifold (M, g) to have *many symmetries*. Among the various notions of highly symmetric manifolds, in this text we focus exclusively on manifolds with *low cohomogeneity*, i.e., admitting an isometric action by a Lie group G so that the orbit space M/G has low dimension. We stress that this is only one particular aspect of the above program, and that we do not address many remarkable results in the area that have been obtained under other interpretations of what it means to have many symmetries. Among these, we highlight manifolds with sec > 0 and large *symmetry rank,*[18] on which there is extensive literature, see [22, 23, 92, 93, 109, 110, 138, 139]. Further details on these other developments can be found in the surveys of Wilking [231] and Ziller [236], which also informed our approach to sec > 0 in low cohomogeneity.

---

[18]By definition, the *symmetry rank* of a manifold (M, g) with an isometric G-action is rank G.

The lowest possible cohomogeneity is zero, which means M/G is a point, i.e., the G-action is transitive. In this case, the only simply-connected manifolds that admit an invariant metric with sec > 0 are listed in Theorem 6.30. Furthermore, optimal pinching constants for homogeneous metrics have also been thoroughly studied, see Sect. 6.2.

The next case is that of manifolds with a cohomogeneity one action, described in Sect. 6.3. In dimensions ≤ 6, Searle [193] showed that a cohomogeneity one manifold carrying an invariant metric with sec > 0 must be diffeomorphic to S^n or $\mathbb{C}P^{n/2}$. The current state-of-the-art result on cohomogeneity one manifolds with sec > 0 is the following almost[19] classification, due to the efforts of Verdiani [218], Grove, Wilking and Ziller [114] and Verdiani and Ziller [220].

Theorem 6.66 (Cohomogeneity one pre-classification of sec > 0**).** *Apart from the CROSS (see Example 6.10), the only simply-connected closed manifolds that can have an invariant cohomogeneity one metric with* sec > 0 *are equivariantly diffeomorphic to*:

- *Berger space $B^7 = SO(5)/SO(3)$;*
- *Eschenburg spaces $E_p^7 = SU(3)/S_p^1$;*
- *Bazaikin spaces $B_p^{13} = SU(5)/S_p^1 Sp(2)$;*
- *One of the candidates P_k^7, or Q_k^7.*

Details on the Eschenburg spaces E_p^7 and Bazaikin spaces B_p^{13} can be found in [236]. These are *biquotients* (another generalization of homogeneous spaces), and were already known to admit metrics with sec > 0. Similarly, many other manifolds in the list (CROSS and the Berger space B^7) were known to have a homogeneous metric with sec > 0 (i.e., they occur in Theorem 6.30). However, note that this does not necessarily imply that they also have invariant metrics with sec > 0 and *cohomogeneity one*. The remaining manifolds P_k^7 and Q_k^7 are infinite families of candidates to have an invariant metric with sec > 0. Their group diagrams are, respectively,

$$H = \Delta Q \subset \left\{ \left(e^{i\theta}, e^{i\theta}\right) \cdot H, \left(e^{j(1-2k)\theta}, e^{j(1+2k)\theta}\right) \cdot H \right\} \subset Sp(1) \times Sp(1), \text{ and}$$

$$H = \{(\pm 1, \pm 1), (\pm i, \pm i)\} \subset \left\{ \left(e^{i\theta}, e^{i\theta}\right) \cdot H, \left(e^{j(k+1)\theta}, e^{-jk\theta}\right) \cdot H \right\}$$
$$\subset Sp(1) \times Sp(1),$$

where ΔQ denotes a diagonal embedding of the group $Q := \{\pm 1, \pm i, \pm j, \pm k\}$. Regarding constructing invariant metrics of positive curvature on P_k^7 and Q_k^7, the case $k = 1$ is trivial, since $P_1^7 = S^7$ and $Q_1^7 = W_{1,1}^7$ are manifolds known to have such metrics. Recently, a construction of an invariant metric with positive curvature on P_2^7 was achieved independently by Dearricott [77] and Grove, Verdiani and Ziller

---

[19]This is not a complete classification, as the *candidate* manifolds P_k^7 and Q_k^7 are not yet known to carry cohomogeneity one metrics with sec > 0, as we explain below.

[112]. In the latter, the manifold P_2^7 was also identified to be an exotic $T_1 S^4$, i.e., homeomorphic (but not diffeomorphic) to the unit tangent bundle $T_1 S^4$ of the sphere S^4. Constructing invariant metrics with positive curvature on P_k^7, $k \geq 3$, and Q_k^7, $k \geq 2$, – or proving that they do not exist – remains a very important and open problem in the area. Finally, we remark that the original classification result in [114] contained one more candidate in dimension 7, denoted R^7, that was very recently ruled out by Verdiani and Ziller [220]. This sequence of results illustrates a successful instance of the above mentioned Grove program, where a new example was discovered in the process of classifying cohomogeneity one positively curved manifolds.

Very little is known regarding the next steps for sec > 0 in low cohomogeneity, even cohomogeneity two. One of the major obstacles is that actions of cohomogeneity ≥ 2 need not be polar, cf. Theorem 5.39. One of the first steps to understand such actions is to recognize the possible orbit spaces. For example, cohomogeneity 2 is still reasonably mild, since all corresponding orbit spaces are orbifolds, see, e.g., [157]. A complete list of possible orbit spaces in cohomogeneity 2 can be found in Yeager [233]. However, checking whether each of these indeed occurs and classifying the corresponding manifolds seems currently elusive in this general framework.

The best result available regarding sec > 0 in higher cohomogeneity is due to Wilking [229], who ingeniously showed that any new examples with large dimension can only occur with large cohomogeneity.

Theorem 6.67. *If M^n admits a metric with sec > 0 invariant under an isometric group action of cohomogeneity $k \geq 1$ with $n > 18(k+1)^2$, then M is homotopy equivalent to a CROSS.*

Instead of increasing the cohomogeneity, one can relax the curvature positivity condition sec > 0 to sec ≥ 0. This substantially enlarges the class of known examples. For instance, in cohomogeneity one, the following result was proved by Grove and Ziller [115]:

Theorem 6.68. *Any cohomogeneity one manifold with codimension two singular orbits (i.e., $\ell_\pm = 1$) admits an invariant metric with sec ≥ 0.*

Proof (Sketch). The construction of such invariant metrics with sec ≥ 0 is through a gluing procedure, inspired by the work of Cheeger [70], where Cheeger deformations were used to prove that the connected sum of any two CROSS admits sec ≥ 0. The key lemma for extending this to the cohomogeneity one setup is showing that if $H \subset K \subset G$ are Lie groups with $K/H = S^1 = \partial D^2$ and Q is a bi-invariant metric on G, then the disk bundle $G \times_K D^2$ has a G-invariant metric with sec ≥ 0 that is a product near the boundary $G \times_K S^1 = G/H$, which is endowed with the normal homogeneous metric induced by Q. Then, if both $\ell_\pm = 1$, one can equip each half $G \times_{K_\pm} D_\pm$ with such a metric and glue them along G/H, following (6.25).

To prove the above key lemma, one needs the crucial fact that if G is a compact Lie group with bi-invariant metric Q and $\mathfrak{a} \subset \mathfrak{g}$ is an abelian Lie subalgebra, then $Q_a := aQ|_{\mathfrak{a}} + Q|_{\mathfrak{a}^\perp}$ has sec ≥ 0 as long as $0 < a < \frac{4}{3}$. This is where the hypotheses $\ell_\pm = 1$ come into play, as the subalgebra corresponding to the tangent space to $K_\pm/H = S^{\ell_\pm}$ is abelian provided it is one-dimensional. With that in mind, consider a product metric on $G \times D$, where G has a metric Q_a and the 2-disk D has a K-invariant metric with sec ≥ 0, e.g., $g_f = dt^2 + f(t)^2 d\theta^2$, where f is concave. Then $Q_a + g_f$ induces a metric $g_{a,f}$ with sec ≥ 0 on the quotient $G \times_K D$. A deformation argument implies that given $1 < a \leq \frac{4}{3}$, one can choose f so that $g_{a,f}$ is product near the boundary G/H, where the induced metric is normal homogeneous.   □

Remark 6.69. Given that the above metrics on the disk bundles $G \times_K D$ are product near the boundary, the resulting metric on the cohomogeneity one manifold M has totally geodesic principal orbits. This is a very strong property, e.g., combined with sec ≥ 0 it implies that each of the disk bundles $G \times_{K_\pm} D_\pm$ are convex subsets of M. Also, it follows that tangent planes to M spanned by one direction tangent to a totally geodesic principal orbit and one direction orthogonal to it are flat. By convexity arguments, these flat planes must then occur at every point of M.

Although the condition $\ell_\pm = 1$ seems somewhat restrictive, the corresponding class of cohomogeneity one manifolds is surprisingly large. In fact, using Theorem 6.68, Grove and Ziller [115] proved the following sequence of results.

Theorem 6.70. *Every principal* $SO(k)$-*bundle* P *over* S^4 *carries a cohomogeneity one action by* $SO(3) \times SO(k)$ *with codimension two singular orbits and hence an invariant metric with* sec ≥ 0.

Thus, every associated bundle $P \times_{SO(k)} M$ (see Theorem 3.51) also carries a metric with sec ≥ 0, where M is a manifold with sec ≥ 0 on which $SO(k)$ acts by isometries. Indeed, since sec$_P \geq 0$, the product $P \times M$ also has sec ≥ 0 and by the Gray-O'Neill formula (2.14), the quotient $P \times_{SO(k)} M$ has sec ≥ 0. This yields:

Corollary 6.71. *The total space of every sphere bundle and vector bundle over* S^4 *carries a metric with* sec ≥ 0.

A particularly interesting consequence regards the *Milnor spheres*, which are S^3-bundles over S^4 and were the first exotic spheres to be discovered.

Corollary 6.72. *At least* 10 *of the* 14 *(unoriented) exotic 7-spheres carry a metric with* sec ≥ 0. *Moreover, on each of the* 4 *(oriented) diffeomorphism types homotopy equivalent to* $\mathbb{R}P^5$ *there exist infinitely many non-isometric metrics with* sec ≥ 0.

Although the above construction gives rise to many metrics of cohomogeneity one and sec ≥ 0, Grove, Verdiani, Wilking and Ziller [111] proved the following negative counterpart result when the singular orbits have larger codimension.

Theorem 6.73. *For any pair of integers* $(\ell_-, \ell_+) \neq (1,1)$, *with* $\ell_\pm \geq 1$, *there is an infinite family of cohomogeneity one manifolds* M *with singular orbits of codimensions* $\ell_\pm + 1$ *and no invariant metric with* sec ≥ 0.

Remark 6.74. Among the above family are the Brieskorn varieties M_d^{2n-1}, $n \geq 4$, $d \geq 3$, including the Kervaire spheres (discussed in Example 6.56), which have $(\ell_-, \ell_+) = (2, n)$ and are shown not to have $SO(2)SO(n)$-invariant metrics with sec ≥ 0 (cf. Exercise 6.57).

Remark 6.75. Although we are not discussing noncompact manifolds with sec ≥ 0 here, we point out that analogous results to Corollary 6.71 regarding vector bundles over the remaining known closed 4-manifolds with sec ≥ 0 were proved in [117].

Let us return to questions related to the Hopf Problem 6.64. Very recently, Grove and Wilking [113] improved the conclusion of Theorem 6.65 from *homeomorphism* to *equivariant diffeomorphism*, in an ingenious proof that uses deep aspects of Alexandrov geometry and the solution of the Poincaré conjecture. Besides Theorem 6.65, Kleiner (in his PhD thesis [141]) proved that the only compact orientable 4-manifolds that admit a metric with sec ≥ 0 invariant under an isometric S^1-action are homeomorphic to S^4, $\mathbb{C}P^2$, $S^2 \times S^2$, $\mathbb{C}P^2 \# \mathbb{C}P^2$ or $\mathbb{C}P^2 \# \overline{\mathbb{C}P}^2$. This conclusion was also improved from homeomorphism to equivariant diffeomorphism in [113].

Altogether, any potential metric with sec > 0 on $S^2 \times S^2$ cannot have continuous symmetry, let alone cohomogeneity one. Nevertheless, cohomogeneity one techniques, combined with conformal methods, can be used to construct metrics on $S^2 \times S^2$ that satisfy a curvature positivity condition stronger than the conditions Ric > 0 and sec ≥ 0 satisfied by the standard product metric. We now describe this construction, following Bettiol [34]. Let (M, g) be a Riemannian manifold and choose a distance (inducing the standard topology) on the Grassmannian bundle $Gr_2 TM$ of 2-planes tangent to M.[20] For each $\theta > 0$ and $\sigma \subset T_p M$, set

$$\sec_g^\theta(\sigma) := \min_{\substack{\sigma' \subset T_p M \\ \text{dist}(\sigma, \sigma') \geq \theta}} \frac{1}{2}\left(\sec_g(\sigma) + \sec_g(\sigma')\right).$$

The condition $\sec_g^\theta > 0$ means that, at every point $p \in M$, the average of sectional curvatures of any two planes $\sigma, \sigma' \subset T_p M$ that are at least $\theta > 0$ apart from each other is positive. Intuitively, one can think of θ as a lower bound for the "angle" or "aperture" between the planes considered. Define the following curvature condition:

$(\sec^{0+} > 0)$ *For all* $\theta > 0$ *there exists a Riemannian metric* g^θ *on M with* $\sec_{g^\theta}^\theta > 0$, *and* g^θ *converges to a metric* g^0 *as* $\theta \searrow 0$.

Any Riemannian manifold with sec > 0 clearly satisfies $\sec^{0+} > 0$. Moreover, routine arguments show that manifolds with $\sec^{0+} > 0$ admit metrics with Ric > 0, proving that $\sec^{0+} > 0$ is an intermediate condition between Ric > 0 and sec > 0:

---

[20]For instance, given $\sigma, \sigma' \subset T_p M$, one can define $\text{dist}(\sigma, \sigma')$ as the Hausdorff distance between the great circles $\sigma \cap S$ and $\sigma' \cap S$, where S is the unit sphere in $T_p M$.

Proposition 6.76. *Let M be a compact manifold satisfying $\sec^{0+} > 0$. Then M admits metrics with $\mathrm{Ric} > 0$, since $\mathrm{Ric}_{g_\theta} > 0$ for $\theta > 0$ sufficiently small. In particular, if $g^\theta \to g^0$ also in the C^2-topology, then $\mathrm{Ric}_{g^0} \geq 0$ and $\sec_{g^0} \geq 0$.*

Proof (Sketch). For any metric g on M, define:

$$
\theta_g := \min_{p \in M} \left(\min_{\substack{v \in T_pM, \\ g(v,v)=1}} \left(\min_{\substack{w_1,w_2 \in T_pM, \\ g(v,w_j)=0, \\ g(w_i,w_j)=\delta_{ij}}} \mathrm{dist}\Big(\mathrm{span}\{v,w_1\}, \mathrm{span}\{v,w_2\}\Big) \right) \right).
$$

The above defines a positive number depending continuously on the metric g, and if $\sec_g^\theta > 0$ for some $0 < \theta \leq \theta_g$, then $\mathrm{Ric}_g > 0$. Given $\mathscr{G} := \{g^\theta : \theta \in [0,1]\}$, let $\theta_* := \min\{\theta_g : g \in \mathscr{G}\}$. It then follows that $\theta_* > 0$ and hence for any $0 < \theta \leq \theta_*$, we have $\mathrm{Ric}_{g_\theta} > 0$. $\square$

Recall that the standard product metric g_0 on $S^2 \times S^2$ satisfies the Einstein equation $\mathrm{Ric}_{g_0} = g_0$, and, in particular, has $\mathrm{Ric} > 0$. However, at every point $p \in (S^2 \times S^2, g_0)$ there exist planes $\sigma, \sigma' \subset T_p(S^2 \times S^2)$ arbitrarily close to each other with $\sec_{g_0}(\sigma) = \sec_{g_0}(\sigma') = 0$. Namely, *mixed planes*, i.e., planes spanned by vectors of the form $(X,0)$ and $(0,Y)$, are g_0-flat and form a continuous family at each point. Following Bettiol [34], we now outline a deformation g^θ, $\theta > 0$, of the product metric g_0 that perturbes these flat planes in such way that $\sec_{g_\theta}^\theta > 0$.

Theorem 6.77. *The manifold $S^2 \times S^2$ satisfies $\sec^{0+} > 0$.*

Proof (Sketch). For each $\theta > 0$, the construction of g^θ has two steps: a Cheeger deformation and a first-order local conformal deformation. The first step of this construction follows the work of Müter [172, Satz 4.26] (see also Ziller [234, p. 5] and Kerin [140, Rem 4.3]).

Let g_0 be the standard product metric on $S^2 \times S^2$. There is an isometric diagonal $SO(3)$-action on $(S^2 \times S^2, g_0)$ given by

$$
A \cdot (p_1, p_2) := (A p_1, A p_2), \quad p = (p_1, p_2) \in S^2 \times S^2 \subset \mathbb{R}^3 \oplus \mathbb{R}^3, \, A \in SO(3).
$$

This is a cohomogeneity one action with trivial principal isotropy, corresponding to $p_1 \neq \pm p_2$; and singular isotropies isomorphic to $SO(2)$, corresponding to $p_1 = \pm p_2$. Thus, the group diagram is $\{1\} \subset \{SO(2), SO(2)\} \subset SO(3)$, cf. Exercise 6.55. The singular orbits of this action are the *diagonal* and *anti-diagonal* submanifolds:

$$
\pm \Delta S^2 := \{(p_1, \pm p_1) : p_1 \in S^2\} \subset S^2 \times S^2.
$$

A careful study of the vertical and horizontal distributions of this cohomogeneity one action reveals the key fact that, at a regular point on $(S^2 \times S^2, g_0)$, there is a *unique* mixed plane that contains a horizontal direction; and, at a singular point, there is a *circle's worth* of such planes. It follows from Corollary 6.7 that the Cheeger deformation $(S^2 \times S^2, g_t)$, $t > 0$, has a unique zero curvature plane at each $p \notin \pm \Delta S^2$ and a circle's worth of zero curvature planes at each $p \in \pm \Delta S^2$, and all other planes have positive curvature. Thus, for any $\theta > 0$, the metric g_t, $t > 0$, satisfies $\sec^\theta_{g_t} \geq 0$ and equality only holds for certain planes tangent at some $p = (p_1, \pm p_1) \in \pm \Delta S^2$. Moreover, these planes are not tangent to $\pm \Delta S^2$.

The second step of the construction consists of perturbing the above Cheeger deformed metric g_t, $t > 0$, with an appropriate symmetric $(0,2)$-tensor h, in a first-order deformation inspired by the work of Strake [199, 200]. Namely, consider the deformation $g_{s,t} := g_t + s\mathrm{h}$ of g_t. For a fixed $\theta > 0$, in order for $\sec^\theta_{g_{s,t}} > 0$ for any $s > 0$ sufficiently small, it suffices that $\frac{d}{ds}\left(\sec_{g_{s,t}}(\sigma_1) + \sec_{g_{s,t}}(\sigma_2)\right)\big|_{s=0} > 0$, for all $\sigma_1, \sigma_2 \subset T_p(S^2 \times S^2)$ with $p \in \pm \Delta S^2$ and $\sec_{g_t}(\sigma_1) = \sec_{g_t}(\sigma_2) = 0$. Indeed, this can be verified using the first-order Taylor expansion of $\sec^\theta_{g_{s,t}}$ in s and the fact that $\sec^\theta_{g_t} \geq 0$ with zeros only along $\pm \Delta S^2$. Choosing $\mathrm{h} = -\phi\, g_t$, we have

$$\tfrac{d}{ds}\sec_{g_{s,t}}(\sigma)\big|_{s=0} = \tfrac{1}{2}\,\mathrm{Hess}\,\phi\,(X,X) + \tfrac{1}{2}\,\mathrm{Hess}\,\phi\,(Y,Y), \qquad (6.45)$$

where the Hessians are with respect to g_t. Since the planes $\sigma \subset T_p(S^2 \times S^2)$ with $\sec_{g_t}(\sigma) = 0$ are never tangent to $\pm \Delta S^2$, one can choose the function ϕ to be the sum of the square distances to $\pm \Delta S^2$, multiplied by an appropriate cutoff function, in order for the above sufficient condition to hold. Since $s > 0$ and $t > 0$ can be chosen arbitrarily small, given any $\theta > 0$ there are $s > 0$ and $t > 0$ such that $g_{s,t}$ is arbitrarily close to g_0 and has $\sec^\theta_{g_{s,t}} > 0$, yielding the desired metrics g^θ that prove that $S^2 \times S^2$ satisfies $\sec^{0+} > 0$. $\square$

Remark 6.78.   The above metrics g^θ on $S^2 \times S^2$ with $\sec^\theta_{g^\theta} > 0$ do not have $\sec \geq 0$, see Bettiol [34, Sec 4.4]. It can also be shown that such metrics can be constructed invariant under the antipodal action of $\mathbb{Z}_2 \oplus \mathbb{Z}_2$, whose quotient is $\mathbb{R}P^2 \times \mathbb{R}P^2$. Recall that from Synge's Theorem (see Petersen [183, p. 172]), $\mathbb{R}P^2 \times \mathbb{R}P^2$ does not admit any metric with $\sec > 0$.

Remark 6.79.   The question of deforming the standard product metric g_0 to improve curvature has been studied in great depth by Bourguignon, Deschamps and Sentenac [50, 51], after Berger [31] elegantly proved that any *first-order* deformation $g_s = g_0 + s\mathrm{h}$ with $\frac{d}{ds}\sec_{g_s}(\sigma)\big|_{s=0} \geq 0$ for all mixed planes σ must satisfy $\frac{d}{ds}\sec_{g_s}(\sigma)\big|_{s=0} = 0$. Despite interesting findings regarding deformations of higher order in [50, 51] and Theorem 6.77, even this *local version* of the Hopf Problem 6.64 (that asks if metrics with $\sec > 0$ on $S^2 \times S^2$ exist arbitrarily close to g_0) remains unanswered.

Remark 6.80.   The standard product metric on $S^2 \times S^2$ can be written as a cohomogeneity one metric (6.27), by using the embedding $S^2 \times S^2 \subset \mathbb{R}^3 \times \mathbb{R}^3$ and choosing as horizontal geodesic, e.g.,

$$\gamma(\tau) = \big((\cos\tau)e_1 + (\sin\tau)e_2, (\cos\tau)e_1 - (\sin\tau)e_2\big), \qquad \tau \in \left[0, \tfrac{\pi}{2}\right]$$

where $\{e_1, e_2, e_3\}$ is the canonical basis of $\mathbb{R}^3$. At each $\gamma(\tau) \in S^2 \times S^2$, $0 < \tau < \frac{\pi}{2}$, we can identify the tangent space to the orbit through $\gamma(\tau)$ as $\mathfrak{m} = \mathfrak{so}(3) \cong \mathbb{R}^3$, and this identification is isometric when $\mathfrak{so}(3)$ is endowed with the standard bi-invariant metric Q and $\mathbb{R}^3$ with the Euclidean metric. Then the standard product metric g_0 is given by $g_0 = d\tau^2 + g_\tau$, and g_τ is related to the Euclidean metric Q via the tensor $P_\tau : \mathfrak{m} \to \mathfrak{m}$, given by

$$P_\tau = \begin{pmatrix} 2\sin\tau & & \\ & 2\cos\tau & \\ & & 2 \end{pmatrix}.$$

In particular, the Cheeger deformation g_t of g_0 can be explicitly written in the form $g_t = d\tau^2 + g_{t,\tau}$, by using Exercise 6.46 and the above formula for P_τ. However, verifying the curvature properties claimed above is much easier through Proposition 6.6, rather than attempting an explicit computation.

We conclude this section with a summary of recent developments regarding another intermediate curvature positivity condition, called *strongly positive curvature*, between sec > 0 and positive-definiteness of the curvature operator. A systematic study of this condition, under the light of the Grove program, was recently initiated by Bettiol and Mendes [36], though this notion is reminiscent of the work of Thorpe [211, 212] in the 1970s and was also used by authors including Pütmann [187], and Grove, Verdiani and Ziller [112], who coined the term. In order to define this condition, recall that the *curvature operator* $R : \Lambda^2 T_p M \to \Lambda^2 T_p M$ of a Riemannian manifold (M, g) is the symmetric endomorphism of $\Lambda^2 T_p M$ given by

$$\langle R(X \wedge Y), Z \wedge W \rangle := R(X, Y, Z, W),$$

where $\langle \cdot, \cdot \rangle$ denotes the inner product on $\Lambda^2 T_p M$ induced by g, and R is defined in (2.3). Any 2-plane $\sigma \subset T_p M$ can be seen as an element[21] $X \wedge Y \in \Lambda^2 T_p M$, by choosing orthonormal vectors $X, Y \in T_p M$ that span σ. From this viewpoint, sectional curvature $\sec(\sigma) = \langle R(\sigma), \sigma \rangle$ is the quadratic form associated to R, restricted to the (oriented) Grassmannian $\mathrm{Gr}_2(T_p M) = \{X \wedge Y \in \Lambda^2 T_p M : \|X \wedge Y\|^2 = 1\}$. Any 4-form $\omega \in \Lambda^4 T_p M$ also induces a symmetric operator on $\Lambda^2 T_p M$, given by

$$\omega : \Lambda^2 T_p M \to \Lambda^2 T_p M, \quad \langle \omega(\alpha), \beta \rangle := \langle \omega, \alpha \wedge \beta \rangle. \tag{6.46}$$

Note that the quadratic form associated to ω clearly vanishes on $\sigma \in \mathrm{Gr}_2(T_p M)$, since $\sigma \wedge \sigma = 0$. Thus, it follows that

$$\sec(\sigma) = \langle R(\sigma), \sigma \rangle = \langle (R + \omega)(\sigma), \sigma \rangle. \tag{6.47}$$

---

[21] For the sake of simplifying notation, here we use the metric g to identify $(0, k)$-tensors and k-forms, that is, $\Lambda^k TM \cong \Lambda^k TM^*$. Furthermore, we omit the parentheses, denoting $\Lambda^k V := \Lambda^k(V)$.

This observation, known as *Thorpe's trick*, implies that if there exists $\omega \in \Lambda^4 T_p M$ such that the so-called *modified curvature operator* $(R + \omega): \Lambda^2 T_p M \to \Lambda^2 T_p M$ is a positive-definite operator, then $\sec(\sigma) > 0$ for all planes $\sigma \subset T_p M$.

Definition 6.81. A Riemannian manifold (M, g) has *strongly positive curvature* if for all $p \in M$, there exists $\omega \in \Lambda^4 T_p M$ such that $(R + \omega): \Lambda^2 T_p M \to \Lambda^2 T_p M$ is positive-definite. Analogously, (M, g) has *strongly nonnegative curvature* if for all $p \in M$, there exists $\omega \in \Lambda^4 T_p M$ such that $(R + \omega)$ is positive-semidefinite.

Remark 6.82. In the above definition, the assignment $M \ni p \mapsto \omega_p \in \Lambda^4 T_p M$ may, in principle, fail to be smooth. However, if (M, g) has strongly positive curvature, then a standard perturbation argument shows that there exists a smooth 4-form $\widetilde{\omega} \in \Omega^4(M)$ such that $R + \widetilde{\omega}$ is positive-definite. The same argument does not work for strongly nonnegative curvature, as it is not an open condition, see [36, Sec 6.4].

In view of (6.47), strongly positive curvature is an intermediate condition between $\sec > 0$ and positive-definiteness of the curvature operator. As mentioned before, manifolds that satisfy the latter are known to be diffeomorphic to spherical space forms by the work of Böhm and Wilking [45]. Thus, the notion of strongly positive curvature might improve our understanding of the gap between finite quotients of spheres and the intriguing class of manifolds that admit $\sec > 0$.

In dimensions ≤ 3, this gap is well-understood, since $\sec > 0$, strongly positive curvature and positive-definiteness of the curvature operator are all equivalent. Remarkably, strongly positive curvature and $\sec > 0$ remain equivalent in dimension 4, providing a different viewpoint on the Hopf Problem 6.64. The proof of this equivalence is originally due to Thorpe [211]; however, it can be greatly simplified using an elegant argument communicated to us by Püttmann (see also [187]), as follows.

Proposition 6.83. *Let (M, g) be a Riemannian manifold with $\dim M \leq 4$. Then (M, g) has strongly positive curvature if and only if it has $\sec > 0$.*

Proof. The only nontrivial implication is that if (M, g) has $\dim M = 4$ and $\sec > 0$, then it also has strongly positive curvature. Fix $p \in M$ and, for each modified curvature operator $(R + \omega): \Lambda^2 T_p M \to \Lambda^2 T_p M$, denote by $\lambda_1(R + \omega) := \min \operatorname{Spec}(R + \omega)$ its smallest eigenvalue. Consider the min-max problem:

$$\lambda := \sup_{\omega \in \Lambda^4 T_p M} \lambda_1(R + \omega). \tag{6.48}$$

It is easy to see that, since the endomorphisms $\omega: \Lambda^2 T_p M \to \Lambda^2 T_p M$ are traceless, the above supremum λ is actually achieved at some $\omega_{max} \in \Lambda^4 T_p M$. Denote by E_λ the subspace of $\Lambda^2 T_p M$ formed by the eigenvectors of $R + \omega_{max}$ with eigenvalue λ. If there exists $\sigma \in E_\lambda \cap \operatorname{Gr}_2(T_p M)$, then $\lambda = \sec(\sigma) > 0$ and hence $R + \omega_{max}$ is positive-definite. Otherwise, we would have $q(\alpha) \neq 0$ for all nonzero $\alpha \in E_\lambda$, where

$$q: \Lambda^2 T_p M \to \Lambda^4 T_p M \cong \mathbb{R}, \quad q(\alpha) := \alpha \wedge \alpha. \tag{6.49}$$

Notice that the image $q(E_\lambda \setminus \{0\}) \subset \mathbb{R} \setminus \{0\}$ is contained in a half-line, say $\mathbb{R}_+ := \{x > 0\}$. Indeed, this is clear if $\dim E_\lambda = 1$, and if $\dim E_\lambda \geq 2$, then $E_\lambda \setminus \{0\}$ is connected and hence so is $q(E_\lambda \setminus \{0\})$. Therefore, for any nonzero $\alpha \in E_\lambda$, we can construct a new modified curvature operator $R + \omega_{max} + \alpha \wedge \alpha$ that satisfies

$$\langle (R + \omega_{max} + \alpha \wedge \alpha)(\beta), \beta \rangle = \langle (R + \omega_{max})(\beta), \beta \rangle + q(\alpha)\|\beta\|^2 > \lambda \|\beta\|^2,$$
(6.50)

for all nonzero $\beta \in \Lambda^2 T_p M$, contradicting the maximality (6.48) of λ. $\qquad \square$

A crucial property of strongly positive curvature is that, similarly to sec > 0, this condition is preserved under Riemannian submersions.[22] This was proved by Bettiol and Mendes [36], using a convenient rearrangement of the Gray-O'Neill formula (2.14) for curvature operators, in terms of the first Bianchi identity.

Theorem 6.84. Let $\pi: (\overline{M}, \overline{g}) \to (M, g)$ be a Riemannian submersion. If $(\overline{M}, \overline{g})$ has strongly positive curvature, then (M, g) also has strongly positive curvature.

Proof. Fix $p \in M$ and $\overline{p} \in \overline{M}$ such that $\pi(\overline{p}) = p$, and identify each $X \in T_p M$ as an element $\overline{X} \in T_{\overline{p}}\overline{M}$ via horizontal lift. On these horizontal vectors, the tensor A is given by $A_X Y = \frac{1}{2}[\overline{X}, \overline{Y}]^{\mathscr{V}}$, see (2.12). As such, it induces a symmetric operator

$$\alpha: \Lambda^2 T_p M \to \Lambda^2 T_p M, \quad \langle \alpha(X \wedge Y), Z \wedge W \rangle := \langle A_X Y, A_Z W \rangle, \qquad (6.51)$$

which is clearly positive-semidefinite. From the Gray-O'Neill formula, see (2.14) and [33, Thm 9.28f], we have

$$\begin{aligned}
\langle R(X \wedge Y), Z \wedge W \rangle &= \langle \overline{R}(\overline{X} \wedge \overline{Y}), \overline{Z} \wedge \overline{W} \rangle + 2\langle A_{\overline{X}}\overline{Y}, A_{\overline{Z}}\overline{W} \rangle \\
&\quad - \langle A_{\overline{Y}}\overline{Z}, A_{\overline{X}}\overline{W} \rangle + \langle A_{\overline{X}}\overline{Z}, A_{\overline{Y}}\overline{W} \rangle \\
&= \langle \overline{R}(\overline{X} \wedge \overline{Y}), \overline{Z} \wedge \overline{W} \rangle + 3\langle \alpha(\overline{X} \wedge \overline{Y}), \overline{Z} \wedge \overline{W} \rangle \\
&\quad - \langle \alpha(\overline{Y} \wedge \overline{Z}), \overline{X} \wedge \overline{W} \rangle - \langle \alpha(\overline{Z} \wedge \overline{X}), \overline{Y} \wedge \overline{W} \rangle \\
&\quad - \langle \alpha(\overline{X} \wedge \overline{Y}), \overline{Z} \wedge \overline{W} \rangle \\
&= \langle \overline{R}(\overline{X} \wedge \overline{Y}), \overline{Z} \wedge \overline{W} \rangle + 3\langle \alpha(\overline{X} \wedge \overline{Y}), \overline{Z} \wedge \overline{W} \rangle \\
&\quad - 3\mathfrak{b}(\alpha)(\overline{X}, \overline{Y}, \overline{Z}, \overline{W}),
\end{aligned}$$

---

[22] Recall that positive-definiteness of the curvature operator is not preserved under Riemannian submersions. For instance, there are Riemannian submersions from the round sphere onto the projective spaces $\mathbb{C}P^n$ and $\mathbb{H}P^n$, but these only have positive-semidefinite curvature operator.

where $\mathfrak{b}$ is the *Bianchi map*, that associates to each symmetric operator $S\colon \Lambda^2 T_p M \to \Lambda^2 T_p M$ the 4-form $\mathfrak{b}(S) \in \Lambda^4 T_p M$ given by[23]:

$$\mathfrak{b}(S)(X,Y,Z,W) := \tfrac{1}{3}\Big(\langle S(X \wedge Y), Z \wedge W \rangle + \langle S(Y \wedge Z), X \wedge W \rangle$$
$$+ \langle S(Z \wedge X), Y \wedge W \rangle \Big).$$

Thus, if there exists $\overline{\omega} \in \Lambda^4 T_{\overline{p}} \overline{M}$ such that $\overline{R} + \overline{\omega}$ is positive-definite, it follows that also $R + \omega$ is positive-definite, by setting $\omega := \overline{\omega} + 3\mathfrak{b}(\alpha) \in \Lambda^4 T_p M$. $\square$

Notice that the above argument implies that strongly nonnegative curvature is also preserved under Riemannian submersions. Furthermore, similar techniques can be used to prove that strongly positive (and nonnegative) curvature are preserved under Cheeger deformations [36, Thm B], cf. Proposition 6.1.

Propelled by the above foundational results, we can analyze the known constructions of metrics with sec > 0 (and sec ≥ 0) to verify whether they yield metrics with strongly positive (and nonnegative) curvature. Regarding the CROSS (recall Example 6.10), it is a direct consequence of Theorem 6.84 that $\mathbb{C}P^n$ and $\mathbb{H}P^n$ have strongly positive curvature, since there are Riemannian submersions from round spheres given by the Hopf bundles (6.20). The same does not hold for the Cayley plane $\mathbb{C}aP^2$; in fact, there are no homogeneous metrics with strongly positive curvature on $\mathbb{C}aP^2$, see [36, Prop. 3.4]. The main reason for this is that $\mathbb{C}aP^2$ admits a unique F_4-invariant metric (see Example 6.21) and any 4-form ω such that $R + \omega$ is positive-definite could be averaged to yield an F_4-invariant (and hence harmonic) 4-form $\widetilde{\omega}$ such that $R + \widetilde{\omega}$ is positive-definite. However, $\mathbb{C}aP^2$ does not admit any nonzero harmonic 4-forms, since $b_4(\mathbb{C}aP^2) = 0$. In particular, this provides an example of a Riemannian manifold that has sec > 0 but does not have strongly positive curvature (algebraic examples were previously constructed by Zoltek [237]).

In order to classify simply-connected homogeneous spaces with strongly positive curvature beyond the CROSS, one may restrict to the list given in Theorem 6.30. By extending previous results of Wallach [221] and Eschenburg [82] with the introduction of *strongly fat* homogeneous bundles, Bettiol and Mendes [36, Thm C] found homogeneous metrics with strongly positive curvature on the Aloff-Wallach spaces $W_{k,l}^7$, the Berger spaces B^7 and B^{13}, and the Wallach flag manifolds W^6 and W^{12}. The only remaining case, the Wallach flag manifold W^{24}, was recently settled

---

[23]The Bianchi map $\mathfrak{b}$ on symmetric operators on $\Lambda^2 T_p M$ can be seen as the orthogonal projection onto the subspace of operators induced by 4-forms via (6.46). The complement $\ker \mathfrak{b}$ of this subspace consists of symmetric operators on $\Lambda^2 T_p M$ that satisfy the first Bianchi identity, which are called *algebraic curvature operators*.

by Bettiol and Mendes [35], using different techniques.[24] Altogether, in terms of the homogeneous classification of sec > 0 (see Theorem 6.30), we have the following:

Theorem 6.85 (Homogeneous classification of strongly positive curvature). *All simply-connected homogeneous spaces that admit a homogeneous metric with sec > 0 admit a homogeneous metric with strongly positive curvature, except for* $\mathbb{C}aP^2$.

Besides homogeneous manifolds, there are currently only a few other known examples of closed manifolds with sec > 0. Namely, these are biquotients (Eschenburg spaces E_p^7 and Bazaikin spaces B_p^{13}) and the exotic $T_1 S^4$, which is the manifold P_2^7 in Theorem 6.66. An infinite family of Eschenburg spaces E_p^7 can be shown to have strongly positive curvature by a limiting argument similar to Escheburg [81, Sec 5], using the fact that all Aloff-Wallach spaces $W_{k,l}^7$ with sec > 0 have strongly positive curvature. Finally, the exotic $T_1 S^4$ was shown to have a cohomogeneity one metric with sec > 0 in [112] by proving that this metric actually has strongly positive curvature. Thus, the task of verifying whether all known examples of closed manifolds with sec > 0 admit a metric with strongly positive curvature is almost complete, and, so far, no exceptions have been found.[25]

Regarding which manifolds with sec ≥ 0 have strongly nonnegative curvature, a very satisfactory start is provided by Theorem 6.84. Since a Lie group (G, Q) with bi-invariant metric has positive-semidefinite curvature operator (recall Proposition 2.26), it follows that every normal homogeneous space G/H has strongly nonnegative curvature. Moreover, in view of [36, Thm B] and other recent work of Bettiol and Mendes, many constructions using Cheeger deformations (including structural results such as Theorem 6.23) carry over to the context of strongly nonnegative curvature, including Theorem 6.68 and its Corollaries 6.71 and 6.72.

---

[24]Moreover, a complete description of the moduli spaces of homogeneous metrics with strongly positive curvature on all Wallach flag manifolds was achieved in [35]. Namely, it is shown that a homogeneous metric on W^6 or W^{12} has strongly positive curvature if and only if it has sec > 0, while a homogeneous metric on W^{24} has strongly positive curvature if and only if it has sec > 0 and does not submerge onto $\mathbb{C}aP^2$ endowed with its standard metric.

[25]We stress that the only topological obstructions to strongly positive curvature currently known are the obstructions for sec > 0.

Appendix A
Rudiments of Smooth Manifolds

In this appendix, we summarize basic definitions and results on smooth manifolds, that are used throughout this book. A more detailed treatment of this subject can be found in Warner [227], Spivak [198] or Lang [149].

A.1 Smooth Manifolds

In general, by smooth or differentiable we mean of class C^∞, unless otherwise specified. In addition, a map is called a *function* if its counter-domain is $\mathbb{R}$.

Definition A.1. A *smooth n-dimensional manifold* M is a second countable para-compact Hausdorff topological space endowed with a *smooth structure*. More precisely, there is a collection of pairs $(U_\alpha, \varphi_\alpha)$ called *charts*, such that U_α are open subsets of M, and, for each α, the map $\varphi_\alpha \colon U_\alpha \subset M \to \varphi_\alpha(U_\alpha) \subset \mathbb{R}^n$ is a homeomorphism between open sets satisfying:

(i) $M = \bigcup_\alpha U_\alpha$;
(ii) If $(U_\alpha, \varphi_\alpha)$ and (U_β, φ_β) are charts, then the *transition map*

$$\varphi_\beta \circ \varphi_\alpha^{-1} \colon \varphi_\alpha(U_\alpha \cap U_\beta) \longrightarrow \varphi_\beta(U_\alpha \cap U_\beta)$$

is smooth. In this case, the charts φ_α and φ_β are said to be C^∞-compatible;
(iii) The collection $\{(U_\alpha, \varphi_\alpha)\}_\alpha$ of charts is maximal with respect to (i) and (ii).

Such collection of charts is called an *atlas* of M. If $(U_\alpha, \varphi_\alpha)$ is a chart and $p \in U_\alpha$, the open set U_α is called a *coordinate neighborhood* of p and φ_α is called a *local coordinate* at p. Furthermore, we write $n = \dim M$ and refer to M as an *n-manifold*.

Similarly, it is possible to define C^k-manifolds and *analytic manifolds* (C^ω-manifolds), by requiring the transition maps to be respectively C^k and analytic.

© Springer International Publishing Switzerland 2015
M.M. Alexandrino, R.G. Bettiol, *Lie Groups and Geometric Aspects of Isometric Actions*, DOI 10.1007/978-3-319-16613-1

Thus, by *analytic structure*, we mean a collection of coordinate systems which overlap analytically, that is, are locally represented by convergent power series.

In this book, we are only interested in C^∞-manifolds, therefore a *manifold* is always considered to be a *smooth manifold*, in the above sense.

Definition A.2. Let M and N be manifolds, and denote their atlases by $\{(U_\alpha, \varphi_\alpha)\}_\alpha$ and $\{(V_\beta, \psi_\beta)\}_\beta$, respectively. A continuous map $f: M \to N$ is said to be *smooth at $p \in M$* if there exist coordinate neighborhoods U_α of p and V_β of $f(p)$, such that $\psi_\beta \circ f \circ \varphi_\alpha^{-1}$ is smooth at $\varphi_\alpha(p)$, as a map between Euclidean spaces.

It is easy to see that if this condition is satisfied by a pair of charts $(U_\alpha, \varphi_\alpha)$ and (V_β, ψ_β), then it holds for any charts. The above definition clearly implies that a function $g: M \to \mathbb{R}$ is *smooth at $p \in M$* if there exists a chart $(U_\alpha, \varphi_\alpha)$ such that $g \circ \varphi_\alpha^{-1}$ is smooth at $g(p)$. Finally, maps (and functions) are called *smooth* if they are smooth at every point $p \in M$.

Let M be a smooth manifold and $p \in M$. The set $C^\infty(M)$ of smooth functions $f: M \to \mathbb{R}$ is an algebra with the usual operations. Consider the subalgebra $C^\infty(p)$ of smooth functions whose domain of definition includes some open neighborhood of p, called the algebra of *germs* of smooth functions at p. We define the *tangent space* to M at p as the vector space T_pM of linear derivations at p, i.e., the set of maps $v: C^\infty(p) \to \mathbb{R}$ satisfying for all $f, g \in C^\infty(p)$,

(i) *Linearity*, i.e., $v(f + g) = v(f) + v(g)$;
(ii) *Leibniz rule*, i.e., $v(fg) = v(f)g(p) + f(p)v(g)$.

The vector space operations are defined for all $v, w \in T_pM$ and $\alpha \in \mathbb{R}$ by

$$(v + w)f := v(f) + w(f), \quad (\alpha v)(f) := \alpha v(f), \quad \text{for all } f \in C^\infty(p).$$

Let (U, φ), where $\varphi = (x_1, \ldots, x_n)$, be a chart around p and $\tilde{f} := f \circ \varphi^{-1}$ the representation of $f \in C^\infty(p)$ in the coordinates given by φ. The *coordinate vectors*

$$\left(\frac{\partial}{\partial x_i}\Big|_p\right) f := \frac{\partial \tilde{f}}{\partial x_i}\Big|_{\varphi(p)}$$

form a basis $\left\{\frac{\partial}{\partial x_i}\big|_p\right\}$ of T_pM. In particular, it follows that $\dim T_pM = \dim M$. In this context, tangent vectors can be explicitly seen as directional derivatives. Indeed, if $v = \sum_{i=1}^n v_i \frac{\partial}{\partial x_i} \in T_pM$, then

$$v(f) = \sum_{i=1}^n v_i \frac{\partial \tilde{f}}{\partial x_i}\Big|_{\varphi(p)}$$

is called the *directional derivative of f in the direction v*. It can also be proved that this definition is independent of the chart φ.

The *tangent vector* to a curve $\alpha\colon (-\varepsilon, \varepsilon) \to M$ at $\alpha(0)$ is defined as $\alpha'(0)(f) = \frac{d}{dt}(f \circ \alpha)\big|_{t=0}$. Indeed, if φ is a local coordinate and $(u_1(t), \ldots, u_n(t)) = \varphi \circ \alpha$, then

$$\alpha'(0) = \sum_{i=1}^{n} u_i'(0) \frac{\partial}{\partial x_i} \in T_p M.$$

Definition A.3. Let $f\colon M \to N$ be a smooth map and let $p \in M$. The *differential*, or *derivative*, of f at p is the linear map $df_p\colon T_p M \to T_{f(p)} N$, such that if $v \in T_p M$, then $df_p v$ is a tangent vector at $f(p)$ satisfying $df_p(v)g = v(g \circ f)$, for all $g \in C^\infty(f(p))$.

In other words, the chain rule is automatically valid on manifolds, as the differential of a map is defined to satisfy it. Equipped with the notion of differential, we can define important classes of maps between manifolds.

Definition A.4. A smooth map $f\colon M \to N$ is said to be an *immersion* if df_p is injective for all $p \in M$, and it is said to be a *submersion* if df_p is surjective for all $p \in M$. An immersion $f\colon M \to N$ is called an *embedding* if $f\colon M \to f(M) \subset N$ is a homeomorphism, considering $f(M)$ endowed with the relative topology.

A smooth bijection f with smooth inverse is called a *diffeomorphism*. From the chain rule, its differential on every point is an isomorphism, and a local converse is given by the *Inverse Function Theorem*. More precisely, this theorem states that if the differential df_p of a smooth map f is an isomorphism, then the restriction of f to some open neighborhood of p is a diffeomorphism onto its image.

Definition A.5. Let P and N be manifolds, with $P \subset N$. Then P is called an *immersed submanifold* of N if the inclusion $i\colon P \hookrightarrow N$ is an immersion. Moreover, if $i\colon P \hookrightarrow N$ is an embedding, then P is called an *(embedded) submanifold*.

Definition A.6. An immersed submanifold $P \subset N$ is *quasi-embedded* if for any smooth map $f\colon M \to N$ with image $f(M)$ lying in P, the induced map $f_0\colon M \to P$ defined by $i \circ f_0 = f$ is smooth, where $i\colon P \hookrightarrow N$ is the inclusion.

Proposition A.7. *If $P \subset N$ is an embedded submanifold, then P is quasi-embedded.*

A.2   Vector Fields

Let M be a manifold and TM be the (total space of its) *tangent bundle*, i.e., $TM := \bigcup_{p \in M} \{p\} \times T_p M$. Denote by $\pi\colon TM \to M$ the *projection map*, defined by $\pi(v) = p$ if $v \in T_p M$. It is possible to prove that TM carries a canonical smooth structure inherited from M, such that π is smooth. In fact, as explained in Sect. 3.1, TM is a fiber bundle; in particular, a vector bundle.

Definition A.8. A *(smooth) vector field* X on M is a (smooth) section of TM, that is, a (smooth) map $X\colon M \to TM$ such that $\pi \circ X = \mathrm{id}$.

Let $p \in M$ and consider a coordinate neighborhood U of p with local coordinates $\varphi = (x_1, \ldots, x_n)$. Given $f \in C^\infty(U)$, the *directional derivative* of f in the direction of X is defined as

$$X(f) \colon U \to \mathbb{R}, \quad X(f)(p) := X_p(f). \tag{A.1}$$

Moreover, if $\left\{ \frac{\partial}{\partial x_i} \big|_q \right\}$ is a basis of $T_q M$ for all $q \in U$, then X can be written as

$$X|_U = \sum_{i=1}^n a_i \frac{\partial}{\partial x_i}, \tag{A.2}$$

where $a_i \colon U \to \mathbb{R}$ are functions.

Proposition A.9. *Let X be a vector field on M. The following are equivalent:*

  (i) *X is smooth;*
 (ii) *For every chart (U, φ), $\varphi = (x_1, \ldots, x_n)$, the functions a_i in (A.2) are smooth;*
(iii) *For every open set V of M and $f \in C^\infty(V)$, we have $X(f) \in C^\infty(V)$.*

The set of smooth vector fields on M is denoted $\mathfrak{X}(M)$, and is clearly a $C^\infty(M)$-module with operations defined pointwise. If $f \colon M \to N$ is a smooth map between two manifolds, $X \in \mathfrak{X}(M)$ and $Y \in \mathfrak{X}(N)$, then X and Y are said to be *f-related* if $df \circ X = Y \circ f$, i.e., if the following diagram commutes.

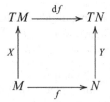

Note that X and Y are f-related if and only if $(Yg) \circ f = X(g \circ f)$ for all $g \in C^\infty(N)$.

Definition A.10. If $X, Y \in \mathfrak{X}(M)$, the *Lie bracket* of the vector fields X and Y is the vector field $[X, Y] \in \mathfrak{X}(M)$ given by

$$[X, Y]f := X(Y(f)) - Y(X(f)), \quad f \in C^\infty(M).$$

Note that the Lie bracket of vector fields is clearly skew-symmetric and satisfies the Jacobi identity (see Definition 1.7). If $X, Y \in \mathfrak{X}(M)$ are such that $[X, Y] = 0$, then X and Y are said to *commute*.

If $f \colon M \to N$ is a smooth map such that $X^1, X^2 \in \mathfrak{X}(M)$ and $Y^1, Y^2 \in \mathfrak{X}(N)$ are f-related, then $[X^1, X^2]$ and $[Y^1, Y^2]$ are also f-related. Indeed, using the above observation, since X^i is f-related to Y^i, it follows that $(Y^i g) \circ f = X^i(g \circ f)$. The conclusion is immediate from the following equation, using again the same observation:

$$\left([Y^1, Y^2]g\right) \circ f = \left(Y^1(Y^2 g)\right) \circ f - \left(Y^2(Y^1 g)\right) \circ f$$
$$= X^1\left((Y^2 g) \circ f\right) - X^2\left((Y^1 g) \circ f\right) \qquad (A.3)$$
$$= X^1\left(X^2(g \circ f)\right) - X^2\left(X^1(g \circ f)\right)$$
$$= [X^1, X^2](g \circ f)$$

In order to obtain a local expression for $[X,Y]$, fix $p \in M$ and consider a chart (U, φ), $\varphi = (x_1, \ldots, x_n)$. According to Proposition A.9, there exist functions $a_i, b_j \in C^\infty(M)$ such that $X = \sum_{i=1}^n a_i \frac{\partial}{\partial x_i}$ and $Y = \sum_{j=1}^n b_j \frac{\partial}{\partial x_j}$. Therefore

$$XY = X\left(\sum_{j=1}^n b_j \frac{\partial}{\partial x_j}\right)$$
$$= \sum_{i=1}^n a_i \frac{\partial}{\partial x_i}\left(\sum_{j=1}^n b_j \frac{\partial}{\partial x_j}\right)$$
$$= \sum_{i=1}^n \sum_{j=1}^n a_i\left(\frac{\partial b_j}{\partial x_i}\frac{\partial}{\partial x_j} + b_j \frac{\partial^2}{\partial x_i \partial x_j}\right)$$
$$= \sum_{i,j=1}^n a_i \frac{\partial b_j}{\partial x_i}\frac{\partial}{\partial x_j} + \sum_{i,j=1}^n a_i b_j \frac{\partial^2}{\partial x_i \partial x_j},$$

and similarly,

$$YX = \sum_{i,j=1}^n b_j \frac{\partial a_i}{\partial x_j}\frac{\partial}{\partial x_i} + \sum_{i,j=1}^n a_i b_j \frac{\partial^2}{\partial x_i \partial x_j}.$$

Applying the definition and expressions above, we recover the well-known local expression for the Lie bracket,

$$[X,Y]|_U = \sum_{i,j=1}^n \left(a_i \frac{\partial b_j}{\partial x_i} - b_i \frac{\partial a_j}{\partial x_i}\right)\frac{\partial}{\partial x_j}. \qquad (A.4)$$

Remark A.11. If $M \subset \mathbb{R}^n$ is an open set, then $X = (a_1, \ldots, a_n)$ and $Y = (b_1, \ldots, b_n)$ are such that $[X,Y] = \overline{\nabla}_X Y - \overline{\nabla}_Y X$, where $\overline{\nabla}$ is the usual derivative of maps on $\mathbb{R}^n$.

A.3   Foliations and the Frobenius Theorem

In this section, we mention some facts about flows, distributions and foliations, including the Frobenius Theorem.

Definition A.12. An *integral curve* of $X \in \mathfrak{X}(M)$ is a smooth curve $\alpha: I \to M$, such that $\alpha'(t) = X(\alpha(t))$ for all $t \in I$.

A classical result from the theory of Ordinary Differential Equations (ODEs) is the existence and uniqueness of integral curves with prescribed initial data.

Theorem A.13. *Given $X \in \mathfrak{X}(M)$ and $p \in M$, there exists a unique maximal integral curve $\alpha_p: I_p \to M$ of X, where I_p is an interval containing 0 and $\alpha_p(0) = p$.*

Maximality in the above result is in the sense that every other integral curve satisfying the same initial condition is a restriction of α_p to an open subinterval of the maximal interval I_p. Let $\mathscr{D}(X)$ be the set of all points $(t,x) \in \mathbb{R} \times M$ such that $t \in I_x$, and define:

$$\varphi^X: \mathscr{D}(X) \to M, \quad \varphi^X(t,x) := \alpha_x(t),$$

where α_x is the maximal integral curve of X with $\alpha_x(0) = x$. The map φ^X is called the *flow* of X, and $\mathscr{D}(X)$ is its *domain of definition*. The following is also a classical result in ODEs.

Theorem A.14. *Given $X \in \mathfrak{X}(M)$, the domain $\mathscr{D}(X)$ is an open subset of $\mathbb{R} \times M$ that contains $\{0\} \times M$ and φ^X is smooth.*

Sometimes, it is also convenient to consider the *local flow* of X, given by the restriction of φ^X to an open subset $(-\delta, \delta) \times U$ of $\mathscr{D}(X)$. For a fixed t, we have local diffeomorphisms φ_t^X, given by $\varphi_t^X(p) := \varphi^X(t, p)$. Note that, for $x \in U$ such that $\varphi_t^X(x) \in U$, if $|s| < \delta$, $|t| < \delta$ and $|s+t| < \delta$, then $\varphi_s^X \circ \varphi_t^X = \varphi_{s+t}^X$.

A smooth vector field X is *complete* if $\mathscr{D}(X) = \mathbb{R} \times M$ and, in this case, φ_t^X is a group of diffeomorphisms parameterized by t, called the *1-parameter group of X*. This is equivalent to saying that there is an $\mathbb{R}$-action on M,

$$\mathbb{R} \times M \ni (t,p) \longmapsto \varphi_t^X(p) \in M.$$

Typical examples of complete vector fields are fields with bounded length. For instance, if M is compact, every smooth vector field on M is complete.

Definition A.15. Let M be an $(n+k)$-manifold. A n-dimensional *distribution*[1] D on M is an assignment of an n-dimensional subspace $D_p \subset T_pM$ for each $p \in M$. The distribution D is said to be *smooth* if, for all $p \in M$, there are n smooth vector fields that span D at all points in a neighborhood of p. A distribution D is *involutive* if, for

---

[1] An n-dimensional distribution D on M is also called a *subbundle* of TM of rank n.

all $p \in M$, given vector fields X, Y in an open neighborhood U of p, with $X_q, Y_q \in D_q$, for all $q \in U$, then $[X, Y]_q \in D_q$, for all $q \in U$. In other words, a distribution is involutive if it is closed with respect to the bracket of vector fields.

Definition A.16. Let M be an $(n+k)$-manifold. A n-*dimensional foliation* of M is a partition $\mathscr{F}$ of M into connected immersed submanifolds of dimension n, called *leaves*, such that, for all $p \in M$ and $v \in T_p L_p$, where $L_p \in \mathscr{F}$ is the leaf that contains p, there is a smooth vector field X on M with $X(p) = v$ and $X(q) \in T_q L_q$, for all $q \in M$.

A trivial example of foliation is $\mathbb{R}^{n+k} = \mathbb{R}^n \times \mathbb{R}^k$, where the leaves are subspaces of the form $\mathbb{R}^n \times \{p\}$, with $p \in \mathbb{R}^k$. More generally, the partition of a manifold into preimages of a smooth submersion is a foliation. Another example of foliation is the partition (3.4) of a manifold into orbits of a group action, provided all the orbits have the same dimension.

Definition A.17. Let $\mathscr{F} = \{L_p : p \in M\}$ be an n-dimensional foliation of an $(n+k)$-manifold M. It is possible to prove that, for all $p \in M$, there exist an open neighborhood U of p, an open neighborhood V of $0 \in \mathbb{R}^{n+k}$ and a diffeomorphism $\psi \colon V \to U$, such that $\psi(V \cap (\mathbb{R}^n \times \{y_0\}))$ is the connected component of $L_{\psi(0,y_0)} \cap U$ that contains $\psi(0, y_0)$. The open set U is called a *simple open neighborhood*, ψ is called a *trivialization*, ψ^{-1} is called a *foliation chart*, and the submanifold $\psi(V \cap (\mathbb{R}^n \times \{y_0\}))$ is called a *plaque*.

Remark A.18. In the general case of singular foliations, we usually do not have trivializations. Nevertheless, plaques can be defined at least for singular Riemannian foliations (see Definition 5.1), as follows. Let P_q be a relatively compact open subset of L_q and z a point in a tubular neighborhood $\mathrm{Tub}(P_q)$. Since $\mathscr{F}$ is a singular Riemannian foliation, each leaf L_z is contained in a stratum (the collection of leaves with the same dimension of L_z). In this stratum, L_z is a regular leaf and hence admits plaques as defined above. A *plaque* P_z of the singular leaf L_z is then defined as a plaque of L_z in its stratum.

If $\mathscr{F} = \{L_p : p \in M\}$ is an n-dimensional foliation, then $D = \{T_p L_p : p \in M\}$ is clearly a smooth involutive n-dimensional distribution. The converse statement is given by the following result:

Frobenius Theorem A.19. *Let D be a smooth n-dimensional involutive distribution on a manifold M. Then D is integrable, i.e., there exists a unique n-dimensional foliation $\mathscr{F} = \{L_p : p \in M\}$ satisfying $D_q = T_q L_q$, for all $q \in M$. Furthermore, each leaf L_p is a quasi-embedded submanifold.*

The above integrability criterion fails for general singular foliations, for which a sufficient integrability condition is given in Sussmann [201]. Further references on foliations are Camacho and Neto [60], Lawson [151] and Molino [165].

A.4   Differential Forms, Integration,
and de Rham Cohomology

In this section, we recall basic notions of tensors, differential forms, integration on manifolds, the Stokes Theorem, and de Rham cohomology.

A $(0,k)$-*tensor* τ on a vector space V is a k-linear functional $\tau \colon V \times \cdots \times V \to \mathbb{R}$. We denote by $T^k(V)$, or $V \otimes \cdots \otimes V$, the vector space of $(0,k)$-tensors on V. The *tensor product* of $\tau_1 \in T^k(V)$ and $\tau_2 \in T^l(V)$ is the element $\tau_1 \otimes \tau_2 \in T^{k+l}(V)$ defined by

$$(\tau_1 \otimes \tau_2)(X_1,\ldots,X_k,X_{k+1},\ldots,X_{k+l}) := \tau_1(X_1,\ldots,X_k)\tau_2(X_{k+1},\ldots,X_{k+l}).$$

A tensor $\tau \in T^k(V)$ is called *symmetric* if, given $X_i \in V$,

$$\tau(X_1,\ldots,X_k) = \tau(X_{\sigma(1)},X_{\sigma(2)},\ldots,X_{\sigma(k)}),$$

for all permutations σ in the group $\mathfrak{S}_k$ of permutations in k letters. Similarly, $\tau \in T^k(V)$ is called *skew-symmetric* if, given $X_i \in V$,

$$\tau(X_1,\ldots,X_k) = \mathrm{sgn}(\sigma)\,\tau(X_{\sigma(1)},X_{\sigma(2)},\ldots,X_{\sigma(k)}),$$

for all $\sigma \in \mathfrak{S}_k$, where $\mathrm{sgn}(\sigma)$ denotes the sign of the permutation σ. We denote by $\Lambda^k(V)$ the vector space of skew-symmetric k-tensors on V. The projection $\mathrm{Alt}_k \colon T^k(V) \to \Lambda^k(V)$, called *alternating projection*, is given by

$$\mathrm{Alt}_k(\tau)(X_1,\ldots,X_k) = \tfrac{1}{k!} \sum_{\sigma \in \mathfrak{S}_k} \mathrm{sgn}(\sigma)\,\tau(X_{\sigma(1)},X_{\sigma(2)},\ldots,X_{\sigma(k)}),$$

for all $X_i \in V$. When k is evident from the context, it is omitted to simplify the notation.

The *wedge product* of $\omega \in \Lambda^k(V)$ and $\eta \in \Lambda^l(V)$ is the element $\omega \wedge \eta \in \Lambda^{k+l}(V)$ defined by

$$\omega \wedge \eta := \tfrac{(k+l)!}{k!l!}\,\mathrm{Alt}_{(k+l)}(\omega \otimes \eta).$$

Remark A.20. If $\omega_1,\ldots,\omega_k \in \Lambda^1(V) = T^1(V)$ are $(0,1)$-tensors, then

$$(\omega_1 \wedge \cdots \wedge \omega_k)(X_1,X_2,\ldots,X_k) = \det \begin{pmatrix} \omega_1(X_1) & \omega_1(X_2) & \cdots & \omega_1(X_k) \\ \omega_2(X_1) & \omega_2(X_2) & \cdots & \omega_2(X_k) \\ \vdots & \vdots & & \vdots \\ \omega_k(X_1) & \omega_k(X_2) & \cdots & \omega_k(X_k) \end{pmatrix}.$$

Some properties of the wedge product are summarized in the following result:

Proposition A.21. *For any* $\omega, \xi \in \Lambda^k(V)$, $\eta \in \Lambda^l(V)$, $\theta \in \Lambda^m(V)$ *and* $\lambda \in \mathbb{R}$,

(i) $(\omega \wedge \eta) \wedge \theta = \omega \wedge (\eta \wedge \theta)$;
(ii) $(\omega + \xi) \wedge \eta = \omega \wedge \eta + \xi \wedge \eta$;
(iii) $(\lambda \omega) \wedge \eta = \lambda (\omega \wedge \eta)$;
(iv) $\omega \wedge \eta = (-1)^{kl} \eta \wedge \omega$.

Moreover, if $\omega_1, \ldots, \omega_n$ *is a basis of* $T^1(V)$, *then* $\omega_{i_1} \wedge \cdots \wedge \omega_{i_r}$, $1 \leq i_1 < \cdots < i_r \leq n$, *form a basis of* $\Lambda^r(V)$, *which hence has dimension* $\binom{n}{r}$, *where* $n = \dim V$. *In particular,* $\Lambda^n(V) \cong \mathbb{R}$.

We now make the transition from tensors and skew-symmetric tensors on a vector space V to tensor fields and differential forms on a manifold M.

Definition A.22. A smooth $(0,k)$-*tensor (field)* on a manifold M is a map τ that assigns to each $p \in M$ an element $\tau_p \in T^k(T_pM)$ with the property that, given smooth vector fields $X_1, \ldots X_k$ on an open subset U of M, then $\tau(X_1, \ldots, X_k)$ is a smooth function on U defined by $\tau(X_1, \ldots, X_k)(p) = \tau(X_1(p), \ldots, X_k(p))$. A smooth *(differential) k-form* on M is a smooth $(0,k)$-tensor ω on M such that $\omega_p \in \Lambda^k(T_pM)$ for all $p \in M$. We denote by $\Omega^k(M)$ the set of all smooth k-forms on M.

Alternatively, using the language of smooth sections of vector bundles, see Definition 3.27 and Example 3.28, $(0,k)$-tensors and k-forms over M are respectively smooth sections of the tensor bundles $TM^* \otimes \cdots \otimes TM^*$ and $\Lambda^k TM^*$ over M.

Remark A.23. A k-form $\omega \in \Omega^k(M)$ is such that

$$\omega|_U = \sum_{i_1 < i_2 < \cdots < i_k} a_{i_1, \ldots, i_k} \, dx_{i_1} \wedge dx_{i_2} \wedge \cdots \wedge dx_{i_k}, \tag{A.5}$$

where (U, φ), $\varphi = (x_1, \ldots, x_n)$, is a chart of M, $a_{i_1, \ldots, i_k} : U \to \mathbb{R}$ are smooth functions and $dx_i\left(\frac{\partial}{\partial x_j}\right) = \delta_{ij}$, cf. (A.2). In particular, 0-forms on M are simply smooth functions $f : M \to \mathbb{R}$.

The differential of a 0-form f is the 1-form $df_p = \sum_{i=1}^n \frac{\partial f}{\partial x_i}\big|_p dx_i$. In what follows, we extend this notion to general k-forms on M.

Proposition A.24. *For all* $k \in \mathbb{N}$, *there exists a unique linear operator* $d : \Omega^k(M) \to \Omega^{k+1}(M)$, *called* exterior derivative, *such that:*

(i) *If* $f \in \Omega^0(M)$, *then* $df \in \Omega^1(M)$ *is the differential of the function* $f \in C^\infty(M)$,

$$df_p = \sum_{i=1}^n \frac{\partial f}{\partial x_i}\bigg|_p dx_i.$$

(ii) *The composition* $\Omega^k(M) \xrightarrow{\mathrm{d}} \Omega^{k+1}(M) \xrightarrow{\mathrm{d}} \Omega^{k+2}(M)$ *is identically null, i.e.* $d^2 = 0$;

(iii) $d(\omega \wedge \eta) = d\omega \wedge \eta + (-1)^j \omega \wedge d\eta$, *for all* $\omega \in \Omega^j(M)$ *and* $\eta \in \Omega^{k-j}(M)$.

The explicit formula for the exterior derivative of a k-form ω applied to $k+1$ smooth vector fields $X_0, \ldots, X_k \in \mathfrak{X}(M)$, is:

$$d\omega(X_0, \ldots, X_k) = \sum_{i=0}^{k} (-1)^i X_i \big(\omega(X_0, \ldots, \widehat{X}_i, \ldots, X_k) \big)$$
$$+ \sum_{i<j} (-1)^{i+j} \omega \big([X_i, X_j], X_0, \ldots, \widehat{X}_i, \ldots, \widehat{X}_j, \ldots, X_k \big),$$

where the *hat* denotes the omission of that element. There is also an explicit formula for d in terms of local charts. If (U, φ), $\varphi = (x_1, \ldots, x_n)$, is a chart of M for which ω has the expression (A.5), then

$$d\omega|_U = \sum_{i_1 < i_2 < \cdots < i_k} da_{i_1, \ldots, i_k} \wedge dx_{i_1} \wedge dx_{i_2} \wedge \cdots \wedge dx_{i_k}.$$

Given a $(0, k)$-tensor τ on N and a smooth map $F: M \to N$, the *pull-back* of τ by F is the $(0, k)$-tensor $F^* \tau$ on M defined by

$$(F^* \tau)_p(X_1, \ldots, X_k) := \tau_{F(p)}(dF_p X_1, \ldots, dF_p X_k), \quad X_i \in T_p N. \tag{A.6}$$

The pull-back of k-forms by F induces a homomorphism $F^*: \Omega^k(N) \to \Omega^k(M)$.

Proposition A.25. F^* *and* d *commute, that is,* $F^* \circ d = d \circ F^*$.

With the above background, we are ready to discuss integration on manifolds, starting from integration in Euclidean space. Let $\omega = f\, dx_1 \wedge dx_2 \wedge \cdots \wedge dx_n$, be an n-form with compact support[2] in $\mathbb{R}^n$. Define

$$\int_U \omega := \int_U f\, dx_1 \ldots dx_n.$$

It $F: V \to U$ is a diffeomorphism between two open sets of $\mathbb{R}^n$ that preserves orientation (i.e., such that $\det dF > 0$), then, by change of variables,

$$\int_U \omega = \int_V F^* \omega. \tag{A.7}$$

Integration on (oriented) manifolds is defined via the above equation. Before giving more details, let us recall the notion of orientation on smooth manifolds.

---

[2] Recall that the support of a function f is defined as $\mathrm{supp}(f) := \overline{\{x \in M \mid f(x) \neq 0\}}$.

Definition A.26. An n-manifold M is *orientable* if it is possible to define a n-form vol $\in \Omega^n(M)$, also denoted $[M]$, which never vanishes, that is, $\mathrm{vol}_p \in \Lambda^n(T_pM) \cong \mathbb{R}$ is nonzero for all $p \in M$. In this case, we say that M is *oriented* by the choice of vol. Moreover, vol is called a *volume form* on M.

Proposition A.27. *A manifold M is orientable if and only if M admits an atlas $\{(U_\alpha, \varphi_\alpha)\}$ of coherently oriented coordinates, i.e., such that $\det \mathrm{d}(\varphi_\alpha \circ \varphi_\beta^{-1}) > 0$.*

Remark A.28. A manifold M of dimension n is called a *Riemannian manifold* if it admits a $(0,2)$-tensor g that is symmetric and positive-definite (called a *Riemannian metric*), see Sect. 2.1. If M is an oriented Riemannian manifold, then there is a uniquely determined n-form vol_g which gives the orientation of M and has the value $+1$ on every oriented orthonormal frame. This form vol_g is the *volume form* of the oriented Riemannian manifold (M, g). In a local chart $\varphi = (x_1, \ldots, x_n)$, we have

$$(\varphi^{-1})^* \mathrm{vol}_g = \sqrt{\det(g_{ij}(x))} \, \mathrm{d}x_1 \wedge \cdots \wedge \mathrm{d}x_n,$$

where $g_{ij} = \mathrm{g}\left(\frac{\partial}{\partial x_i}, \frac{\partial}{\partial x_j}\right)$.

The first step to define integration of n-forms on an oriented n-manifold M is to define integration of n-forms with small support. Let $\{(U_\alpha, \varphi_\alpha)\}$ be a coherent atlas for M and let ω be an n-form with compact support contained in U_α. We define

$$\int_{U_\alpha} \omega := \int_{\varphi(U_\alpha)} (\varphi^{-1})^* \omega.$$

Note that this is well-defined, i.e., if we consider another chart (U_β, φ_β) such that the support of ω is contained in U_β, then we get the same number. This follows from (A.7) and Proposition A.27, using that $(\varphi_\beta \circ \varphi_\alpha^{-1})^* = (\varphi_\alpha^{-1})^* \circ \varphi_\beta^*$.

In order to define the integration of any n-form with compact support (not necessarily contained in a coordinate neighborhood) we need to use partitions of unity. A *partition of unity* on M is a collection of functions $\{f_\alpha : M \to [0,1]\}$ such that:

(i) $\{\mathrm{supp} f_\alpha\}$ is a locally finite cover of M;
(ii) $\sum_\alpha f_\alpha(x) = 1$ for every $x \in M$.

Given any manifold M (orientable or not), any open cover $\{U_\alpha\}$ of M admits a *subordinate* partition of unity, that is, a partition of unity satisfying $\mathrm{supp} f_\alpha \subset U_\alpha$.

Let $K = \mathrm{supp}\, \omega$ and consider a finite cover of K by coordinate neighborhoods U_α such that $\overline{U_\alpha}$ is compact. Consider $\{f_\alpha\}$ a partition of unity subordinate to $\{U_\alpha\}$. Note that $f_\alpha \omega$ has compact support contained in U_α, and define

$$\int_M \omega := \sum_\alpha \int_{U_\alpha} f_\alpha \omega.$$

One can check that in fact this definition does not depend on the cover or on the partition of unity. Here, the fact that K is compact was very important. For example, it guarantees that the above sum is finite and hence well-defined.

With the concept of integration of an n-form with compact support, we can generalize (A.7) to a change of variables formula for integration of forms.

Proposition A.29. *If M and N are oriented n-manifolds, $F: M \to N$ is a diffeomorphism that preserves orientation, and ω is an n-form with compact support on N, then*

$$\int_N \omega = \int_M F^* \omega.$$

In order to state the Stokes Theorem, which is a fundamental result of integration of differential forms, we first discuss the definition of manifolds with boundary. A *smooth n-manifold M with boundary* is a smooth n-manifold whose smooth atlas $\{(U_\alpha, \varphi_\alpha)\}$ is formed by open subsets U_α that are homeomorphic either to the half space $H^n = \{(x_1, \ldots, x_n), x_n \geq 0\}$ or to $\mathbb{R}^n$. This class extends that of n-manifolds (without boundary), for which all U_α are homeomorphic to $\mathbb{R}^n$. Compact manifolds (without boundary) are called *closed manifolds*.

Consider the upper half space H^n in $\mathbb{R}^n$ with the orientation $dx_1 \wedge \ldots \wedge dx_n$. The *induced orientation* on its boundary $\partial H^n = \{x_n = 0\}$ is $(-1)^n dx_1 \wedge \ldots \wedge dx_{n-1}$.[3] In general, if M is an oriented manifold with boundary, we define the *induced orientation* $[\partial M]$ on ∂M as follows. If φ is an orientation-preserving diffeomorphism from some open subset U of M to the upper half space H^n, then we set $[\partial M]|_{\partial U} = \varphi^*[\partial H^n]$. We are now ready to state the *Stokes Theorem*:

Theorem A.30. *Let ω be an $(n-1)$-form with compact support on an oriented n-manifold M, and consider ∂M with the induced orientation. Then*

$$\int_M d\omega = \int_{\partial M} \omega.$$

In particular, if M is an oriented closed manifold, $\int_M d\omega = 0$ for all $\omega \in \Omega^{n-1}(M)$.

We conclude this section with some comments on the *de Rham cohomology* of a manifold. A k-form ω is called *closed* if $d\omega = 0$, and *exact* if there exists a $(k-1)$-form η such that $\omega = d\eta$. Since $d^2 = 0$, see Proposition A.24, every exact form is closed. The converse is not true, i.e., there exists closed forms that are not exact.

Example A.31. Consider $\omega = \frac{1}{x^2+y^2}(-y\,dx + x\,dy)$ defined on $M = \mathbb{R}^2 \setminus \{(0,0)\}$. One can check that $d\omega = 0$, i.e., ω is a closed 1-form. If there was a 0-form (i.e., a function) f such that $df = \omega$, the Stokes Theorem would imply that $\int_{S^1} \omega = 0$,

---

[3]The sign $(-1)^n$ is needed to keep the Stokes Theorem sign-free.

however, by the definition of integration of forms, one can check that $\int_{S^1} \omega = 2\pi$.
Therefore, $\omega = \frac{1}{x^2+y^2}(-y\,dx + x\,dy)$ is closed, but not exact, on $\mathbb{R}^2 \setminus \{(0,0)\}$.

Note that on each open connected subset S of $\mathbb{R}^2 \setminus \{(0,0)\}$ that does not intersect the half line[4] $[0,+\infty) \times \{0\}$, the 1-form ω above is exact. In fact, it is given by $\omega = d\theta$, where θ is the angle element of $S \subset \mathbb{R}^2 \cong \mathbb{C}$, such that $x + iy = re^{i\theta}$ with $r = \sqrt{x^2+y^2}$, i.e., $\theta(x,y) = \arctan \frac{y}{x}$. In particular, θ can only be continuously defined on such subsets S, because otherwise θ would assume values that differ by 2π on points arbitrarily close to each other. This example shows the strong dependence of the notions of closedness and exactness on the topology of the underlying manifold.

The *k*th *de Rham cohomology group* of M is defined to be the vector space

$$H^k(M,\mathbb{R}) := \frac{\ker d^k}{\operatorname{Im} d^{k-1}} = \frac{\{\text{closed } k\text{-forms on } M\}}{\{\text{exact } k\text{-forms on } M\}}, \qquad (A.8)$$

where $\Omega^{k-1}(M) \xrightarrow{d^{k-1}} \Omega^k(M) \xrightarrow{d^k} \Omega^{k+1}(M)$ are the operators from Proposition A.24.

The above is a measure of how far closed k-forms on M are from being exact. Clearly, the de Rham cohomology groups are invariant under diffeomorphisms. This differentiable invariant is closely related to the topology of M, since it can be proved that $H^k(M,\mathbb{R})$ is isomorphic to the kth singular cohomology of M.

Regarding Example A.31, $H^1\big(\mathbb{R}^2 \setminus \{(0,0)\}, \mathbb{R}\big) \cong \mathbb{R}$ is actually spanned by the class of ω. On the other hand, if M is diffeomorphic to $\mathbb{R}^n$, then $H^k(M,\mathbb{R}) \cong \mathbb{R}$ if $k = 0$ and $H^k(M,\mathbb{R}) \cong 0$ if $k > 0$, by the so-called *Poincaré Lemma*.

If M is a closed n-manifold, then M is orientable if and only if $H^n(M,\mathbb{R}) \cong \mathbb{R}$, in which case the latter is spanned by the class of a volume form of M, and the isomorphism $H^n(M,\mathbb{R}) \cong \mathbb{R}$ is given by integration. The *Betti numbers* of a closed n-manifold M are defined as $b_k(M) := \dim H^k(M,\mathbb{R})$, and the Euler characteristic of M is given by their alternate sum:

$$\chi(M) = \sum_{k=0}^{n} (-1)^k b_k(M).$$

A fundamental result on the de Rham cohomology of closed oriented manifolds is *Poincaré duality*, which states that if M is a closed oriented n-manifold, then $b_k(M) = b_{n-k}(M)$, for all $0 \le k \le n$.

Example A.32. The Betti numbers of the n-sphere S^n are $b_0(S^n) = b_n(S^n) = 1$ and $b_k(S^n) = 0$ for all $1 \le k \le n-1$. In particular, $\chi(S^n) = 2$ if n is even and $\chi(S^n) = 0$ if n is odd.

For more about de Rham cohomology, see Bott and Tu [48] or Warner [227].

---

[4]It suffices to require that S does not intersect a fixed unbounded curve in $\mathbb{R}^2$ starting at the origin.

References

1. A.V. Alekseevsky, D.V. Alekseevsky, Riemannian G-manifold with one-dimensional orbit space. Ann. Glob. Anal. Geom. **11**(3), 197–211 (1993)
2. D.V. Alekseevsky, F. Podestà, Compact cohomogeneity one Riemannian manifolds of positive Euler characteristic and quaternionic Kähler manifolds, in *Geometry, Topology and Physics*, Campinas, 1996 (de Gruyter, Berlin, 1997), pp. 1–33
3. M.M. Alexandrino, Integrable Riemannian submersion with singularities. Geom. Dedicata **108**, 141–152 (2004)
4. M.M. Alexandrino, Singular Riemannian foliations with sections. Ill. J. Math. **48**(4), 1163–1182 (2004)
5. M.M. Alexandrino, Generalizations of isoparametric foliations. Mat. Contemp. **28**, 29–50 (2005)
6. M.M. Alexandrino, Proofs of conjectures about singular Riemannian foliations. Geom. Dedicata **119**(1), 219–234 (2006)
7. M.M. Alexandrino, Singular holonomy of singular Riemannian foliations with sections. Mat. Contemp. **33**, 23–55 (2007)
8. M.M. Alexandrino, Desingularization of singular Riemannian foliation. Geom. Dedicata **149**, 397–416 (2010)
9. M.M. Alexandrino, On polar foliations and fundamental group. Results Math. **60**(1–4), 213–223 (2011)
10. M.M. Alexandrino, Polar foliations and isoparametric maps. Ann. Glob. Anal. Geom. **41**(2), 187–198 (2012)
11. M.M. Alexandrino, L. Biliotti, R. Pedrosa, *Lectures on Isometric Actions*. XV Escola de Geometria Diferencial, Publicações matemáticas (IMPA, Fortaleza, 2008)
12. M.M. Alexandrino, R. Briquet, D. Töben, Progress in the theory of singular Riemannian foliations. Differ. Geom. Appl. **31**, 248–267 (2013)
13. M.M. Alexandrino, C. Gorodski, Singular Riemannian foliations with sections, transnormal maps and basic forms. Ann. Glob. Anal. Geom. **32**(3), 209–223 (2007)
14. M.M. Alexandrino, M.A. Javaloyes, On closed geodesics in the leaf space of singular Riemannian foliations. Glasg. Math. J. **53**, 555–568 (2011)
15. M.M. Alexandrino, A. Lytchak, On smoothness of isometries between orbit spaces, in *Riemannian Geometry and Applications, Proceedings RIGA* (Ed. Univ. Bucureşti, 2011), pp. 17–28. arXiv:1107.2563
16. M.M. Alexandrino, M. Radeschi, Smoothness of isometric flows on orbit spaces and applications to the theory of foliations (2014, preprint). arXiv:1301.2735

© Springer International Publishing Switzerland 2015

M.M. Alexandrino, R.G. Bettiol, *Lie Groups and Geometric Aspects of Isometric Actions*, DOI 10.1007/978-3-319-16613-1

17. M.M. Alexandrino, M. Radeschi, Mean curvature flow of singular Riemannian foliations (2014, preprint). arXiv:1408.543
18. M.M. Alexandrino, M. Radeschi, Isometries between leaf spaces. Geom. Dedicata **174**, 193–201 (2015)
19. M.M. Alexandrino, T. Töben, Singular Riemannian foliations on simply connected spaces. Differ. Geom. Appl. **24**, 383–397 (2006)
20. M.M. Alexandrino, T. Töben, Equifocality of singular Riemannian foliation. Proc. Am. Math. Soc. **136**, 3271–3280 (2008)
21. S. Aloff, N.R. Wallach, An infinite family of distinct 7-manifolds admitting positively curved Riemannian structures. Bull. Am. Math. Soc. **81**, 93–97 (1975)
22. M. Amann, L. Kennard, On a generalized conjecture of Hopf with symmetry. arXiv:1402.7255
23. M. Amann, L. Kennard, Topological properties of positively curved manifolds with symmetry. Geom. Funct. Anal. **24**(5), 1377–1405 (2014)
24. A. Arvanitogeorgos, *An Introduction to Lie Groups and the Geometry of Homogeneous Spaces*, Providence (AMS, 1999)
25. A. Back, W.-Y. Hsiang, Equivariant geometry and Kervaire spheres. Trans. Am. Math. Soc. **304**, 207–227 (1987)
26. J. Baez, The octonions. Bull. Am. Math. Soc. **39**(2), 145–205 (2002). Errata: Bull. Am. Math. Soc. (N.S.) **42**(2), 213 (2005). arXiv:math/0105155
27. L. Berard-Bergery, Les variétés riemanniennes homogènes simplement connexes de dimension impaire à courbure strictement positive. J. Math. Pures Appl. **55**, 47–67 (1976)
28. V.N. Berestovskii, Homogeneous Riemannian manifolds of positive Ricci curvature. Math. Notes **58**, 905–909 (1995)
29. M. Berger, Les variétés Riemanniennes 1/4-pincées. Ann. Sc. Norm. Super. Pisa Cl. Sci. **14**(2), 161–170 (1960)
30. M. Berger, Les variétés riemanniennes homogènes normales simplement connexes à courbure strictement positive. Ann. Scuola Norm. Sup. Pisa **15**, 179–246 (1961)
31. M. Berger, Trois remarques sur les variétés riemanniennes à courbure positive. C. R. Acad. Sci. Paris Sér. **A-B 263**, A76–A78 (1966)
32. J. Berndt, S. Console, C. Olmos, *Submanifolds and Holonomy*. Research Notes in Mathematics, vol. 434 (Chapman and Hall/CRC, Boca Raton, 2003)
33. A.L. Besse, *Einstein Manifolds*. Series of Modern Surveys in Mathematics (Springer, Berlin, 2002)
34. R.G. Bettiol, Positive biorthogonal curvature on $S^2 \times S^2$. Proc. Am. Math. Soc. **142**(12), 4341–4353 (2014)
35. R.G. Bettiol, R.A.E. Mendes, Flag manifolds with strongly positive curvature. Math. Z. (to appear)
36. R.G. Bettiol, R.A.E. Mendes, Strongly positive curvature (2014, preprint). arXiv:1403.2117
37. R.G. Bettiol, P. Piccione, Bifurcation and local rigidity of homogeneous solutions to the Yamabe problem on spheres. Calc. Var. Partial Differ. Equ. **47**(3–4), 789–807 (2013)
38. R.G. Bettiol, P. Piccione, Multiplicity of solutions to the Yamabe problem on collapsing Riemannian submersions. Pac. J. Math. **266**(1), 1–21 (2013)
39. R.G. Bettiol, P. Piccione, Delaunay type hypersurfaces in cohomogeneity one manifolds. arXiv:1306.6043
40. L. Biliotti, Coisotropic and polar actions on compact irreducible Hermitian symmetric spaces. Trans. Am. Math. Soc. **358**, 3003–3022 (2006)
41. C. Böhm, Inhomogeneous Einstein metrics on low-dimensional spheres and other low-dimensional spaces. Invent. Math. **134**(1), 145–176 (1998)
42. C. Böhm, Non-compact cohomogeneity one Einstein manifolds. Bull. Soc. Math. Fr. **127**(1), 135–177 (1999)
43. C. Böhm, Non-existence of cohomogeneity one Einstein metrics. Math. Ann. **314**(1), 109–125 (1999)

44. C. Böhm, M. Wang, W. Ziller, A variational approach for compact homogeneous Einstein manifolds. Geom. Funct. Anal. **14**(4), 681–733 (2004)
45. C. Böhm, B. Wilking, Manifolds with positive curvature operators are space forms. Ann. Math. **167**(2), 1079–1097 (2008)
46. W.M. Boothby, *An Introduction to Differentiable Manifolds and Riemannian Geometry*, 2nd edn. (Academic, Amsterdam/New York, 2003)
47. A. Borel, Some remarks about Lie groups transitive on spheres and tori. Bull. Am. Math. Soc. **55**, 580–587 (1949)
48. R. Bott, L. Tu, *Differential Forms in Algebraic Topology*. Graduate Texts in Mathematics, vol. 82 (Springer, New York, 1982)
49. H. Boualem, Feuilletages riemanniens singuliers transversalement integrables. Compositio Mathematica **95**, 101–125 (1995)
50. J.-P. Bourguignon, A. Deschamps, P. Sentenac, Conjecture de H. Hopf sur les produits de variétés. Ann. Sci. École Norm. Sup. **5**(4), 277–302 (1972)
51. J.-P. Bourguignon, A. Deschamps, P. Sentenac, Quelques variations particuliéres d'un produit de métriques. Ann. Sci. École Norm. Sup. (4) **6**, 1–16 (1973)
52. J.-P. Bourguignon, H. Karcher, Curvature operators: pinching estimates and geometric examples. Ann. Sci. École Norm. Sup. (4) **11**(1), 71–92 (1978)
53. G. Bredon, *Introduction to Compact Transformation Groups*. Pure and Applied Mathematics, vol. 46 (Academic, New York/London, 1972)
54. S. Brendle, Embedded minimal tori in S^3 and the Lawson conjecture. Acta Math. **211**(2), 177–190 (2013)
55. S. Brendle, R. Schoen, Manifolds with $1/4$-pinched curvature are space forms. J. Am. Math. Soc. **22**, 287–307 (2009)
56. T. Bröken, T. Tom Dieck, *Representations of Compact Lie Group*. Graduate Texts in Mathematics, vol. 98 (Springer, New York, 1995)
57. R.L. Bryant, *An Introduction to Lie Groups and Symplectic Geometry*. Geometry and Quantum Field Theory, IAS/Park City, Mathematics Series, vol. 1 (American Mathematical Society, Providence, 1995)
58. D. Bump, *Lie Groups*. Graduate Texts in Mathematics, vol. 225, 2nd edn. (Springer, New York, 2013)
59. D. Burago, Y. Burago, S. Ivanov, *A Course in Metric Geometry*. Graduate Studies in Mathematics, vol. 33 (American Mathematical Society, Providence, 2001)
60. C. Camacho, A.L. Neto, *Teoria geométrica das folheações*. Projeto Euclides (IMPA, Brasilia, 1979)
61. M.P. do Carmo, *Geometria Riemanniana*. Projeto Euclides, 3rd edn. (IMPA, Rio de Janeiro, 2005)
62. É. Cartan, Familles de surfaces isoparamétriques dans les espaces á courbure constante. Ann. di Mat. **17**, 177–191 (1938)
63. É. Cartan, Sur des familles remarquables d'hypersurfaces isoparamétriques dans les espaces shériques. Math. Z. **45**, 335–367 (1939)
64. É. Cartan, Sur quelques familles remarquables d'hypersurfaces. C. R. Congrès Math. Liège 30–41 (1939)
65. É. Cartan, Sur des familles d'hypersurfaces isoparamétriques des espaces sphériques à 5 et à 9 dimensions. Revista Univ. Tucumán **1**, 5–22 (1940)
66. R. Carter, G. Segal, I. MacDonald, *Lectures on Lie Groups and Lie Algebras*. London Mathematical Society, Student Texts, vol. 32 (Cambridge University Press, 1995)
67. S. Carter, A. West, Isoparametric systems and transnormality. Proc. Lond. Math. Soc. **51**, 520–542 (1985)
68. S. Carter, A. West, Generalised Cartan polynomials. J. Lond. Math. Soc. **32**, 305–316 (1985)
69. T.E. Cecil, Q.-S. Chi, G.R. Jensen, Isoparametric hypersurfaces with four principal curvatures. Ann. Math. (2) **166**(1), 1–76 (2007)
70. J. Cheeger, Some examples of manifolds of nonnegative curvature. J. Differ. Geom. **8**, 623–628 (1973)

71. J. Cheeger, D. Ebin, *Comparison Theorems in Riemannian Geometry* (North-Holland Publishing, Amsterdam, 1975)
72. J. Dadok, Polar coordinates induced by actions of compact Lie groups. Trans. Am. Math. Soc. **288**, 125–137 (1985)
73. A. Dancer, S. Hall, M. Wang, Cohomogeneity one shrinking Ricci solitons: an analytic and numerical study. Asian J. Math. **17**(1), 33–61 (2013)
74. A. Dancer, M. Wang, The cohomogeneity one Einstein equations from the Hamiltonian viewpoint. J. Reine Angew. Math. **524**, 97–128 (2000)
75. A. Dancer, M. Wang, On Ricci solitons of cohomogeneity one. Ann. Glob. Anal. Geom. **39**(3), 259–292 (2011)
76. M.W. Davis, Groups generated by reflections and aspherical manifolds not covered by Euclidean space. Ann. Math. **117**, 293–324 (1983)
77. O. Dearricott, A 7-manifold with positive curvature. Duke Math. J. **158**(2), 307–346 (2011)
78. A. Deitmar, *A First Course in Harmonic Analysis*. Universitext, 2nd edn. (Springer, New York, 2005)
79. J.J. Duistermaat, J.A.C. Kolk, *Lie Groups*. Universitext (Springer, Berlin/New York, 2000)
80. H. Elíasson, Die Krümmung des Raumes Sp(2)/SU(2) von Berger. Math. Ann. **164**, 317–323 (1966)
81. J.-H. Eschenburg, New examples of manifolds with strictly positive curvature. Invent. Math. **66**(3), 469–480 (1982)
82. J.-H. Eschenburg, *Freie isometrische Aktionen auf kompakten Lie-Gruppen mit positiv gekrümmten Orbiträumen*. Schriftenreihe des Mathematischen Instituts der Universität Münster, 2. Serie, 32 (1984)
83. J.-H. Eschenburg, M. Wang, The initial value problem for cohomogeneity one Einstein metrics. J. Geom. Anal. **10**(1), 109–137 (2000)
84. F. Fang, K. Grove, G. Thorbergsson, *Tits Geometry and Positive Curvature* (2012, preprint). arXiv:1205.6222 [math.DG]
85. H.D. Fegan, *Introduction to Compact Lie Groups*. Series in Pure Mathematics, vol. 13 (World Scientific, Singapore, 1998)
86. L. Fehér, B.G. Pusztai, Twisted spin Sutherland models from quantum Hamiltonian reduction. Phys. A: Math. Theor. **41**, 194009 (2008)
87. L. Fehér, B.G. Pusztai, Hamiltonian reductions of free particles under polar actions of compact Lie groups. Theor. Math. Phys. **155**, 646–658 (2008)
88. B.L. Feigin, D.B. Fuchs, V.V. Gorbatsevich, O.V. Schvartsman, E.B. Vinberg, *Lie Groups and Lie Algebras II*. Encyclopaedia of Mathematical Sciences, vol. 21 (Springer, Moscow, 1988)
89. D. Ferus, H. Karcher, H.F. Münzner, Cliffordalgebren und neue isoparametrische Hyperflächen. Math. Z. **177**, 479–502 (1981)
90. W. Fulton, J. Harris, *Representation Theory. A First Course*. Graduate Texts in Mathematics, vol. 129, Readings in Mathematics (Springer, New York, 1991)
91. F. Galaz-Garcia, C. Searle, Cohomogeneity one Alexandrov spaces. Transform. Groups **16**(1), 91–107 (2011)
92. F. Galaz-Garcia, C. Searle, Low-dimensional manifolds with non-negative curvature and maximal symmetry rank. Proc. Am. Math. Soc. **139**(7), 2559–2564 (2011)
93. F. Galaz-Garcia, C. Searle, Nonnegatively curved 5-manifolds with almost maximal symmetry rank. Geom. Topol. **18**(3), 1397–1435 (2014)
94. R. Gangolli, V.S. Varadarajan, *Harmonic Analysis of Spherical Functions on Real Reductive Groups*. Ergebnisse der Mathematik und ihrer Grenzgebiete, vol. 101 (Springer, Berlin/New York, 1988)
95. J. Ge, Z. Tang, Isoparametric functions on exotic spheres. J. Reine Angew. Math. **683**, 161–180 (2013)
96. J. Ge, Z. Tang, Geometry of isoparametric hypersurfaces in Riemannian manifolds. Asian J. Math. **18**(1), 117–125 (2014)
97. Robert E. Krieger Publishing Co., Inc. (Malabar, FL, 1994)

98. V.V. Gorbatsevich, A.L. Onishchik, E.B. Vinberg, *Foundations of Lie Theory and Lie Transformation Groups* (Springer, Moscow, 1988)

99. V.V. Gorbatsevich, A.L. Onishchik, E.B. Vinberg, *Lie Groups and Lie Algebras I*. Encyclopaedia of Mathematical Sciences, vol. 20 (Springer, Moscow, 1988)

100. V.V. Gorbatsevich, A.L. Onishchik, E.B. Vinberg, *Lie Groups and Lie Algebras III*. Encyclopaedia of Mathematical Sciences, vol. 41 (Springer, Moscow, 1990)

101. C. Gorodski, *A Metric Approach to Representations of Compact Lie Groups* (2014, preprint), http://www.ime.usp.br/~gorodski/publications.html, or http://math.osu.edu/file/metric-approach-representations-compact-lie-groups.pdf

102. C. Gorodski, E. Heintze, Homogenous structures and rigidity of isoparametric submanifolds in Hilbert spaces. J. Fixed Point Theory Appl. **11**(1), 93–136 (2012)

103. F. Gozzi, Low dimensional polar actions. Geom. Dedicata (to appear). arXiv:1407.0638

104. A. Gray, Pseudo-Riemannian almost product manifolds and submersions. J. Math. Mech. **16**, 715–737 (1967)

105. D. Gromoll, G. Walschap, *Metric Foliations and Curvatures*. Progress in Mathematics, vol. 268 (Birkhäuser, 2009)

106. K. Grove, *Riemannian Geometry: A Metric Entrance*. Lecture Notes Series (Aarhus), vol. 65 (University of Aarhus, Department of Mathematics, Aarhus, 1999)

107. K. Grove, Geometry of, and via, symmetries, in *Conformal, Riemannian and Lagrangian Geometry*, Knoxville, 2000. University Lecture Series, vol. 27 (American Mathematical Society, Providence, 2002), pp. 31–53

108. K. Grove, D. Gromoll, The low-dimensional metric foliations of Euclidean spheres. J. Differ. Geom. **28**(1), 143–156 (1988)

109. K. Grove, C. Searle, Positively curved manifolds with maximal symmetry-rank. J. Pure Appl. Algebra **91**(1–3), 137–142 (1994)

110. K. Grove, C. Searle, Differential topological restrictions by curvature and symmetry. J. Differ. Geom. **47**(3), 530–559 (1997)

111. K. Grove, L. Verdiani, B. Wilking, W. Ziller, Non-negative curvature obstructions in cohomogeneity one and the Kervaire spheres. Ann. Sc. Norm. Super. Pisa Cl. Sci. (5) **5**(2), 159–170 (2006)

112. K. Grove, L. Verdiani, W. Ziller, An exotic $T_1 S^4$ with positive curvature. Geom. Funct. Anal. **21**(3), 499–524 (2011)

113. K. Grove, B. Wilking, A knot characterization and 1-connected nonnegatively curved 4-manifolds with circle symmetry. Geom. Topol. **18**(5), 3091–3110 (2014)

114. K. Grove, B. Wilking, W. Ziller, Positively curved cohomogeneity one manifolds and 3-Sasakian geometry. J. Differ. Geom. **78**(1), 33–111 (2008)

115. K. Grove, W. Ziller, Curvature and symmetry of Milnor spheres. Ann. Math. (2) **152**(1), 331–367 (2000)

116. K. Grove, W. Ziller, Cohomogeneity one manifolds with positive Ricci curvature. Invent. Math. **149**(3), 619–646 (2002)

117. K. Grove, W. Ziller, Lifting group actions and nonnegative curvature. Trans. Am. Math. Soc. **363**(6), 2865–2890 (2011)

118. K. Grove, W. Ziller, Polar manifolds and actions. J. Fixed Point Theory Appl. **11**(2), 279–313 (2012)

119. M.A. Guest, *Harmonic Maps, Loop Groups and Integrable Systems*. London Mathematical Society, Student Texts, vol. 38 (Cambridge University Press, Cambridge/New York, 1997)

120. L. Guijarro, G. Walschap, When is a Riemannian submersion homogeneous? Geom. Dedicata **125**, 47–52 (2007)

121. B.C. Hall, *Lie Groups, Lie Algebras and Representations, an Elementary Introduction*. Graduate Texts in Mathematics (Springer, New York, 2004)

122. C.E. Harle, Isoparametric families of submanifolds. Bol. Soc. Brasil. Mat. **13**, 491–513 (1982)

123. F.R. Harvey, *Spinors and Calibrations*. Vol. 9 of Perspectives in Mathematics (Academic, Boston, 1990)

124. E. Heintze, X. Liu, C. Olmos, Isoparametric submanifolds and a Chevalley–type restriction theorem, in *Integrable Systems, Geometry, and Topology*. AMS/IP Studies in Advanced Mathematics, vol. 36 (American Mathematical Society, Providence, 2006), pp. 151–190

125. S. Helgason, *Groups and Geometric Analysis. Integral Geometry, Invariant Differential Operators, and Spherical Functions*. Corrected reprint of the 1984 original. Mathematical Surveys and Monographs, vol. 83 (American Mathematical Society, Providence, 2000)

126. S. Helgason, *Differential Geometry, Lie Groups and Symmetric Spaces*. Graduate Studies in Mathematics, vol. 34 (American Mathematical Society, Providence, 2001)

127. R. Hermann, Variational completeness for compact symmetric spaces. Proc. Am. Math. Soc. **11**, 544–546 (1960)

128. J. Hilgert, K.-H. Neeb, *Structure and Geometry of Lie Groups*. Springer Monographs in Mathematics (Springer, New York, 2012)

129. W.-Y. Hsiang, *Lectures on Lie Groups*. Series on University Mathematics, vol. 2 (World Scientific Publishing, River Edge, 2000)

130. W.C. Hsiang, W.Y. Hsiang, On compact subgroups of the diffeomorphism groups of Kervaire spheres. Ann. Math. **85**, 359–369 (1967)

131. W.Y. Hsiang, B. Kleiner, On the topology of positively curved 4-manifolds with symmetry. J. Differ. Geom. **29**(3), 615–621 (1989)

132. G.A. Hunt, A theorem of Elie Cartan. Proc. Am. Math. Soc. **7**, 307–308 (1956)

133. M. Ise, M. Takeuchi, *Lie Groups I*. Translations of Mathematical Monographs, vol. 85 (American Mathematical Society, Providence, 1991)

134. G. Jensen, Einstein metrics on principal fibre bundles. J. Differ. Geom. **8**, 599–614 (1973)

135. J. Jost, *Riemannian Geometry and Geometric Analysis*. Universitext, 2nd edn. (Springer, Berlin/New York, 1998)

136. Y. Katznelson, *An Introduction to Harmonic Analysis* (Dover, New York, 1968)

137. K. Kawakubo, *The Theory of Transformation Groups* (Oxford University Press, Oxford/New York, 1991)

138. L. Kennard, On the Hopf conjecture with symmetry. Geom. Topol. **161**(1), 563–593 (2013)

139. L. Kennard, Positively curved Riemannian metrics with logarithmic symmetry rank bounds. Comment. Math. Helv. **89**, 937–962 (2014)

140. M. Kerin, On the curvature of biquotients. Math. Ann. **352**(1), 155–178 (2012)

141. B. Kleiner, Riemannian four-manifolds with nonnegative curvature and continuous symmetry, Ph.D. thesis, University of California, Berkeley (1990)

142. W. Klingenberg, Contributions to Riemannian geometry in the large. Ann. Math. **69**(3), 654–666 (1959)

143. W. Klingenberg, Über Riemannsche Mannigfaltigkeiten mit positiver Krümmung. Comment. Math. Helv. **35**(1), 47–54 (1961)

144. A.W. Knapp, *Representation Theory of Semisimple Groups* (Princeton University Press, Princeton, 1986)

145. A.W. Knapp, *Lie Groups: Beyond an Introduction*. Progress in Mathematics, vol. 140, 2nd edn. (Birkhäuser, Boston, 2002)

146. A. Kollross, A classification of hyperpolar and cohomogeneity one actions. Trans. Am. Math. Soc. **354**(2), 571–612 (2002)

147. R. Lafuente, J. Lauret, On homogeneous Ricci solitons. Q. J. Math. **65**(2), 399–419 (2014)

148. R. Lafuente, J. Lauret, Structure of homogeneous Ricci solitons and the Alekseevskii conjecture. J. Differ. Geom. **98**(2), 315–347 (2014)

149. S. Lang, *Differential and Riemannian Manifolds*. Graduate Texts in Mathematics (Springer, New York, 1995)

150. J. Lauret, Ricci flow of homogeneous manifolds. Math. Z. **274**(1–2), 373–403 (2013)

151. H.B. Lawson, Foliations. Bull. Am. Math. Soc. **80**, 3 (1974)

152. J.M. Lee, *Riemannian Manifolds: An Introduction to Curvature*. Graduate Texts in Mathematics (Springer, New York, 1997)

153. A. Lytchak, *Singular Riemannian Foliations on Space Without Conjugate Points*. Differential Geometry (World Scientific Publishing, Hackensack, 2009), pp. 75–82

154. A. Lytchak, Geometric resolution of singular Riemannian foliations. Geom. Dedicata **149**, 379–395 (2010)
155. A. Lytchak, Polar foliations on symmetric spaces. Geom. Funct. Anal. **24**, 1298–1315 (2014)
156. A. Lytchak, G. Thorbergsson, Variationally complete actions on nonnegatively curved manifolds. Ill. J. Math. **51**(2), 605–615 (2007)
157. A. Lytchak, G. Thorbergsson, Curvature explosion in quotients and applications. J. Differ. Geom. **85**(1), 117–139 (2010)
158. J. Milnor, *Morse Theory*. Annals of Mathematics Studies, vol. 51 (Princeton University Press, Princeton, 1969)
159. J. Milnor, Curvatures of left-invariant metrics on Lie groups. Adv. Math. **21**(3), 293–329 (1976)
160. W. Misner, K.S. Thorne, J.A. Wheeler, *Gravitation* (Freeman and Company, New York, 1973)
161. R. Miyaoka, Transnormal functions on a Riemannian manifold. Differ. Geom. Appl. **31**, 130–139 (2013)
162. R. Miyaoka, Isoparametric hypersurfaces with $(g, m) = (6, 2)$. Ann. Math. (2) **177**(1), 53–110 (2013). Errata: Ann. Math. (2) **180**(3), 1221 (2014)
163. I. Moerdijk, J. Mrčun, *Introduction to Foliations and Lie Groupoids*. Cambridge Studies in Advanced Mathematics (Cambridge University Press, Cambridge/New York, 2003)
164. P. Molino, Desingularisation Des Feuilletages Riemanniens. Am. J. Math. **106**(5), 1091–1106 (1984)
165. P. Molino, *Riemannian Foliations*. Progress in Mathematics, vol. 73 (Birkhäuser, Boston, 1988)
166. P. Molino, M. Pierrot, Théorèmes de slice et holonomie des feuilletages Riemanniens singuliers. Ann. Inst. Fourier (Grenoble) **37**(4), 207–223 (1987)
167. D. Montgomery, H. Samelson, Transformation groups of spheres. Ann. Math. (2) **44**, 454–470 (1943)
168. D. Montgomery, L. Zippin, *Topological Transformation Groups* (Interscience Publishers, New York, 1955)
169. M. Mucha, Polar actions on certain principal bundles over symmetric spaces of compact type. Proc. Am. Math. Soc. **139**(6), 2249–2255 (2011)
170. H.F. Münzner, Isoparametrische Hyperflächen in Sphären I. Math. Ann. **251**, 57–71 (1980)
171. H.F. Münzner, Isoparametrische Hyperflächen in Sphären II. Math. Ann. **256**, 215–232 (1981)
172. M. Müter, Krümmungserhöhende Deformationen mittels Gruppenaktionen, PhD thesis, University of Münster (1987)
173. S. Myers, N. Steenrod, The group of isometries of a Riemannian manifold. Ann. Math. **40**(2), 400–416 (1939)
174. L. Ni, *Notes on Transnormal Functions on Riemannian Manifolds*, unpublished (1997), available at: http://math.ucsd.edu/~lni/academic/isopara.pdf
175. M. Noumi, *Painlevé Equations Through Symmetry*. Translations of Mathematical Monographs, vol. 223 (American Mathematical Society, Providence, 2004)
176. P.J. Olver, *Applications of Lie Groups to Differential Equations*. Graduate Texts in Mathematics, vol. 107, 2nd edn. (Springer, New York, 2000)
177. B. O'Neill, The fundamental equations of a submersion. Mich. Math. J. **13**, 459–469 (1966)
178. A.L. Onishchik, Transitive compact transformation groups. Mat. Sb. **60**, 447–485 (1963); translated in English in Am. Math. Soc. Transl. **55**, 153–194 (1966)
179. A.L. Onishchik, Topology of transitive transformation groups. Johann Ambrosius Barth Verlag GmbH, Leipzig, 1994. 300 pp. ISBN: 3-335-00355-1
180. R.S. Palais, On the existence of slices for actions of noncompact groups. Ann. Math. **73**, 295–323 (1961)
181. R.S. Palais, C.L. Terng, A general theory of canonical forms. Trans. Am. Math. Soc. **300**, 236–238 (1987)
182. R.S. Palais, C.L. Terng, *Critical Point Theory and Submanifold Geometry*. Lecture Notes in Mathematics, vol. 1353 (Springer, Berlin/New York, 1988)

183. P. Petersen, *Riemannian Geometry*. Graduate Texts in Mathematics, **171**, 2nd edn. New York, NY, USA (Springer, 2006)

184. P. Petersen, F. Wilhelm, An exotic sphere with positive sectional curvature. arXiv:0805.0812

185. M.J. Pflaum, H. Posthuma, X. Tang, Geometry of orbit spaces of proper Lie groupoids. J. Reine Angew. Math. **694**, 49–84 (2014)

186. F. Podestà, G. Thorbergsson, Polar actions on rank one symmetric spaces. J. Differ. Geom. **53**(1), 131–175 (1999)

187. T. Püttmann, Optimal pinching constants of odd dimensional homogeneous spaces. Invent. Math. **138**, 631–684 (1999)

188. M. Radeschi, Low dimensional singular Riemannian foliations in spheres (2012, preprint). arXiv:1203.6113

189. M. Radeschi, Clifford algebras and new singular Riemannian foliations in spheres. Geom. Funct. Anal. **24**, 1660–1682 (2014)

190. H. Rauch, A contribution to differential geometry in the large. Ann. Math. **54**(1), 38–55 (1951)

191. L.A.B. San Martin, *Álgebras de Lie* (Editora da Unicamp, 1999)

192. L. Schwachhöfer, K. Tapp, Homogeneous metrics with nonnegative curvature. J. Geom. Anal. **19**(4), 929–943 (2009)

193. C. Searle, Cohomogeneity and positive curvature in low dimensions. Math. Z. **214**, 491–498 (1993): Err. ibet. **226**, 165–167 (1997)

194. C. Searle, F. Wilhelm, How to lift positive Ricci curvature (2013, preprint). arXiv:1311.1809

195. J.P. Serre, *Complex Semisimple Lie Algebras*. Monographs in Mathematics (Springer, Berlin/New York, 2001)

196. A. Siffert, A new structural approach to isoparametric hypersurfaces in spheres (2014, preprint). arXiv:1410.6206

197. F.L.N. Spindola, Grupos de Lie, ações próprias e a conjectura de Palais–Terng, MSc. dissertation, Universidade de São Paulo (2008)

198. M. Spivak, *A Comprehensive Introduction to Differential Geometry*, vol. 1 (Publish or Perish, Boston, 1975)

199. M. Strake, Variationen von Metriken nichtnegativer Schnittkrümmung. PhD thesis, University of Münster (1986)

200. M. Strake, Curvature increasing metric variations. Math. Ann. **276**(4), 633–641 (1987)

201. H. Sussmann, Orbits of families of vector fields and integrability of distributions. Trans. Am. Math. Soc. **180**, 171–188 (1973)

202. J. Szenthe, *A generalization of the Weyl group*. Acta Math. Hungar. **41**(3–4), 347–357 (1983)

203. J. Szenthe, Orthogonally transversal submanifolds and the generalizations of the Weyl group. Periodica Mathematica Hungarica **15**(4), 281–299 (1984)

204. C.L. Terng, Isoparametric submanifolds and their Coxeter groups. J. Differ. Geom. **21**, 79–107 (1985)

205. C.L. Terng, G. Thorbergsson, Submanifold geometry in symmetric spaces. J. Differ. Geom. **42**, 665–718 (1995)

206. G. Thorbergsson, Isoparametric foliations and their buildings. Ann. Math. **133**, 429–446 (1991)

207. G. Thorbergsson, *Differentialgeometrie I und II*. Lecture Notes (University of Cologne, 1997)

208. G. Thorbergsson, A survey on isoparametric hypersurfaces and their generalizations, in *Handbook of Differential Geometry*, ed. by F. Dillen, L. Verstraelen, vol. 1 (Elsevier, Amsterdam/New York, 2000)

209. G. Thorbergsson, Transformation groups and submanifold geometry. Rendiconti di Matematica, Serie VII **25**, 1–16 (2005)

210. G. Thorbergsson, Singular Riemannian foliations and isoparametric submanifolds. Milan J. Math. **78**(1), 355–370 (2010)

211. J.A. Thorpe, On the curvature tensor of a positively curved 4-manifold, in *Proceedings of the Thirteenth Biennial Seminar of the Canadian Mathematical Congress*, Dalhousie University, Halifax, 1971, vol. 2 (Canadian Mathematical Congress, Montreal, 1972), pp. 156–159

212. J.A. Thorpe, The zeros of nonnegative curvature operators. J. Differ. Geom. **5**, 113–125 (1971). Erratum: J. Differ. Geom. **11**(2), 315 (1976)

213. D. Töben, Parallel focal structure and singular Riemannian foliations. Trans. Am. Math. Soc. **358**, 1677–1704 (2006)

214. D. Töben, Singular Riemannian foliations on nonpositively curved manifolds. Math. Z. **255**(2), 427–436 (2007)

215. F.M. Valiev, Precise estimates for the sectional curvatures of homogeneous Riemannian metrics on Wallach spaces. Sib. Mat. Zhurn. **20**, 248–262 (1979)

216. V.S. Varadarajan, *Lie Groups, Lie Algebras, and Their Representations*. Graduate Texts in Mathematics (Springer, New York, 1984)

217. V.S. Varadarajan, *An Introduction to Harmonic Analysis on Semisimple Lie Groups*. Cambridge Studies in Advanced Mathematics, vol. 16 (Cambridge University Press, Cambridge/New York, 1999)

218. L. Verdiani, Cohomogeneity one manifolds of even dimension with strictly positive sectional curvature. J. Differ. Geom. **68**, 31–72 (2004)

219. L. Verdiani, W. Ziller, Positively curved homogeneous metrics on spheres. Math. Z. **261**(3), 473–488 (2009)

220. L. Verdiani, W. Ziller, Concavity and rigidity in non-negative curvature. J. Differ. Geom. **97**, 349–375 (2014)

221. N.R. Wallach, Compact homogeneous Riemannian manifolds with strictly positive curvature. Ann. Math. **96**, 277–295 (1972)

222. G. Walschap, *Metric Structures in Differential Geometry*. Graduate Texts in Mathematics (Springer, New York, 2004)

223. Q.M. Wang, Isoparametric hypersurfaces in complex projective spaces, in *Proceedings of the 1980 Beijing Symposium on Differential Geometry and Differential Equations*, Beijing, 1980 (Science Press, Beijing, 1982), pp. 1509–1523

224. Q.M. Wang, Isoparametric functions on Riemannian manifolds. I. Math. Ann. **277**, 639–646 (1987)

225. M. Wang, W. Ziller, On normal homogeneous Einstein manifolds. Ann. Sci. École Norm. Sup. (4) **18**(4), 563–633 (1985)

226. M. Wang, W. Ziller, Existence and nonexistence of homogeneous Einstein metrics. Invent. Math. **84**(1), 177–194 (1986)

227. F.W. Warner, *Foundations of Differentiable Manifolds and Lie Groups*. Graduate Texts in Mathematics (Springer, New York, 1983)

228. S. Wiesendorf, Taut submanifolds and foliations. J. Diff. Geom. (to appear). arXiv:1112.5965

229. B. Wilking, Positively curved manifolds with symmetry. Ann. Math. **163**, 607–668 (2006)

230. B. Wilking, A duality theorem for Riemannian foliations in nonnegative sectional curvature. Geom. Funct. Anal. **17**(4), 1297–1320 (2007)

231. B. Wilking, Nonnegatively and positively curved manifolds, in *Surveys in Differential Geometry*, ed. by J. Cheeger, K. Grove, vol. XI (International Press, Somerville, 2007), pp. 25–62

232. B. Wilking and W. Ziller. Revisiting homogeneous spaces with positive curvature. (2015, preprint) arXiv:1503.06256

233. J. Yeager, Geometric and topological ellipticity in cohomogeneity two. Ph.D. thesis, University of Maryland (2012)

234. W. Ziller, On M. Mueter's Ph.D. thesis on Cheeger deformations. arXiv:0909.0161

235. W. Ziller, Homogeneous Einstein metrics on spheres and projective spaces. Math. Ann. **259**(3), 351–358 (1982)

236. W. Ziller, Examples of Riemannian manifolds with non-negative sectional curvature, in *Surveys in Differential Geometry*, ed. by J. Cheeger, K. Grove, vol. XI (International Press, Somerville, 2007), pp. 63–102

237. S. Zoltek, Nonnegative curvature operators: some nontrivial examples. J. Differ. Geom. **14**, 303–315 (1979)

Index

Symbols
G-equivariant map, 53
$G(x)$, 52
G_x, 52
M_{princ}, 75
$M_x^{\approx}, M_x^{\sim}$, 79
S_x, 64, 120
$Z(G), Z(\mathfrak{g})$, 23
$\mathscr{S}_\xi$, 36
Ad, 16, 52
$\text{Iso}(M, \mathbf{g})$, 32, 69
Tub, 68
ad, 16
exp, 13, 15
$\mathbb{C}P^n$, 59, 63, 146, 166
$\mathbb{H}P^n$, 59, 63, 146
II, 35
$\mathfrak{X}(M)$, 188
$\mathfrak{gl}(n, \mathbb{R}), \mathfrak{gl}(n, \mathbb{C})$, 5, 14, 21
$\mathfrak{sl}(n, \mathbb{R}), \mathfrak{sl}(n, \mathbb{C})$, 14, 21
$\mathfrak{so}(n), \mathfrak{o}(n)$, 21, 104
$\mathfrak{sp}(n)$, 21
$\mathfrak{su}(n), \mathfrak{u}(n)$, 21, 39, 47, 98, 105
$\text{GL}(n, \mathbb{R}), \text{GL}(n, \mathbb{C})$, 14
$\text{SL}(n, \mathbb{R}), \text{SL}(n, \mathbb{C})$, 4, 14
$\text{SU}(n)$, 21, 47
$\text{Sp}(n)$, 21, 63
$\text{U}(n)$, 21
$\text{GL}(n, \mathbb{R}), \text{GL}(n, \mathbb{C})$, 21
$\text{GL}(n, \mathbb{R}), \text{GL}(n, \mathbb{C})$, 4
$\text{Gr}_k(V)$, 64
$\text{O}(n)$, 4, 21
$\text{SL}(n, \mathbb{R}), \text{SL}(n, \mathbb{C})$, 21
$\text{SO}(n)$, 4, 7, 21, 23, 56, 63

$\text{SU}(n)$, 4, 21–24, 39, 63, 77
$\text{Sp}(n)$, 5, 22
$\text{U}(n)$, 4, 24, 45
μ^G, 52
$\mu^\mathbf{g}$, 52
μ_x, 52
∇, 28
Ric, 34
scal, 34
H, 36
k-Grassmannian, 64
$\mathscr{X}_\mathscr{F}$, 109
$\text{SO}(n)$, 104
$\text{SU}(n)$, 88, 98, 105
$\text{U}(n)$, 88
1-parameter subgroup, 13

A
Action, 51
    adjoint, 16, 52, 85
    by isometries, 52
    cohomogeneity, 89, 157
    commutes, 62
    diagonal, 53
    effective, 52
    field, 55
    free, 52
    hyperpolar, 88
    orbit-equivalent, 90
    polar, 78, 88, 117
    product, 53
    proper, 56
    properly discontinuous, 56

© Springer International Publishing Switzerland 2015
M.M. Alexandrino, R.G. Bettiol, *Lie Groups and Geometric Aspects
of Isometric Actions*, DOI 10.1007/978-3-319-16613-1

Action (*cont.*)
   section, 88
   subaction, 53
   transitive, 52
Adjoint
   action, 16, 52, 85
   representation, 16
   representation in $GL(n, \mathbb{R})$, 17

B
Berger sphere, 151
Betti number, 197
Bi-invariant
   k-form, 38
   metric, 38
   metric of $SU(n)$, 39
   vector field, 6
Bianchi identity, 33, 181
Brieskhorn variety, 168

C
Campbell formulas, 15
Centralizer, 24
Cheeger deformation, 140, 151
   cohomogeneity one, 163
Cheeger reparametrization, 144
Cohomogeneity, 89, 139
Compact-open topology, 32
Complex flag manifold, 64
Connection, 28
   basic, 112
   Bott, 112
   compatible with metric, 28
   formula, 29
   Levi-Civita, 29
   symmetric, 28
Covariant derivative, 29
Covering map, 10
CROSS, 146
Curvature, 32
   Gaussian, 36
   mean, 36, 162
   operator, 33, 179
   principal, 36
   Ricci, 34
   scalar, 34
   sectional, 33
   strongly nonnegative, 180
   strongly positive, 180

D
Delaunay surfaces, 162
Differential form, 193

Distribution, 190
   horizontal, 37
   integrable, 191
   involutive, 190
   vertical, 37
Dynkin diagram, 103

E
Einstein
   constant, 35
   equation (in vacuum), 35
   manifold, 35, 152, 153
Endpoint map, 110
Equifocal submanifold, 110
Equifocality, 112
Equivariant
   map, 53
   normal field, 74
Exponential map, 13
Exterior derivative, 193

F
Fiber bundle, 57
   associated to a principal bundle, 66
   base, 57
   bundle charts, 57
   fiber, 57
   principal, 58
   projection, 57
   structure group, 57
   total space, 57
   transition function, 57
   trivial, 57
   vector bundle, 57
Fixed point, 52
   set, 79
Flag, 64
Foliation, 191
   homogeneous, 55
   isoparametric, 91
   leaf, 191
   polar, 110
   Riemannian, 109
   section, 110
   singular, 109
   singular Riemannian, 109
   singular Riemannian, with sections, 110
Frame bundle, 58

G
Gauss equation, 36
Geodesic, 29
   flow, 30

segment, 30
   minimal, 31
   unit speed, 29
Grassmannian, 64
Gray-O'Neill formula, 38, 181

H
Holonomy
   group, 69, 113
   map, 112
   pseudogroup, 114
   singular map, 121
Homogeneous bundle, 145, 151
Homogeneous metric, 148, 149
   normal, 153
   on projective spaces, 152
   on spheres, 149, 156
Hopf action, 54, 59, 164
Hopf bundle, 54, 59, 150, 151, 156
Hopf Problem, 172

I
Ideal, 43
   simple, 43
Ineffective kernel, 52
Isometry, 32, 52
   group, 32
      full, 146
   pseudogroup, 113
Isoparametric
   function, 127
   map, 131
   submanifold, 90, 91, 127
Isoparametric hypersurface, 130
Isotropy
   group, 52
   representation, 53, 72, 89
Isotypic component, 149

J
Jacobi equation, 33
Jacobi field, 33
Jacobi identity, 5

K
Kervaire sphere, 168
Killing
   form, 41
   form of $SU(n)$, 47
   vector field, 32

Kleiner's lemma, 72
Koszul formula, 29

L
Lawson conjecture, 165
Leaf, 191
   regular, 109
   singular, 109
Left-invariant
   k-form, 38
   metric, 38
   vector field, 6
Length, 31
Lie algebra, 5
   bracket, 5
   center, 23
   homomorphism, 6, 8
   ideal, 43
   of $SO(3)$, 7
   of $SU(n)$, 39
   of a Lie group, 7
   radical, 44
   semisimple, 41
   simple, 43
   solvable, 43
   subalgebra, 8
Lie group, 3
   action, 51
   center, 23, 45
   classical, 4
   exceptional, 106
   homomorphism, 7
   normal subgroup, 43
   rank, 87
   root, 94
   semisimple, 41
   subgroup, 8
Local section, 120

M
Manifold, 185
   C^k, 185
   analytic, 185
   atlas, 185
   chart, 185
   closed, 196
   geodesically complete, 31
   tangent space, 186
   with boundary, 196
Metric, 27
   G-invariant, 52
   bi-invariant, 38

Metric (*cont.*)
    Einstein, 35
    left-invariant, 38
    right-invariant, 38
Molino Conjecture, 125

N

Normal bundle, 58, 90
    flat, 90
    globally flat, 90
Normal slice, 72
Normal space, 35
Normalizer, 24

O

Orbifold, 114
    good, 114
Orbit, 52
    exceptional, 77, 160
    larger type, 77
    local type, 79
    principal, 73, 75
    regular, 77
    same local type, 79
    same type, 77
    singular, 77
    space, 54
    type, 79

P

Parallel translation, 30, 112
Partition of unity, 195
Pinching constant, 156
Plaque, 111, 191
Polar foliation, 110
Principal bundle, 58
Principal direction, 36
Principal orbit, 73, 75

R

Representation, 15, 53
    s-representation, 90
    isotropy, 53, 89
    slice, 72
Riemannian
    curvature, 32
    distance, 31
    exponential map, 31
    good orbifold, 114
    immersion, 35
    isometry, 32

length, 31
    manifold, 27
    metric, 27
    orbifold, 114
    submersion, 37
        integrable, 37
Right-invariant
    k-form, 38
    metric, 38
    vector field, 6
Root, 94
    base, 103
    coroot, 94
    diagram, 103
    of $SU(n)$, 98
    positive, 97
    simple, 103
    system, 94

S

Second fundamental form, 35
Semisimple
    Lie algebra, 41
    Lie group, 41
Shape operator, 36
Simple
    ideal, 43
    Lie algebra, 43
Simple open subset, 111, 191
Singular Riemannian foliation with sections,
    110
Singular stratification of local section, 121
Slice, 64, 120
    normal, 72
    representation, 72
Space form, 34
    spherical, 156, 180
Sphere theorem, 156
Stabilizer, 52
Strata, 82
Stratification, 82
Strongly positive curvature, 180
Submanifold, 187
    embedded, 187
    equifocal, 110
    immersed, 187
    isoparametric, 90
    minimal, 36, 83, 162, 165
    quasi-embedded, 187, 191
    totally geodesic, 36, 41
Submersion, 37
Surface, 28, 30
    of revolution, 30

Symmetric space, 41, 89, 157
    pair, 89, 157
    rank, 90

T

Tensor product, 192
Theorem
    Bonnet-Myers, 35
    Cartan, 44
    Constant Rank, 20
    Frobenius, 191
    Hopf-Rinow, 31
    Inverse Function, 187
    Levi-Civita, 29
    Lie's Third, 7
    Maximal Torus, 86
    Myers-Steenrod, 32
    Principal orbit, 75
    Stokes, 196
    Tubular Neighborhood, 68
Thorpe's trick, 180
Torsion, 28
Torus, 4, 18, 85
    infinitesimal generator, 85
    maximal, 86
Totally geodesic submanifold, 36, 41
Transnormal map, 131

Tubular neighborhood, 68
Twisted space, 66

V

Vector bundle, 57
    cotangent, 57
    frame bundle, 58
    normal bundle, 58
    tangent, 57
Vector field, 187
    1-parameter group, 190
    f-related, 188
    action field, 55
    flow, 190
    Killing, 32
    Lie bracket, 188
    normal foliated, 112
    normal parallel, 90
    parallel, 30
Volume form, 195

W

Wedge product, 192
Weyl
    chamber, 97, 121
    group, 99, 102, 169
    pseudogroup, 122

Printed in the United States
By Bookmasters